Artificial General Intelligence-Based Drones for Climate Change

Inam Ullah Khan
Multimedia University, Cyberjaya, Malaysia

Oroos Arshi
Sharda University, Greater Noida, India

Keshav Kaushik
University of Petroleum and Energy Studies, India

Hamed Taherdoost
University Canada West, Canada & GUS Institute, Global University Systems, UK & Hamta Business Corporation, Canada & Q Minded, Quark Minded Technology Inc., Canada & Westcliff University, USA & Gisma University of Applied Sciences, Germany

Khadija Slimani
University of Technology of Belfort Montbeliard, France

IGI Global
Scientific Publishing
Publishing Tomorrow's Research Today

Vice President of Editorial Melissa Wagner
Director of Acquisitions Mikaela Felty
Director of Book Development Jocelynn Hessler
Production Manager Mike Brehm
Cover Design Phillip Shickler

Published in the United States of America by
IGI Global Scientific Publishing
701 East Chocolate Avenue
Hershey, PA, 17033, USA
Tel: 717-533-8845
Fax: 717-533-7115
Website: https://www.igi-global.com E-mail: cust@igi-global.com

Library of Congress Cataloging-in-Publication Data
Library of Congress Cataloging-in-Publication Data

Names: Khan, Inam Ullah, editor. | Arshi, Oroos, 1999- editor. | Kaushik, Keshav, editor. | Taherdoost, Hamed, 1983- editor. | Slimani, Khadija, 1991- editor.
Title: Artificial general intelligence-based drones for climate change / edited by Inam Khan, Oroos Arshi, Keshav Kaushik, Hamed Taherdoost, Khadija Slimani.
Description: Hershey PA : IGI Global, [2025] | Includes bibliographical references. | Summary: "The book addresses the challenges and opportunities inherent in deploying AGI-enabled drones for climate-related missions. It explores topics such as cybersecurity, cloud computing, machine learning, IoT integration, and deep learning applications in the context of climate change"-- Provided by publisher.
Identifiers: LCCN 2024045712 (print) | LCCN 2024045713 (ebook) | ISBN 9798369364574 (hardcover) | ISBN 9798369364628 (paperback) | ISBN 9798369364581 (ebook)
Subjects: LCSH: Climatic changes--Remote sensing. | Climatic changes--Research--Methodology. | Drone aircraft in remote sensing. | Artificial intelligence--Environmental applications.
Classification: LCC QC903 .A7853 2025 (print) | LCC QC903 (ebook) | DDC 551.6028/4--dc23/eng/20250128
LC record available at https://lccn.loc.gov/2024045712
LC ebook record available at https://lccn.loc.gov/2024045713

British Cataloguing in Publication Data
A Cataloguing in Publication record for this book is available from the British Library.

Editorial Advisory Board

Table of Contents

Detailed Table of Contents

Drones have changed dramatically over the last several decades, from being a notion limited to science fiction and military uses to being an essential component of many businesses and daily life. This change started when the military started using Unmanned Aerial Vehicles (UAVs) in the early 20th century for combat and surveillance operations. Drone technology has developed over time, demonstrating human creativity and the never-ending quest of innovation that has resulted in the development of complex devices that are more compact, effective, and competent than before. The drone has a long history. These include public acceptability, technological constraints, integration of airspace, and legal frameworks. Unlocking the full potential of drones will need persistent communication, investment in research and development, and cooperation across stakeholders to overcome these obstacles. This chapter discusses the drone technology evolution, its application and future work.

hybrid approaches, and graph neural networks for swarm security. The discussion encompasses cutting-edge paradigms such as Zero Trust-Architecture, post-quantum-cryptography, and AGI safety considerations essential for autonomous systems. It concludes with a research agenda addressing quantum-resistant, federated learning, self-adaptive architectures, and formal verification methods providing a framework for addressing security challenges at the critical intersection of AGI, drone tech, and climate applications.

AGI-enabled drones are vital yet vulnerable systems, especially in climate monitoring and autonomous aerial missions, making cybersecurity a top priority. Dynamic, decentralized UAVs face cyber threats like DoS, spoofing, replay, and FDI attacks. This research proposes and evaluates lightweight ML-based Intrusion Detection Systems (IDS) tailored for UAVs. Five supervised models—Naïve Bayes, Decision Tree, K-Nearest Neighbors, Random Forest, and LightGBM—were tested across three datasets: a CTGAN-augmented mix of CIC-IDS2017 and UNSW-NB15, a physical-cyber UAV testbed dataset, and a T-ITS behavioral dataset. The experimental design included preprocessing, normalization, class balancing, and performance evaluation using accuracy, precision, recall, F1-score, and confusion matrix. LightGBM achieved the highest accuracy and generalization, while Decision Tree and Naïve Bayes offered optimal efficiency and speed, making them ideal for onboard IDS in AGI-driven drones.

Unmanned aerial drones (UAVs) or drones are becoming more and more common as people use them for surveillance and aerial photography among other purposes. In this new era, artificial intelligence (AGI) is a versatile AI technology that can be applied to various fields, including logistics, defence farming, and agriculture. AGI can collect and analyse real-time data, fostering communication with IOT devices and ensuring seamless operations. Finally, only by addressing safety related issues,

coastal environments which would immensely help the Indian subcontinent. The chapter will also include international best practices for monitoring and assessing sea level rise.

Muhammad Hammad Nawaz, University of Mianwali, Pakistan
Faisal Rehman, University of Mianwali, Pakistan

The purpose of this chapter is to review the dispersed uses of drone technologies within the built environment and as an instrument of climate change study. Unmanned aerial vehicles (or drones) have become very useful for environmental scientists to collect data, monitor, and analyse in the urban environment. The first section provides extensive details about drones operating. The systems showcase complex capabilities because they incorporate high-quality cameras together with LiDAR sensors and thermal imaging equipment. Modern tools developed for drones allow researchers to obtain precise measurements of space and environment information needed for urban climate analysis. Furthermore, the chapter investigates the integration of drone technology into environmental research and urban analysis. The paper describes essential applications of drone systems beginning with urban heat island mapping followed by building energy efficiency assessment and air quality management and finishing with greenhouse gas emission location work.

Bodhisattwa Baidya, Ramakrishna Mission Vidyamandira, India
Atanu Mondal, Ramakrishna Mission Vidyamandira, India

This chapter explores the cutting-edge integration of Artificial General Intelligence (AGI) powered drones with Software-Defined Networking (SDN), quantum computing, edge computing, LIDAR technology, and satellite systems for advanced climate change monitoring and mitigation. We present a comprehensive framework for long-term autonomous monitoring that leverages these technologies to revolutionize environmental data collection, analysis, and response strategies. The chapter discusses the synergistic potential of AGI drones equipped with LIDAR, connected through SDN architectures, and supported by quantum and edge computing capabilities. We examine how these systems can interact with Low Earth Orbit (LEO) and Geostationary Earth Orbit (GEO) satellites to create a global, real-time environmental monitoring network. Through detailed case studies, methodologies, and algorithms, we illustrate the transformative impact of this integrated approach on climate change research and mitigation efforts.

Preface

Artificial General Intelligence-Based Drones for Climate Change is a book that explores the use of advanced technologies, particularly Artificial General Intelligence (AGI), to combat climate change. The book explains how AGI can enhance drones' capabilities in monitoring environmental parameters, predicting weather patterns, and assessing biodiversity. It also discusses the causes, effects, and implications of climate change, emphasizing the need for innovative solutions. The book also covers challenges and opportunities in deploying AGI-enabled drones, including cybersecurity, cloud computing, machine learning, IoT integration, and deep learning applications. It emphasizes the importance of responsible and ethical deployment of these technologies.The book explores the use of Artificial Intelligence (AI)-powered drones in disaster management, highlighting their potential in detecting wildfires, monitoring floods, and tracking hurricanes. It also discusses their integration with edge computing, 5G/6G networks, and satellite communications for seamless data transmission. The book also discusses the societal and policy implications of AGI-based drones for climate change solutions, addressing privacy, regulatory frameworks, AI ethics, and global governance. It serves as a valuable resource for AI researchers, environmental scientists, and policymakers.

- Artificial General Intelligence-Based Drones are revolutionizing environmental monitoring, mitigating climate change challenges, and providing autonomous, adaptive solutions for real-time data collection, disaster management, and ecological preservation.
- The integration of advanced artificial intelligence (AGI) into UAVs could revolutionize environmental monitoring by enabling autonomous decision-making, optimizing routes, identifying environmental anomalies, predicting climate patterns, and collaborating in swarm formations to tackle large-scale crises.
- AGI-based drones offer real-time damage assessment, disaster prevention, carbon sequestration, forest health monitoring, precision agriculture, and

marine ecosystem conservation, making them crucial in combating climate change

- AGI-based UAVs are instrumental in wildlife and biodiversity conservation by tracking endangered species, detecting illegal deforestation, and monitoring habitat changes globally.
- The integration of AGI with drone-based precision agriculture can enhance farmers' irrigation, pest control, and crop health monitoring, thereby reducing resource wastage and enhancing food security.

A Description of the Target Audience

- **Researchers & Academics:** Experts in AI, robotics, environmental science, and UAV technology are exploring novel applications of AGI in climate monitoring and disaster management, seeking insights for environmental sustainability.
- **Industry Professionals & Engineers:** Professionals in UAV development, AI automation, and climate technology are exploring the integration of AGI into drone frameworks, while engineers and data scientists are working on autonomous aerial solutions.
- **Climate Change & Sustainability Advocates**: NGOs, environmental activists, and sustainability leaders are seeking technological innovations to combat climate change and promote eco-friendly solutions, while also advocating for AI-driven climate initiatives.
- **AI & UAVs enthusiasts:** AI enthusiasts explore the convergence of machine learning, UAV advancements, and AGI in tackling climate issues, exploring autonomous decision-making and adaptive drone behavior.

The book "Artificial General Intelligence-Based Drones for Climate Change" explores the use of AI in autonomously monitoring and responding to environmental challenges. It aims to revolutionize climate monitoring by providing real-time decisions, optimizing resource utilization, and providing predictive insights. The book encourages interdisciplinary collaboration among AI researchers, UAV engineers, climate scientists, and policymakers, addressing regulatory challenges and ethical considerations. It presents real-world case studies and future research directions, envisioning a future where autonomous drones become essential tools for mitigating climate change. This book explores the use of Artificial General Intelligence (AGI)-powered drones for climate resilience. It highlights their ability to adapt to real-time environmental data, making them efficient in unpredictable conditions. The book encourages researchers, engineers, and policymakers to embrace AI-driven innovation for sustainable development and environmental protection in the 21st century.

Acknowledgements

I want to acknowledge my dear friend Keshav Kaushik.

- Dr. Inam Ullah Khan

I dedicate this book to my parents Mr. Anzar Ahmad Tauseef and Mrs. Mumtaz Arshi.

- Ms. Oroos Arshi

Dedicated to my beloved parents (Sh. Vijay Kaushik and Smt. Saroj Kaushik); my wife (Priyanka); my daughter and son (Kashvi and Harshiv). May god bless everyone.

Har Har Mahadev!!

- Keshav Kaushik

I would like to express my sincere gratitude to all the contributors, authors, reviewers, and editors of this book for their valuable contributions and dedication. Special thanks to my colleagues at University Canada West, Q Minded | Quark Minded Technology, and Hamta Business Corporation for their continued encouragement and support throughout this journey.

I am deeply grateful to my family for their unwavering support and inspiration.

- Hamed Taherdoost

Chapter 1
Introduction to Artificial General Intelligence (AGI) in Drones

Roheen Qamar
https://orcid.org/0000-0003-4169-9455
Quaid-e-Awam University of Engineering, Science, and Technology, Pakistan

Baqar Ali Zardari
https://orcid.org/0000-0002-3573-1191
Quaid-e-Awam University of Engineering, Science, and Technology, Pakistan

ABSTRACT

Drones have changed dramatically over the last several decades, from being a notion limited to science fiction and military uses to being an essential component of many businesses and daily life. This change started when the military started using Unmanned Aerial Vehicles (UAVs) in the early 20th century for combat and surveillance operations. Drone technology has developed over time, demonstrating human creativity and the never-ending quest of innovation that has resulted in the development of complex devices that are more compact, effective, and competent than before. The drone has a long history. These include public acceptability, technological constraints, integration of airspace, and legal frameworks. Unlocking the full potential of drones will need persistent communication, investment in research and development, and cooperation across stakeholders to overcome these obstacles. This chapter discusses the drone technology evolution, its application and future work.

DOI: 10.4018/979-8-3693-6457-4.ch001

I. INTRODUCTION

Drones, also known as unmanned aerial vehicles (UAVs), have advanced significantly since their introduction in the early 1900s. Drone technology has evolved from its humble origins as military reconnaissance equipment to its current position in agriculture, filmmaking, and infrastructure assessment, which has been interesting to see. Drone technology has expanded from military uses to daily use in a variety of businesses. Consumer drones are now widely used for photography, racing, and other purposes. Drone photography and videography provide new viewpoints that have transformed visual storytelling. Drones have come a long way since their inception. What started as a military tool has now become a popular technology among hobbyists, filmmakers, and even businesses. The use of drones has increased significantly in recent years, and with that comes an evolution in drone technology. In this article, we will explore the past, present, and future of drone technology. Recent research and industry publications illustrate how drone technology has advanced swiftly, enabling a wide range of applications from agriculture to healthcare. Businesses are increasingly using drones for purposes like aerial mapping, surveillance, and deliveries, reflecting a shift toward automation and efficiency (Mairaj et al., 2019).

Previously, the term "drone" meant an unmanned, radio-controlled military or target-towing aircraft. A drone is now described as an aircraft that flies without a pilot at the controls, instead relying on the help of a ground operator or automated flying with no human participation. Drones have a wide range of uses, including crime scene surveillance, marine species detection, habitat degradation assessment, agricultural monitoring, and vegetation mapping. Drone mapping also has several uses in industries such as building, agriculture, mining, and infrastructure inspection. Drones have recently expanded their use in disaster or humanitarian aid to encompass search and rescue operations, catastrophe prevention, and disaster management (Hassanalian et al., 2018) drone shown in fig 1.

Figure 1. Drone architecture

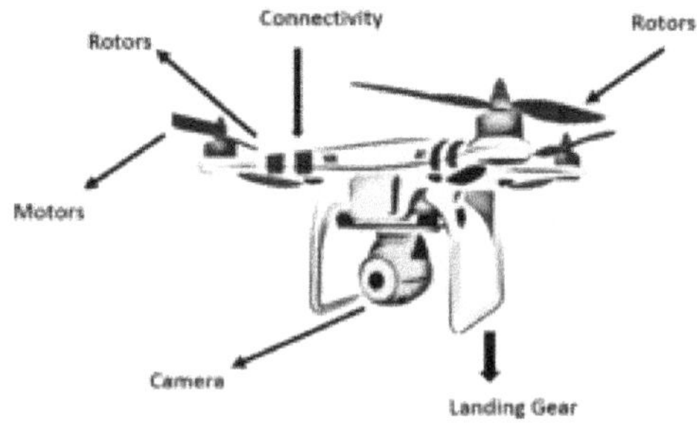

The integration of modern sensors and AI algorithms has improved drone capabilities, enabling for more precise data collecting and analysis across several sectors.

Regulatory frameworks are constantly adapting to the growing usage of drones, defining their deployment in fields such as transportation, infrastructure inspection, and emergency response. As drone technology advances, its revolutionary potential across industries becomes more apparent, opening the door for novel solutions and new commercial possibilities.

Drones were initially conceived as a basic gadget, but their complexity has increased as stated objectives have become more complicated. Drone variety is a key aspect in establishing their operating capabilities, which are influenced by size, power, and application circumstances. Drones are grouped into nine types: fixed-wing, flapping-wing, rotary-wing, tilt-rotor, ducted fan, helicopter, ornithopter, and unusual (Azar et al., 2021).

Catastrophes of any kind are harmful to the health and welfare of the affected people, but mass catastrophes disproportionately damage a huge number of victims. Natural and man-made catastrophes are classed based on whether they are caused by an environmental, medical, industrial, or terrorist incident. As a result, when a disaster happens, a quick and effective reaction is vital for supporting the population, lowering the number of casualties, and lessening economic damage. Regardless of the nature of the catastrophe or emergency, disaster management is described as a process or practice that encompasses mitigation, readiness, response, and recovery.

Consequently, the use of drones in a disaster has the following advantages: they reduce the time required to locate victims and the time required for subsequent intervention by searching a large area in a short period of time, as well as providing critical information to rescuers about the route that must be taken during search and rescue missions. Drones may also seek for living persons trapped beneath debris utilizing sensors such as noise, binary, vibration, and heat detection. These highlight the benefits of having drones on-site during disasters, as well as drones' capacity to be crucial instruments for gathering aerial photos (Adamopoulos & Rinaudo, 2020).

During World War I, both the United States and the United Kingdom utilized unmanned aircraft for observation and reconnaissance. The Royal Navy's Air Department created the "Aerial Target," a remotely controlled aircraft intended to train anti-aircraft gunners. During the conflict, the United States Army employed drones to photograph enemy locations from the air.

After World War II, drone technology continued to develop, and drones became increasingly sophisticated. In the 1950s, the U.S. Air Force began developing the first long-range reconnaissance drone, called the Q-2C Firebee. This drone could fly at high altitudes and was used for intelligence gathering during the Cold War.

In the 1960s, the U.S. military began developing a new generation of drones, called unmanned combat aerial vehicles (UCAVs). These drones were designed to be able to engage in combat without risking human pilots' lives. However, it wasn't until the 1990s that UCAVs began to be deployed in real-world combat situations.

The development and growth of drones is a fascinating adventure that has transformed numerous sectors and caught the imaginations of people all around the world. Drones have progressed from their humble origins as remote-controlled toys to sophisticated flying vehicles capable of recording stunning footage. In this essay, we will look at how drones came to be, how they have evolved throughout time, and how they have become vital tools in a variety of sectors. Whether you are a newbie hoping to learn the fundamentals or a professional searching for the most recent updates, this article will provide you useful insights into the history and growth of drones. So tighten your seatbelts and join us on this amazing voyage through time (Elghaish et al., 2021).

1.1 Early Development

Drones, also known as unmanned aerial vehicles (UAVs), have a rich and intriguing history that spans decades. The idea of employing unmanned flying aircraft may be traced back to the concepts and sketches of Leonardo da Vinci, a famous artist and inventor. In the late 15th century, da Vinci imagined vehicles that resembled birds and bats and could fly without a pilot (Alfeo et al., 2018) types are shown in table 1.

Table 1. Early development

1.	**Leonardo Da Vinci's Designs**	Leonardo da Vinci's plans for flying aircraft were groundbreaking at the time. He sketched out many concepts, including a glider and an ornithopter propelled by flapping wings to resemble bird flight. While da Vinci's plans were never built or flown, they provided the groundwork for subsequent advances in unmanned flying.
2.	**World War I and II**	The actual progress of drone technology occurred in the early twentieth century, namely during World Wars I and II. Both sides of the battle recognized the possibilities of utilizing unmanned aircraft for reconnaissance. The first significant usage of drones happened during World War I, when the British military created the "Aerial Target" to distract and confound German defenses. During WWII, the United States and Germany achieved tremendous advances in drone technology. The US military created the "Radio plane OQ-2" as a target for anti-aircraft gunnery practice. Meanwhile, the German military created the "V-1" flying bomb, an early prototype of a cruise missile that could be launched from the ground and directed to its target (Magistretti & Dell'Era, 2019).
3.	**Early Experimental Drones**	After WWII, the development of drones halted dramatically. However, in the 1960s, the US military started experimenting with remotely piloted aircraft for surveillance. The "Ryan Firebee" was one of the first successful remotely controlled drones, capable of flying at great heights and capturing images (Chou, 2018).

1.2 Military Applications

While drones were initially designed for reconnaissance, their military uses swiftly extended to incorporate a variety of other functions. Technological advancements enabled the development of armed drones and enhanced surveillance capabilities. Discuss in Table 2.

Table 2. Military drone

1.	**Surveillance and Reconnaissance**	Drones are still used mostly for surveillance and reconnaissance in the military. The capacity to collect real-time data and conduct aerial surveillance without endangering human lives is vital to armed forces throughout the world. Drones may be outfitted with high-resolution cameras and infrared imaging sensors, allowing them to record precise footage while providing critical intelligence to military commanders.
2.	**Weaponries Drones**	The introduction of armed drones has revolutionized modern combat. Armed drones, also known as unmanned combat aerial vehicles (UCAVs), can hit targets precisely and accurately. The US military's employment of armed drones, such as the Predator and Reaper, has helped it perform targeted attacks on high-value targets in combat zones.
3.	**Developments in Military Drone Technology**	Military drone technology continues to transform modern combat. Drones have become a vital tool for armed forces due to their improved flight capabilities, longer durability, and increased payload capacity. The integration of artificial intelligence and autonomous flying systems has increased the capabilities of military drones, allowing more complicated missions to be completed independently (Sharma, 2019).

1.3 Commercial and Civilian Drones

Although drones were originally designed for military objectives, their use in commercial and civilian sectors has increased dramatically in recent years. Drones are popular among enthusiasts, photographers, and filmmakers because to their ease of use and low cost.as shown in Table 3

Table 3. Commercial and civilian drones

1.	**Rise in Consumer Drones**	Consumer drones are becoming increasingly popular due to technological breakthroughs and lower production costs. Consumer drones are now widely available and reasonably priced, making them accessible to a diverse variety of consumers. These drones are intended for recreational use, allowing people to shoot amazing aerial footage, discover new viewpoints, and enjoy the thrill of flight.
2.	**Regulatory and Legal Frameworks**	With the fast rise in the usage of drones, governments throughout the world have created rules and legal frameworks to ensure safe and responsible drone use. These rules include topics including flying limitations, registration procedures, and privacy concerns. To prevent legal and safety difficulties, drone enthusiasts should become aware with and follow these restrictions (AL-Dosari et al., 2023).
3.	**Applications for Photography and Videography**	Photography and videography are two of the most important applications of drones in both business and consumer settings. Drones outfitted with high-quality cameras and stabilization technologies enable photographers and filmmakers to capture breathtaking aerial pictures that were previously only feasible with costly helicopter or crane setups. Drones have transformed the way we record and photograph the world around us, offering a new viewpoint and limitless creative opportunities (Vujičić et al., 2022).

1.4 Technological Advancements

Technological developments have driven the growth of drones. Miniaturization, innovations in flight control systems, and advances in image technology have all led to the creation of more competent and adaptable drones. Categories are given in table 4.

Table 4. Technological advancements

1.	**Rise in Consumer Drones**	Consumer drones are becoming increasingly popular due to technological breakthroughs and lower production costs. Consumer drones are now widely available and reasonably priced, making them accessible to a diverse variety of consumers. These drones are intended for recreational use, allowing people to shoot amazing aerial footage, discover new viewpoints, and enjoy the thrill of flight.
2.	**Regulatory and Legal Frameworks**	With the fast rise in the usage of drones, governments throughout the world have created rules and legal frameworks to ensure safe and responsible drone use. These rules include topics including flying limitations, registration procedures, and privacy concerns. To prevent legal and safety difficulties, drone enthusiasts should become aware with and follow these restrictions (AL-Dosari et al., 2023).
3.	**Camera and Imaging Innovations**	The developments in camera and image technology have dramatically increased the possibilities of drones for photography and filming. High-resolution cameras, picture stabilization technology, and powerful image processing techniques have enabled the collection of professional-grade aerial film and still photos (Hermanus et al., 2023).

2. TYPES OF DRONES

Drones come in a variety of forms, sizes, and combinations, each intended to fulfill a unique function. Understanding the many sorts of drones is critical to appreciate the adaptability of this technology.

A. Quadcopters and Multirotor

Quadcopters, sometimes known as quadrotors, are among the most popular and most identifiable forms of drones. They have four rotors stacked in a square configuration, which offers stability and agility. Multirotors include more than four rotors, such as hexacopters (six rotors) and octocopters (eight rotors). Quadcopters are popular for jobs such as aerial photography, surveillance, and recreational flight because they are easy to operate and maneuver.

B. Modern-Day Drones

Drones are now employed in a number of civilian purposes, as opposed to just military reconnaissance. Drones have gotten smaller, lighter, and more inexpensive, making them available to both corporations and people. Some of the most common applications for drones nowadays. Shown in fig 2

Figure 2. Modern drone

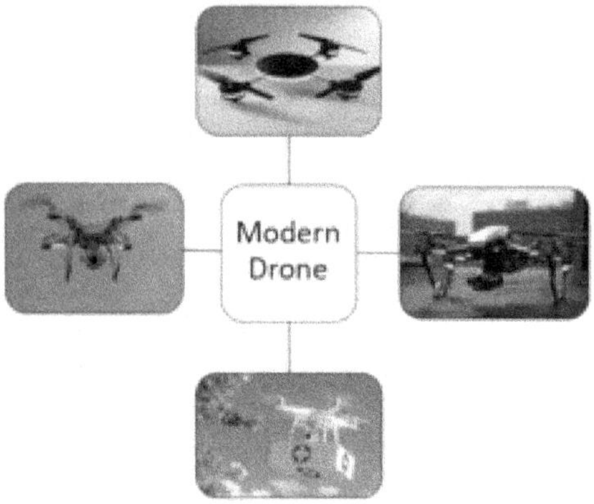

C. Small Drone

Unlike the first four drone kinds described below, which may potentially cost tens or even hundreds of thousands of dollars, drones typically cost under $100. In this usage, "small" often refers to lengths ranging from 20 to 80 inches.

These drones are exclusively recreational and cannot do many of the business activities that other models can. Cameras onboard a very stable multi-rotor, for example, may record beautiful aerial images and video when correctly installed.

Small drones, on the other hand, are typically too light and lack the steadiness essential for precise image capture. Nonetheless, these solutions might be a terrific economical introduction to the world of drones (Maghazei & Netland, 2020).

D. Micro Drone

While smaller drones may represent enjoyment in the perspective of customers, micro drones are strictly business for military that utilize them. The most well-known example of this sort of drone in use today is the Black Hornet, which was built for the British military.

Since their introduction in 2013, the British military has employed these tiny 1" x 4" drones to inspect walls and other installations in Afghanistan. While cameras may be overkill for recreational tiny drones, the particular micro cameras on these drones can give valuable intelligence. When not in use, Black Hornets can be kept in a specific belt.

E. Tactical Drones

These drones are large enough to not be pocket-sized, but they are still significantly smaller than those employed for general combat and larger jobs. The Raven is the US military's favorite tactical drone, measuring 4.5 feet and weighing pounds (Guvenc et al., 2018).

These sorts of drones are frequently employed for surveillance purposes. The Ravens, like the Black Hornets, may be mounted with unique infrared cameras, allowing soldiers to see the surroundings well even at night.

The devices include GPS technology. While they are basic and do not have many bells and whistles, this makes them very accessible and easy for soldiers to operate without the need for specific training.

F. Reconnaissance Drones

With another military drone class, we go up in size slightly to drones that are not intended to be handed. Instead, these drones, which are around 16 feet long and launched from the ground, are referred to as Medium Altitude Long Endurance (MALE) or High Altitude Long Endurance (HALE).

These drones are among the most widely used by military throughout the globe. The Heron, created by Israeli Aerospace Industries, has produced drones of this type for military reconnaissance in the United States, Canada, Turkey, India, Morocco, and Australia.

G. Large Combat Drones

When you think of “drones” in a military context, these are most likely the sorts that come to mind. The US uses variants such as the Predator and Reaper, which are around 36 feet long and can fire on targets with air-to-surface missiles and laser-guided bombs.

These devices have a 1,000-mile range and can run for 14 hours. These drones have been employed in activities such as military attacks against Pakistan and other nations with which the United States is not formally at war.

The United States may be the most renowned (or notorious) user of drones, but it is far from alone. Fellow NATO nations like as the UK, Spain, and France also employ them, while China has produced its own version, the CH-4, which possesses (Iost Filho et al., 2020).

H. Non-Combat Large Drones

In contrast, there are big drones that are not designed to be used in conflict. These may perform a number of tasks, including reconnaissance, and are utilized for larger-scale recon operations than their smaller Black Hornet and Raven equivalents.

For example, Northrop Grumman's Global Hawk is predominantly utilized above war zones but is not designed for fighting. Instead, it is utilized for surveillance purposes, such as monitoring cell phone calls.

The fact that these drones do not engage in warfare does not make them any less costly. The Global Hawk, for example, may cost up to $131 million, which excludes ground infrastructure.

I. TARGET AND DECOY DRONES

One of the most essential things to remember about military drones is that they may serve several purposes depending on the scenario. For example, certain drones may be employed for observation, while others are designed for striking capability; these serve as decoys.

However, the appearance of those decoy missions is likely to vary based on the mission's specific nature. As a result, these military drones must be prepared to serve as decoys in a variety of scenarios.

Some target and decoy drones, for example, can complete their objective by imitating an oncoming missile. This can draw fire from ground anti-aircraft troops, diverting them from any real combat drones or oncoming missiles.

J. GPS Drones

Now that we've covered the basic sorts of drones for commercial and military applications, let's take a look at some specialist varieties, beginning with drones intended exclusively for GPS purposes. These drones communicate with satellites via a GPS signal. They may then use this to plan out the rest of their journey, generating data that can be extracted and used for other purposes. For example, this can be an excellent method for mapping out huge topographies. Naturally, this requires a lot of battery power. As a result, the finest GPS drones are programmed to autonomously return to their base when their battery runs low or they reach the end of their working range.

K. Photography Drones

We've already discussed drones' photo-taking capabilities. While a few lesser choices can take limited images, and there are certainly multiple drones capable of doing so for armies, if you want to shoot professional photos, you'll need a commercial drone specifically developed for the purpose.

As previously said, drones with numerous rotors will provide the steadiness required for shooting decent photos or videos.

If you want to shoot professional-quality photos, be sure your drone is equipped with a professional-grade camera. 4K camera drones are capable of taking high-resolution photos. Some photography drones employ precision stability and autonomous flying modes to take photographs.

L. Racing Drones

Racing drones may achieve speeds of up to 60 mph. However, because so much force is concentrated on a single goal, they frequently do not have the same flying time as their counterparts.

Whether you want to construct your own racing drone from start or buy one, you should consult firms like Walkera for advice, since their units are among the quickest and may cost less than $300 (Jiménez López & Mulero-Pázmány, 2019).

3. APPLICATIONS OF DRONE TECHNOLOGY

1) Military Drone Technology

The military is perhaps the oldest, best-known, and most contentious deployment of drones. In the early 1940s, British and American forces began using primitive types of drones to spy on the Axis powers. Drones nowadays are significantly more advanced than UAVs of the past, with thermal imaging, laser rangefinders, and even bombing capabilities. The MQ-9 Reaper is a widely recognized military drone. The aircraft is 36 feet long, capable of flying unnoticed at 50,000 feet, and equipped with a variety of weapons and intelligence collecting equipment.

2) Delivery Drone Technology

Delivery drones are generally unmanned aerial vehicles (UAVs) that deliver food, goods, or commodities to your doorstep. These flying vehicles are referred to as “last mile” delivery drones since they transport goods from local stores or warehouses. Instead of relying on inefficient delivery drivers and trucks, businesses and supermarket chains throughout the country are turning to drones for a more efficient delivery solution. These drones can transport 55 pounds of stuff to your front door without you having to leave the house. Amazon, Walmart, Google, FedEx, UPS, and several larger firms are all exploring different sorts of delivery drones (Roldán-Gómez et al., 2021).

3) Drone for Emergency Public Rescue

Because of the magnitude or severity of the disaster, it is not always safe to deploy humans into a rescue situation. Here's where drones come into play. In the case of a capsized boat or a drowning individual, officials can use an Autonomous Underwater Vehicle (AUV) to help with the recovery.

4) Drone For Agriculture

Drones have also proved to be beneficial in the agriculture industry, giving farmers a number of alternatives for improving their crops to maximum productivity while minimizing physical strain. UAVs simplify field surveys, seeding across fields, tracking animals, and estimating crop yields, all while saving farm personnel valuable time.

5) Drone For Outer Space

NASA and the US Air Force have been testing unmanned aircraft intended for space travel. The Air Force's ultra-secretive X-37B UAV resembles a tiny spaceship. It has quietly circled the Earth for the last two years, breaking the record for the longest unmanned aircraft flight (781 days and counting). Despite its ambiguity, the Air Force has claimed that "the primary objectives of the X-37B are twofold: reusable spacecraft technologies for America's future in space and operating experiments that can be returned to, and examined on Earth." When it comes to the future of space exploration and innovation, drones appear to be the top priority. Other categories are shown in fig 3.

Figure 3. Drone router

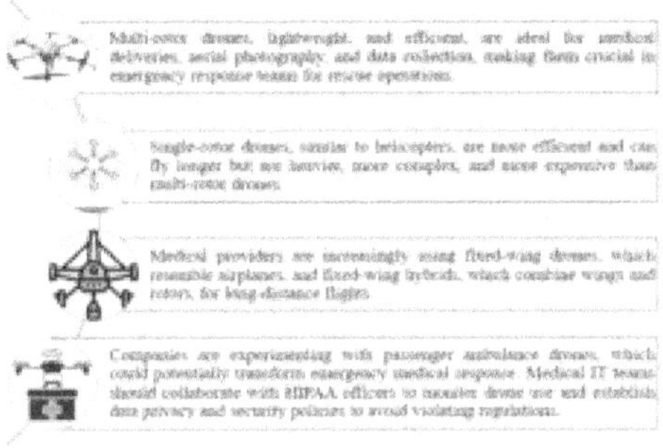

6) Drone for Environmental and Historical Protection

Drones are a less expensive and more effective approach to protect animals. It is difficult to track wildlife numbers while people are around. The capacity to follow moving groups of animals, such as orangutans in Borneo and bison on the Great Plains, helps wildlife conservationists to acquire a better knowledge of the health of their species and ecosystems. Conservation drones are also important in combating poaching in Asia and Africa.

7) Drones in Medicine

How do you get medical supplies to people living in rural areas? What device may be used to convey organs to transplant recipients? Drones can answer both of these questions. Unmanned aircraft vehicles are increasingly being used to transport emergency medical supplies and products to isolated settlements in Alaska. Instead, then relying on dog sleds, snowmobiles, or ambulances that can't handle the snow, Alaskans are turning to drones to deliver life-saving medical supplies quickly.

Drones are being utilized to deliver donated organs to transplant recipients. A kidney was recently transferred from one hospital in Maryland to another in less than five minutes using a specially developed drone. This has the potential to lower the excruciatingly slow rate at which donations usually arrive (if at all). Organs are often delivered via chartered or commercial planes. Delays and errors in judgment cause hazardous two-hour or longer delays in 4% of all organ deliveries. Drones can drastically decrease time while providing a safer and more secure way of organ transfer (Zailani et al., 2020).

8) Drone for 3D Modelling

LiDAR drones are outfitted with LiDAR sensors, which survey landscapes and collect comprehensive data that may be used to generate 3D models. Drones using LiDAR technology can deliver far more precise data than drones without it. Aside from making it simpler for drones to navigate different surfaces, LiDAR enables them to detect targets in search and rescue operations, analyze crops in agriculture, and perform a variety of other tasks.

9) Drone For Photography

Drones have proved beneficial to aerial photographers who use UAVs to get wide pictures. Have you ever longed to see a bird's-eye perspective of your favorite city, beach, or building? There are drones made specifically for photography, which provide a unique viewpoint on some of your favorite locales (Mishra et al., 2022) shown in fig 4.

Figure 4. Drone photography

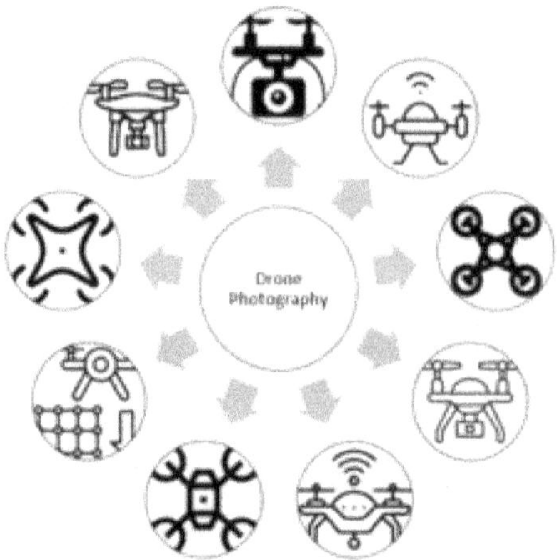

4. RELATED WORK

Boccadoro et al. (2021) Drones are increasingly being used for wildlife and historical conservation, as they offer a cost-effective and efficient method for tracking animal populations and fighting poaching. They are also being used in global forestry projects, helping to reforest deforested land. This paper aims to categorize the multifaceted aspects of Internet of Things (IoT) and propose a classification approach based on two main directions: the Internet protocol stack structure and the challenges faced by each layer. The paper also examines privacy and security considerations, and discusses research challenges and future directions in the IoT field. The goal is to provide a deeper understanding of the potential benefits of drones in wildlife and historical conservation.

Hanover et al. (2024) The increasing use of autonomous drones for surveying, search and rescue, and last-mile delivery necessitates robust, safety-critical algorithms for complex and uncertain environments. Autonomous drone racing (ADR) is a high-speed navigation task that requires real-time algorithms robust to motion blur, high dynamic range, model uncertainties, aerodynamic disturbances, and unpredictable opponents. This survey covers the evolution of ADR across model-based and learning-based approaches and highlights challenges and open questions for future development.

Park et al. (2021) this paper surveys anti-drone systems, focusing on non-military drone usage and analyzing detection, identification, and neutralization technologies. It proposes a hypothetical system for adaptable defense, discusses potential drone-side safety schemes, and proposes future solutions to address challenges in constructing military-grade anti-drone systems due to installation and operation costs.

Labib et al. (2021) integration of Unmanned Aerial Vehicles (UAVs) into the Internet of Things (IoT) has led to new applications and challenges in privacy, security, and safe operation of UAVs. This study explores the current state of UAVs and low-altitude airspace traffic management, highlighting synergies between scientific research and standardization efforts to enable safe UAV operations while considering IoT inherent challenges such as security, data protection, and privacy.

Tsiamis et al. (2019) This study analyzes OECD countries' national legislations on drone use, focusing on similarities and differences. The analysis reveals a wide range of drone classifications, with some countries implementing legislation during recent drone applications. The study highlights the ethical dilemmas arising from drone use, suggesting a proposal for basic legislation based on existing criteria and limitations. The findings suggest a need for future policymakers to address these issues.

Sliusar et al. (2022) The paper explores the use of unmanned aerial vehicles (UAV) in managing municipal solid waste landfills and dumpsites. Despite a ten-year history, the use of UAVs has increased in the last 3-4 years. The research mainly studies spatial and volumetric characteristics of landfills, with a quarter focusing on monitoring landfill gas emissions. Current technologies, such as UAVs, sensors, and GIS, enable the identification and assessment of environmental effects and compliance with standards, potentially improving waste disposal management.

Nayyar et al. (2020) The Internet of Drone Things (IoDT) is envisioned as the future of drone backend via Internet of Things, Smart Computer Vision, Cloud Computing, enhanced wireless communication, big data, and cutting-edge security approaches. Drones are increasingly being used in a variety of industries, including agriculture, industry, government, and commercial organizations, as well as Smart Cities and rural surveillance. With IoDT-based implementations, all existing sectors will become intelligent and smart in terms of monitoring, surveillance, search and rescue, and other functions. In this work, we give a conceptual presentation of new nomenclature, i.e., Internet of Drone Things (IoDT), as well as related technologies, applications, security challenges, and real-time IoDT implementation using case studies from Agriculture and Smart Cities.

Sibanda et al. (2021) Accurate and timely information on surface water quality and quantity is required for a variety of uses, including irrigated agriculture. In-field water quality and quantity data from unmanned aerial vehicle systems (UAVs) might help close spatial data gaps by generating near-real-time, fine resolution, and spatially explicit information needed for water resource accounting. This study analyzed the accomplishments, prospects, and problems of using UAV data to map and simulate water quality and quantity. To accomplish this study purpose, a systematic review was used. The findings indicate limited growth in the utility of UAVs, particularly in the global south. This might be due in part to high expenses, a lack of required skills, and rules governing drone purchase and operations costs.

Ausonio et al. (2021) The study proposes an innovative forest firefighting system using hundreds of unmanned aerial vehicles (UAVs) to generate a continuous flow of extinguishing liquid on the fire front, simulating rain. The system is validated in Mediterranean scrub by computing critical water flow rate and estimating active fire front extinguishment. Simulation results show the system can support current forest firefighting techniques.

5. THE FUTURE OF DRONE TECHNOLOGY

As with any new technology, there are also challenges and concerns surrounding the use of drones. One of the biggest concerns is privacy. Drones equipped with cameras can be used to spy on individuals without their knowledge or consent. There are also concerns about drones being used to deliver weapons or explosives.

Regulation is another challenge facing the drone industry. The Federal Aviation Administration (FAA) in the United States and other regulatory bodies around the world have established rules and guidelines for the use of drones, including restrictions on where drones can be flown and how high they can go.

Safety is also a concern, as drones can collide with other objects in the air or on the ground, causing damage or injury. As drone technology continues to evolve, safety regulations and guidelines will need to be updated to ensure the safety of people and property (Johnson et al., 2021).

The future of drone technology seems promising, with longer flight periods, higher autonomy, enhanced sensors, swarm technology, AI integration, urban air mobility, environmental monitoring, and space exploration. These improvements will have a tremendous influence on a variety of sectors, perhaps transforming the way we live our everyday lives.

To overcome these obstacles and concerns, a clear regulatory framework is required. Governments and regulatory organizations must collaborate closely with the drone sector to guarantee that safety norms and regulations are continually updated to reflect technological improvements. It's also critical to address privacy issues and guarantee that drones are utilized safely (Chen et al., 2020).

6. CHALLENGES AND CONCERNS

While drones have enormous potential advantages, various obstacles and issues must be addressed to ensure that this technology is used safely and responsibly.

Figure 5. Challenges of drone

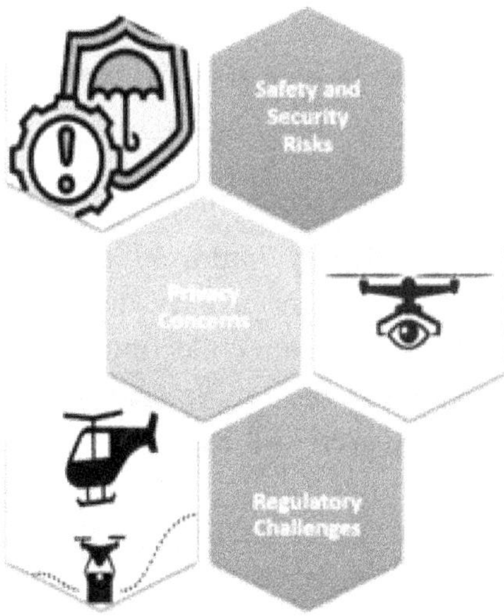

1. Safety and Security Risks

One of the biggest worries about drones is the possibility of mishaps and crashes, particularly in busy skies. The rise of drones raises worries about airspace security and the risk of uncontrolled drone activities in sensitive locations.

2. Privacy Concerns

As drones grow increasingly common, privacy concerns have surfaced. Drones outfitted with cameras can record sensitive pictures, infringing on an individual's private rights. Strict restrictions and standards are required to achieve a balance between the benefits of drone technology and the preservation of personal privacy.

3. Regulatory Challenges

The fast growth of drone technology has created regulatory issues for governments all around the world. Balancing the requirement for safety and responsible drone operations with the goal to promote innovation and economic growth is a difficult challenge. Governments must constantly adjust and alter their policies to keep up with the changing nature of this technology (Ayamga et al., 2021).

7. EFFECT OF DRONES

Drones have had a profound influence on many elements of society, with uses that go beyond the military and commercial sectors.

A. Ecological Applications

Drones have significantly improved environmental monitoring and conservation efforts. Drones give researchers and scientists with a cost-effective and efficient tool for collecting crucial data, including tracking wildlife populations, monitoring deforestation, and mapping ecosystems.

B. **Disaster Response and Search and Rescue**

Drones have shown to be quite useful in disaster response and search-and-rescue missions. Drones outfitted with thermal imaging sensors and capable of scanning huge regions fast can help locate survivors, assess damage, and deliver critical supplies in disaster circumstances.

C. **Economic Impact**

The drone industry has boosted economic growth and employment creation in a variety of areas. Drones have a considerable economic influence, ranging from production and sales to the creation of support services. Furthermore, the use of drones in industries such as agriculture, construction, and cinematography has increased efficiency, production, and income creation (Kunertova, 2023).

CONCLUSION

Drone technology has evolved dramatically over the years, from its early usage in military surveillance to its current employment in a wide range of civilian applications. As drone technology advances, the possibilities for their application are limitless, and we may anticipate drones to play an increasingly vital part in our lives. However, like with any new technology, obstacles and concerns must be addressed to ensure that drones are used safely and responsibly.

In this chapter discuss about the drone technology evolution and its challenges. Drones have come a long way since their initial use as military surveillance instruments. Today, they are utilized for a variety of purposes, including photography and package delivery. As drone technology advances, we should expect even more intriguing uses in the future.

There are hundreds of various drone models on the market, and more are being developed all the time. With a little expertise and the correct equipment, you can snap breathtaking aerial photos that were previously difficult to obtain without renting a helicopter or plane. Drone technology has evolved from military use to civilian applications, catering to specific tasks and industries. Its future is bright, with a focus on agriculture, conservation, healthcare, urban air mobility, AI, and sustainability efforts, showcasing human creativity and potential.

REFERENCES

Adamopoulos, E., & Rinaudo, F. (2020). UAS-based archaeological remote sensing: Review, meta-analysis and state-of-the-art. *Drones (Basel)*, *4*(3), 46. DOI: 10.3390/drones4030046

AL-Dosari, K., Hunaiti, Z., & Balachandran, W.AL-Dosari. (2023). Systematic review on civilian drones in safety and security applications. *Drones (Basel)*, *7*(3), 210. DOI: 10.3390/drones7030210

Alfeo, A. L., Cimino, M. G. C. A., De Francesco, N., Lega, M., & Vaglini, G. (2018). Design and simulation of the emergent behavior of small drones swarming for distributed target localization. *Journal of Computational Science*, *29*, 19–33. DOI: 10.1016/j.jocs.2018.09.014

Ausonio, E., Bagnerini, P., & Ghio, M. (2021). Drone swarms in fire suppression activities: A conceptual framework. *Drones (Basel)*, *5*(1), 17. DOI: 10.3390/drones5010017

Ayamga, M., Tekinerdogan, B., & Kassahun, A. (2021). Exploring the challenges posed by regulations for the use of drones in agriculture in the African context. *Land (Basel)*, *10*(2), 164. DOI: 10.3390/land10020164

Azar, A. T., Koubaa, A., Mohamed, N. A., Ibrahim, H. A., Ibrahim, Z. F., Kazim, M., & Ammar, A.. (2021). Drone deep reinforcement learning: A review. *Electronics (Basel)*, *10*(9), 999. DOI: 10.3390/electronics10090999

Boccadoro, P., Striccoli, D., & Grieco, L. A. (2021). An extensive survey on the Internet of Drones. *Ad Hoc Networks*, *122*, 102600. DOI: 10.1016/j.adhoc.2021.102600

Chen, W., Liu, J., Guo, H., & Kato, N. (2020). Toward robust and intelligent drone swarm: Challenges and future directions. *IEEE Network*, *34*(4), 278–283. DOI: 10.1109/MNET.001.1900521

Chou, P.-N. (2018). Smart technology for sustainable curriculum: Using drone to support young students' learning. *Sustainability (Basel)*, *10*(10), 3819. DOI: 10.3390/su10103819

Elghaish, F., Matarneh, S., Talebi, S., Kagioglou, M., Hosseini, M. R., & Abrishami, S. (2021). Toward digitalization in the construction industry with immersive and drones technologies: A critical literature review. *Smart and Sustainable Built Environment*, *10*(3), 345–363. DOI: 10.1108/SASBE-06-2020-0077

Guvenc, I., Koohifar, F., Singh, S., Sichitiu, M. L., & Matolak, D. (2018). Detection, tracking, and interdiction for amateur drones. *IEEE Communications Magazine*, *56*(4), 75–81. DOI: 10.1109/MCOM.2018.1700455

Hanover, D., Loquercio, A., Bauersfeld, L., Romero, A., Penicka, R., Song, Y., Cioffi, G., Kaufmann, E., & Scaramuzza, D. (2024). Autonomous drone racing: A survey. *IEEE Transactions on Robotics*, *40*, 3044–3067. DOI: 10.1109/TRO.2024.3400838

Hassanalian, M., Rice, D., & Abdelkefi, A. (2018). Evolution of space drones for planetary exploration: A review. *Progress in Aerospace Sciences*, *97*, 61–105. DOI: 10.1016/j.paerosci.2018.01.003

Hermanus, D. R., Supangkat, S. H., & Hidayat, F. (2023). Web-based anomaly detection for smart urban living: Drone photography and videography. *Jurnal Sistem Cerdas*, *6*(2), 144–152. DOI: 10.37396/jsc.v6i2.330

Iost Filho, F. H., Heldens, W. B., Kong, Z., & de Lange, E. S. (2020). Drones: Innovative technology for use in precision pest management. *Journal of Economic Entomology*, *113*(1), 1–25. DOI: 10.1093/jee/toz268 PMID: 31811713

Jiménez López, J., & Mulero-Pázmány, M. (2019). Drones for conservation in protected areas: Present and future. *Drones (Basel)*, *3*(1), 10. DOI: 10.3390/drones3010010

Johnson, A. M., Cunningham, C. J., Arnold, E., Rosamond, W. D., & Zègre-Hemsey, J. K. (2021). Impact of using drones in emergency medicine: What does the future hold? *Open Access Emergency Medicine : OAEM*, *13*, 487–498. DOI: 10.2147/OAEM.S247020 PMID: 34815722

Kunertova, D. (2023). The war in Ukraine shows the game-changing effect of drones depends on the game. *Bulletin of the Atomic Scientists*, *79*(2), 95–102. DOI: 10.1080/00963402.2023.2178180

Labib, N. S., Brust, M. R., Danoy, G., & Bouvry, P. (2021). The rise of drones in internet of things: A survey on the evolution, prospects and challenges of unmanned aerial vehicles. *IEEE Access : Practical Innovations, Open Solutions*, *9*, 115466–115487. DOI: 10.1109/ACCESS.2021.3104963

Maghazei, O., & Netland, T. (2020). Drones in manufacturing: Exploring opportunities for research and practice. *Journal of Manufacturing Technology Management*, *31*(6), 1237–1259. DOI: 10.1108/JMTM-03-2019-0099

Magistretti, S., & Dell'Era, C. (2019). Unveiling opportunities afforded by emerging technologies: Evidences from the drone industry. *Technology Analysis and Strategic Management*, *31*(5), 606–623. DOI: 10.1080/09537325.2018.1538497

Mairaj, A., Baba, A. I., & Javaid, A. Y. (2019). Application specific drone simulators: Recent advances and challenges. *Simulation Modelling Practice and Theory, 94*, 100–117. DOI: 10.1016/j.simpat.2019.01.004

Merkert, R., & Bushell, J. (2020). Managing the drone revolution: A systematic literature review into the current use of airborne drones and future strategic directions for their effective control. *Journal of Air Transport Management, 89*, 101929. DOI: 10.1016/j.jairtraman.2020.101929 PMID: 32952321

Mishra, D., Vegni, A. M., Loscrí, V., & Natalizio, E. (2022). Drone networking in the 6G era: A technology overview. *IEEE Communications Standards Magazine, 5*(4), 88–95. DOI: 10.1109/MCOMSTD.0001.2100016

Nayyar, A., Nguyen, B. L., & Nguyen, N. G. (2020). The internet of drone things (IoDT): future envision of smart drones. In *First International Conference on Sustainable Technologies* DOI: 10.1007/978-981-15-0029-9_45

Park, S., Kim, H. T., Lee, S., Joo, H., & Kim, H. (2021). Survey on anti-drone systems: Components, designs, and challenges. *IEEE Access : Practical Innovations, Open Solutions, 9*, 42635–42659. DOI: 10.1109/ACCESS.2021.3065926

Roldán-Gómez, J. J., González-Gironda, E., & Barrientos, A. (2021). A survey on robotic technologies for forest firefighting: Applying drone swarms to improve firefighters' efficiency and safety. *Applied Sciences (Basel, Switzerland), 11*(1), 363. DOI: 10.3390/app11010363 PMID: 36660242

Shahmoradi, J., Talebi, E., Roghanchi, P., & Hassanalian, M. (2020). A comprehensive review of applications of drone technology in the mining industry. *Drones (Basel), 4*(3), 34. DOI: 10.3390/drones4030034

Sharma, V. (2019). Advances in drone communications, state-of-the-art and architectures. *Drones (Basel), 3*(1), 21. DOI: 10.3390/drones3010021

Sibanda, M., Mutanga, O., Chimonyo, V. G., Clulow, A. D., Shoko, C., Mazvimavi, D., Dube, T., & Mabhaudhi, T. (2021). Application of drone technologies in surface water resources monitoring and assessment: A systematic review of progress, challenges, and opportunities in the global south. *Drones (Basel), 5*(3), 84. DOI: 10.3390/drones5030084

Sliusar, N., Filkin, T., Huber-Humer, M., & Ritzkowski, M. (2022). Drone technology in municipal solid waste management and landfilling: A comprehensive review. *Waste Management (New York, N.Y.), 139*, 1–16. DOI: 10.1016/j.wasman.2021.12.006 PMID: 34923184

Tsiamis, N., Efthymiou, L., & Tsagarakis, K. P. (2019). A comparative analysis of the legislation evolution for drone use in OECD countries. *Drones (Basel)*, *3*(4), 75. DOI: 10.3390/drones3040075

Vujičić, M. D., Kennell, J., Stankov, U., Gretzel, U., Vasiljević, Đ. A., & Morrison, A. M. (2022). Keeping up with the drones! Techno-social dimensions of tourist drone videography. *Technology in Society*, *68*, 101838. DOI: 10.1016/j.techsoc.2021.101838

Yaacoub, J. P., Noura, H., Salman, O., & Chehab, A. (2020). Security analysis of drones systems: Attacks, limitations, and recommendations. *Internet of Things : Engineering Cyber Physical Human Systems*, *11*, 100218. DOI: 10.1016/j.iot.2020.100218 PMID: 38620271

Zailani, M. A. H., Sabudin, R. Z. A. R., Rahman, R. A., Saiboon, I. M., Ismail, A., & Mahdy, Z. A. (2020). Drone for medical products transportation in maternal healthcare: A systematic review and framework for future research. *Medicine*, *99*(36), e21967. DOI: 10.1097/MD.0000000000021967 PMID: 32899033

Chapter 2

Artificial General Intelligence-Powered Drones for Climate-Responsive Infrastructure

Muhammad Ameer Hamza
COMSATS University Islamabad, Abbottabad, Pakistan

Saveera Saveera
Rhine-Waal University of Applied Sciences, Kleve, Germany

Muhammad Abbas
University of Wah, Pakistan

Ubaid Ullah
University of Wah, Pakistan

Zaid Muhammad
Abdulwali Khan University, Pakistan

Muhammad Usama
https://orcid.org/0000-0002-6029-8828
Hafencity University, Hamburg, Germany

ABSTRACT

This chapter explores the integration of autonomous, intelligent, and adaptable solutions to address the challenges climate change poses to infrastructure sustainability. Traditional infrastructure maintenance and monitoring approaches are often reactive, costly, and inefficient, particularly in the face of rising global temperatures and environmental unpredictability. The chapter focuses on the potential of drone technology combined with artificial general intelligence (AGI) to create climate-responsive infrastructure solutions. AGI-powered drones can enhance real-time monitoring, predictive maintenance, and climate adaptation strategies, with the ability to evolve and improve over time. This chapter discusses the design, applications, advantages, and limitations of AGI-powered drones for infrastructure

DOI: 10.4018/979-8-3693-6457-4.ch002

management. It also examines how these drones can support and augment traditional human-led infrastructure efforts, creating a hybrid approach that enhances resilience and efficiency.

1.INTRODUCTION

Rising frequency of occurrences connected to climate demands innovative concepts for infrastructure design, maintenance, and monitoring. Among the ongoing climate processes threatening infrastructure stability are rising sea levels, more frequent droughts, and more violent storms (Panarelli, 2015; Upadhyaya et al., n.d.). Although they could ignore new problems, conventional infrastructure management mostly depends on regular inspections: Delayed interventions and insufficient predictive maintenance therefore lead to many infrastructure breakdowns (I. U. Khan et al., 2022; Luu et al., 2025; Ojo, 2024) One amazing approach to improve infrastructure resilience is drone technology paired with artificial intelligence. Unlike traditional drones dependent on human input and preprogramming, AGIs-powered drones function with quite great degree of autonomy (Gardezi et al., 2024; Usama et al., 2023). Perfect for challenging and irregular illnesses, they can evaluate, diagnose, and suggest therapies devoid of personal contact (Carlucci et al., 2024; Dharmarathne et al., 2024; Mhlanga & Dzingirai, 2024) This part discusses artificial intelligence (AGI) drones and its uses in the maintenance of sensitive infrastructure for the environment The growing demand for environmentally friendly infrastructure solutions highlights the need of automation motivated by artificial intelligence to control infrastructure related with climate change (Koranteng et al., 2019; Osman et al., 2025).

2. DRONES: AGI COMPONENT

Beyond current artificial intelligence, artificial intelligence (AGI) offers context-specific adaptability, autonomous decision-making, and self-learning (Nguyen et al., 2011; Seelam et al., 2025) While some artificial intelligence systems concentrate on certain objectives, artificial general intelligence (AGI) enables drones to assess a spectrum of meteorological and infrastructure-related data (I. U. Khan, Abdollahi, et al., 2021; I. U. Khan, Shah, et al., 2021; M. A. Khan et al., 2018) There should be use for this in many other disciplines. Artificial general intelligence (AGI) has the ability to generalise knowledge, unlike rule-based artificial intelligence (which follows pre-programmed instructions), which lets drones adapt to changing surroundings and make decisions free from human control (Irshad et al., 2023; I. U. Khan,

Qureshi, et al., 2020; A. Kumar & Jain, 2021). Particularly infrastructure monitoring rely on this skill as unexpected events and environmental circumstances quickly change the surroundings. Using real-time data, artificial general intelligence (AGI) lets drones maximise their operations, identify patterns, and predict probable infrastructure issues. With minimal human participation, autonomous technologies made feasible by drones and artificial intelligence (AGI) can solve difficult infrastructure challenges. AGI-powered drones would also be able to make better decisions by assessing massive amounts of data from many sources—such as sensor networks, satellite images, and historical climate records. Great instruments for character sustainability and infrastructure resilience are self-improving AGIs driven drones (Alsamhi et al., 2019; Mahar, 2021).

3. DRONE OPERATIONS UNDER ARTIFICIAL GENERAL INTELLIGENCE

Good AGI-powered drone deployment calls a uniform networking design, smart software, and robust hardware. Among other technological instruments of these drones, high-end cameras, LiDAR sensors, thermal imaging, and multispectral sensors help to allow the collecting of massive quantities of data (Cortesão et al., 2020). These sensors allow artificial intelligence (AGI) equipped drones to find structural flaws like microcracks and material degradation unseen to the human eye (Dou et al., 2023). Together defining the software architecture managing real-time obtained data are deep learning algorithms, artificial intelligence-driven analytics systems, and neural networks. AGI-powered drones might explore data locally instead of dependent on far-off cloud processing using edge computing technologies, therefore reducing latency and increasing efficiency. Artificial intelligence powered drone operations largely rely on connection, hence 5G technology offers quick data transmission and a user-friendly interface with smart infrastructure systems. By use of securely distributed data, blockchain technology provides infrastructure evaluations openness and integrity. Furthermore very essential for artificial intelligence-driven drone design is the combination of predictive analytics systems with autonomous navigation. These technologies have let drones fly in difficult surroundings, avoid obstructions, and give areas needing immediate attention high priority. Making decisions based on previous as well as current data considerably improves the effectiveness of infrastructure maintenance and monitoring (St. Clair & Hyde, 2009).

Figure 1. AGI-powered drone architecture

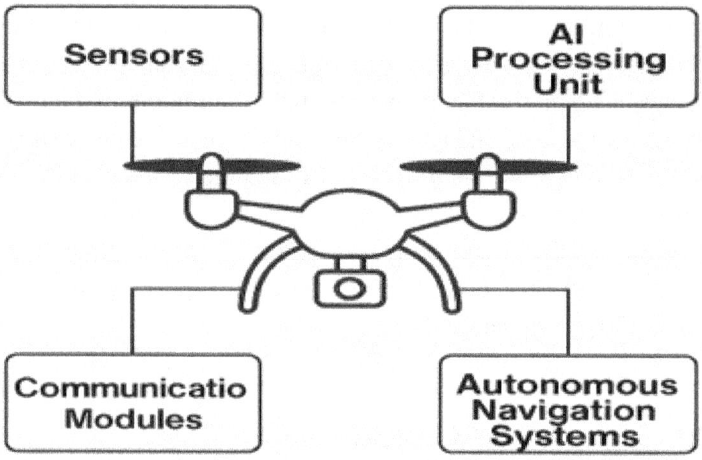

Figure 2. AGI-powered drone systems

4. PROGRAM OF REVERSAL OF INFRASTRUCTURE FOR CLIMATE CHANGE

Drone driven by artificial intelligence have a lot of applications in infrastructure. One of the most crucial applications is structural health monitoring, when drones hunt for evidence of climate-related degradation in highways, bridges, tunnels, and buildings (Henrion et al., 2021; Webber et al., 2022) These drones provide quick repairs and reduce the possibility of catastrophic failures as frequent aerial inspections allow for early wear and tear diagnosis. During floods, earthquakes, and storms artificial intelligence (AGI) drones provide rapid assessment and mitigating reaction capability. By rapidly scanning affected areas, spotting possible hazards, and providing emergency staff with essential data, these drones help to handle disasters (De Schiller et al., 2016). Monitoring the infrastructure connected with solar fields and wind farms, which are components of renewable energy sources, drones are also being employed very widely (J. Neumann, 2009). The efficiency of the energy infrastructure determines sustainability objectives mostly; artificial intelligence (AI) drones improve the dependability and applicability of the systems. Artificial intelligence (AGI) drones monitoring reservoirs, dams, and flood-prone areas might help with water resource management. These drones monitor water levels, avoid silky structures, and check structural integrity of buildings, thereby supporting efficient water management initiatives (S. Kumar & Kanwar, 2022; J. E. Neumann & Price, 2009) Artificial intelligence (AI) equipped drones will optimise urban growth in smart cities and climate change-responsive urban design. To let cities to handle the problems brought on by climate change, they support traffic monitoring, air quality assessment, and the building of green infrastructure (Proag & Proag, 2021; Vatan Kaptan, 2020).

Figure 3. Program of reversal of infrastructure for climate change: Mind map

5.ARTIFICIAL INTELLIGENCE DRIVEN DRONE TECHNOLOGY'S BENEFITS

Artificial intelligence (AGI) drone repairs of infrastructure have various advantages. The primary benefit is higher efficiency as real-time data collecting and processing accelerates infrastructure monitoring reaction times (Lamoureux et al., 2015). By reducing running and personnel expenses, automating maintenance and inspection activities may save a lot of money (Gabrielle Huet, 2020). By lowering human presence in dangerous locations such tall buildings, unstable bridges, and disaster-torn regions, AGI-powered drones also increase safety. Preventive maintenance—what predictive analytics stops before it starts—has several advantages as well. By always learning from fresh data, artificial intelligence (AGI) drones increase forecast accuracy, thereby accelerating and enhancing the efficiency of maintenance chores.

These drones are very important instruments for many infrastructure uses as AGI's adaptability allows it to function in a range of conditions. Their scalability allows AGI-powered drone solutions to be deployed in both urban and rural areas to solve infrastructural issues globally, hence increasing their effect (Kim et al., 2017; La Roche & Liggett, 2001; Yang et al., 2020)

6.ETHICAL CONUNDRUMS AND CONFLICTS

Though they have numerous benefits, artificial intelligence (AGI) drones generate major ethical questions. Technical issues include computer constraints, energy consumption, and data processing bottlenecks greatly restrict their general use (Bikkasani, 2024; EASTAFF & PRIYA, n.d.) Regulatory limitations are especially challenging as responsible implementation depends on respect of privacy rules, cybersecurity policies, and aviation requirements. If we are to guarantee the responsible use of AGI-powered drones, then the ethical questions of artificial intelligence bias, decision-making responsibility, and the environmental effect of drone manufacture must be sufficiently answered (Sonko et al., 2024). Moreover, efficient integration of AGI-powered drones into infrastructure management depends on public confidence and permission. Encouragement of moral AI research and transparent policies including stakeholders could help to tackle these issues.

Figure 4. Conceptual overview of AI-Driven drone technology: benefits and ethical challenges

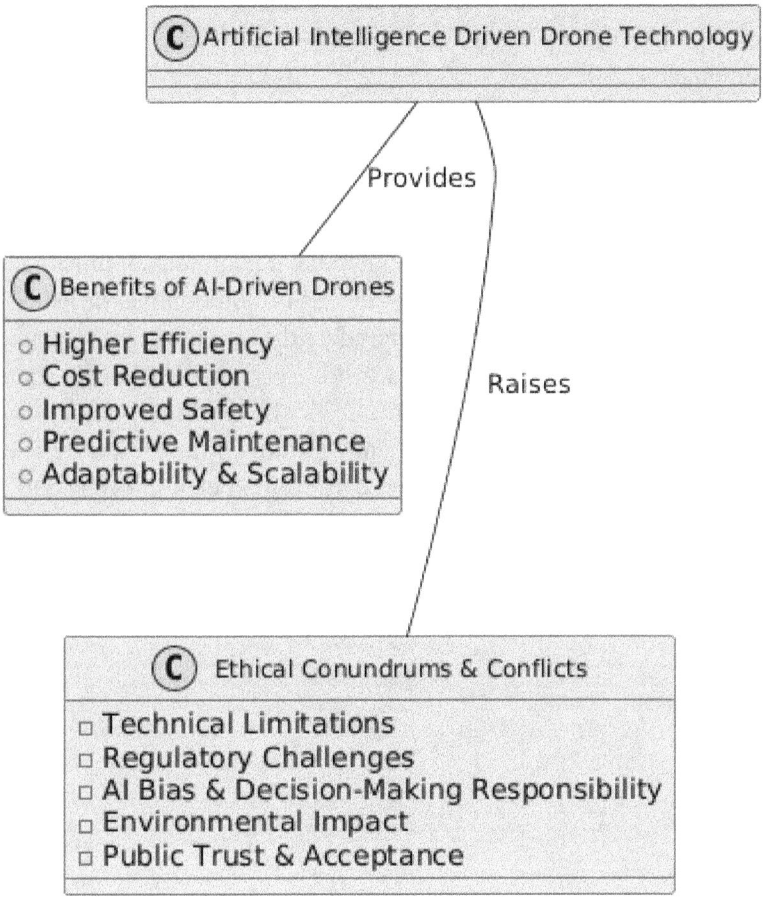

7.THOUGHTS AND FUTURE PROSPECTS

Constant developments in robotics, artificial intelligence, and climate research will define the direction artificial intelligent drones go. To really fulfil their potential, politicians, scholars, and engineers have to go across disciplinary lines (Brandsma et al., 2024; Looman, 2017; Obaid, 2023) Future advancements should mostly be focused on improving artificial intelligence openness, guaranteeing regulatory compliance, and optimising computer speed. Looking at solar-powered drones also helps to address energy consumption problems. Moreover, reducing the threats related to data security depends on investing in artificial intelligence-powered cybersecurity solutions. This part offers doable suggestions for the design, operation, and development of AGI-powered drones for infrastructure sensitive to climate change (Caballero-Martin et al., 2024; I. U. Khan, Nain Zukhraf, et al., 2020; Salmon et al., 2024)

8. CONCLUSION

Artificial intelligence (AGI) drones constitute a major development in climate-sensitive infrastructure management. By means of autonomous decision-making, real-time monitoring, and predictive analytics, these drones greatly enhance the resilience and sustainability of important infrastructure. Although strategic development and ethical issues largely define its effective integration into current infrastructure systems, this is still difficult. Future infrastructure resilience and climate adaption may be much enhanced by developments in drone and artificial intelligence technologies.

REFERENCES

Alsamhi, S. H., Ma, O., Ansari, M. S., & Almalki, F. A. (2019). Survey on collaborative smart drones and internet of things for improving smartness of smart cities. *IEEE Access : Practical Innovations, Open Solutions*, *7*, 128125–128152. DOI: 10.1109/ACCESS.2019.2934998

Bikkasani, D. C. (2024). Navigating artificial general intelligence (AGI): Societal implications, ethical considerations, and governance strategies. *AI and Ethics*, *•••*, 1–16. DOI: 10.1007/s43681-024-00642-z

Brandsma, S., Lenzholzer, S., Carsjens, G. J., Brown, R. D., & Tavares, S. (2024). Implementation of urban climate-responsive design strategies: An international overview. *Journal of Urban Design*, *29*(5), 598–623. DOI: 10.1080/13574809.2024.2314760

Caballero-Martin, D., Lopez-Guede, J. M., Estevez, J., & Graña, M. (2024). Artificial intelligence applied to drone control: A state of the art. *Drones (Basel)*, *8*(7), 296. DOI: 10.3390/drones8070296

Carlucci, F., Campagna, L. M., & Fiorito, F. (2024). *Responsive Envelopes and Climate Change*. Springer. DOI: 10.1007/978-3-031-58101-4

Cortesão, J., Lenzholzer, S., Mülder, J., Klok, L., Jacobs, C., & Kluck, J. (2020). Visual guidelines for climate-responsive urban design. *Sustainable Cities and Society*, *60*, 102245. DOI: 10.1016/j.scs.2020.102245

De Schiller, S., Bentley, I., & Watson, G. B. (2016). Sustainable urban form: environment and climate responsive design. In *Designing Sustainable Cities in the Developing World* (pp. 163–178). Routledge.

Dharmarathne, G., Waduge, A. O., Bogahawaththa, M., Rathnayake, U., & Meddage, D. P. P. (2024). Adapting cities to the surge: A comprehensive review of climate-induced urban flooding. *Results in Engineering*, *22*, 102123. DOI: 10.1016/j.rineng.2024.102123

Dou, F., Ye, J., Yuan, G., Lu, Q., Niu, W., Sun, H., Guan, L., Lu, G., Mai, G., & Liu, N. (2023). Towards artificial general intelligence (agi) in the internet of things (iot): Opportunities and challenges. *ArXiv Preprint ArXiv:2309.07438*.

Eastaff, M. S., & Priya, M. A. S. (n.d.). Artificial General Intelligence (AGI)–Balancing Promise and Peril. Advances in Future Trends and Technology, 186.

Gabrielle Huet, V. (2020). Infrastructure Projects and Climate Change Adaption in the Era of Grassroots Movement Resurgence: Suggestions fro Transformational Actions.

Gardezi, H., Ikrama, M., Usama, M., Iqbal, M., Jalal, F. E., Hussain, A., & Li, X. (2024). Predictive modeling of rutting depth in modified asphalt mixes using gene-expression programming (GEP): A sustainable use of RAP, fly ash, and plastic waste. *Construction & Building Materials*, *443*, 137809. DOI: 10.1016/j.conbuildmat.2024.137809

Henrion, L., Zhang, D., Li, V., & Sick, V. (2021). Built infrastructure renewal and climate change mitigation can both find solutions in CO2. *Frontiers in Sustainability*, *2*, 733133. DOI: 10.3389/frsus.2021.733133

Irshad, M. A., Sattar, S., Nawaz, R., Al-Hussain, S. A., Rizwan, M., Bukhari, A., Waseem, M., Irfan, A., Inam, A., & Zaki, M. E. A. (2023). Enhancing chromium removal and recovery from industrial wastewater using sustainable and efficient nanomaterial: A review. *Ecotoxicology and Environmental Safety*, *263*, 115231. DOI: 10.1016/j.ecoenv.2023.115231 PMID: 37429088

Khan, I. U., Abdollahi, A., Alturki, R., Alshehri, M. D., Ikram, M. A., Alyamani, H. J., & Khan, S. (2021). Intelligent detection system enabled attack probability using Markov chain in aerial networks. *Wireless Communications and Mobile Computing*, *2021*(1), 1–9. DOI: 10.1155/2021/1542657

Khan, I. U., Abdollahi, A., Jamil, A., Baig, B., Aziz, M. A., & Subhan, F. (2022). A Novel Design of FANET Routing Protocol Aided 5G Communication Using IoT. J. *Mobile Multimedia*, *18*(5), 1333–1354.

Khan, I. U., Nain Zukhraf, S. Z., Abdollahi, A., Imran, S. A., Qureshi, I. M., Aziz, M. A., & Hussian Shah, S. B. (2020, November). Reinforce based optimization in wireless communication technologies and routing techniques using internet of flying vehicles. In *Proceedings of the 4th International Conference on Future Networks and Distributed Systems* (pp. 1-6).

Khan, I. U., Qureshi, I. M., Aziz, M. A., Cheema, T. A., & Shah, S. B. H. (2020). Smart IoT control-based nature inspired energy efficient routing protocol for flying ad hoc network (FANET). *IEEE Access : Practical Innovations, Open Solutions*, *8*, 56371–56378. DOI: 10.1109/ACCESS.2020.2981531

Khan, I. U., Shah, S. B. H., Wang, L., Aziz, M. A., Stephan, T., & Kumar, N. (2021). Routing protocols & unmanned aerial vehicles autonomous localization in flying networks. *International Journal of Communication Systems*, •••, 4885.

Khan, M. A., Alvi, B. A., Safi, A., & Khan, I. U. (2018). Drones for good in smart cities: a review. *Proc. Int. Conf. Elect., Electron., Comput., Commun., Mech. Comput.(EECCMC)*, 1–6.

Kim, Y., Eisenberg, D. A., Bondank, E. N., Chester, M. V., Mascaro, G., & Underwood, B. S. (2017). Fail-safe and safe-to-fail adaptation: Decision-making for urban flooding under climate change. *Climatic Change, 145*(3-4), 397–412. DOI: 10.1007/s10584-017-2090-1

Koranteng, C., Simons, B., & Essel, C. (2019). Climate responsive buildings: A comfort assessment of buildings on KNUST campus, Kumasi. *Journal of Engineering. Defence Technology, 17*(5), 862–877.

Kumar, A., & Jain, S. (2021). Drone-based monitoring and redirecting system. In *Development and Future of Internet of Drones (IoD): Insights, Trends and Road Ahead* (pp. 163–183). Springer International Publishing.

Kumar, S., & Kanwar, S. R. (2022). Community Participatory Approach to Reverse Climate Change.

La Roche, P., & Liggett, R. (2001). Very simple design tools: A web based assistant for the design of climate responsive buildings. *Architectural Science Review, 44*(4), 437–448. DOI: 10.1080/00038628.2001.9696924

Lamoureux, S., Forbes, D. L., Bell, T., Manson, G., Rudy, A. C. A., Lalonde, J., & Fraser, P. R. (2015). In Stern, G. A., & Gaden, A. (Eds.), *The impact of climate change on infrastructure in the western and central Canadian Arctic. From Science to Policy in the Western and Central Canadian Arctic: An Integrated Regional Impact Study (IRIS) of climate change and modernization* (pp. 301–341). ArcticNet.

Looman, R. (2017). Climate-responsive design: A framework for an energy concept design-decision support tool for architects using principles of climate-responsive design. A+ BE| Architecture and the Built Environment, (1), 1-282.

Luu, T. M. N., Yadav, M., Srivastava, A. P., & Gopal, K. (2025). Innovation and Technology to Address the Challenges of Climate Change. In *Effects of Climate Change on Social and Economic Factors* (pp. 423–464). IGI Global.

Mahar, W. A. (2021). *Methodology for the design of climate-responsive houses for improved thermal comfort in cold semi-arid climates.* Universite de Liege.

Mhlanga, D., & Dzingirai, M. (2024). Smart technologies in agriculture: Advancements and benefits for sustainable development. In *Responsible Business and Sustainable Development* (pp. 57–72). Routledge.

Neumann, J. (2009). Adaptation to climate change: Revisiting infrastructure norms. *Issue Brief*, •••, 9–15.

Neumann, J. E., & Price, J. C. (2009). Adapting to Climate Change: The Public Policy Response–Public Infrastructure.

Nguyen, A.-T., Tran, Q.-B., Tran, D.-Q., & Reiter, S. (2011). An investigation on climate responsive design strategies of vernacular housing in Vietnam. *Building and Environment*, *46*(10), 2088–2106. DOI: 10.1016/j.buildenv.2011.04.019

Obaid, O. I. (2023). From machine learning to artificial general intelligence: A roadmap and implications. *Mesopotamian Journal of Big Data*, *2023*, 81–91. DOI: 10.58496/MJBD/2023/012

Ojo, B. (2024). Strategies for the optimization of critical infrastructure projects to enhance urban resilience to climate change. *The Journal of Scientific and Engineering Research*, *11*, 107–123.

Osman, O., Yehya, A., & Maalouf, E. (2025). New perspective on designing climate-responsive latent energy storage systems. *Renewable & Sustainable Energy Reviews*, *211*, 115286. DOI: 10.1016/j.rser.2024.115286

Panarelli, G. (2015). Climate responsive design. Research, strategies and assessment criteria. Technological innovation and Climate Responsive Architecture Approach in the project. *Housing Policies and Urban Economics*, *3*, 92–94.

Proag, V., & Proag, V. (2021). Climate change and infrastructure. Infrastructure Planning and Management: An Integrated Approach, 279-304.

Salmon, P. M., McLean, S., Carden, T., King, B. J., Thompson, J., Baber, C., Stanton, N. A., & Read, G. J. M. (2024). When tomorrow comes: A prospective risk assessment of a future artificial general intelligence-based uncrewed combat aerial vehicle system. *Applied Ergonomics*, *117*, 104245. DOI: 10.1016/j.apergo.2024.104245 PMID: 38320386

Seelam, D. R., Kidiyur, M. D., Whig, P., Gupta, S. K., & Balantrapu, S. S. (2025). Integrating Artificial Intelligence in Blue-Green Infrastructure: Enhancing Sustainability and Resilience. In Integrating Blue-Green Infrastructure Into Urban Development (pp. 347-372). IGI Global Scientific Publishing.

Sonko, S., Adewusi, A. O., Obi, O. C., Onwusinkwue, S., & Atadoga, A. (2024). A critical review towards artificial general intelligence: Challenges, ethical considerations, and the path forward. *World Journal of Advanced Research and Reviews*, *21*(3), 1262–1268. DOI: 10.30574/wjarr.2024.21.3.0817

St. Clair, P., & Hyde, R. (2009). Towards a new model for climate responsive design at the university of the sunshine coast chancellery. *Journal of Green Building*, *4*(3), 3–20. DOI: 10.3992/jgb.4.3.3

Upadhyaya, S., Gyawali, P., Shrestha, S., & Bhandari, S. (n.d.). *KC, S.* The Fundamental Role of GNSS in Modern Surveying and Mapping to Support Climate Responsive Land Governance and to Enhance Disaster Resilience.

Usama, M. M., Gardezi, H., Jalal, M. F. E., Rehman, M. M. A., Javed, M. N., Janjua, D. S., & Iqbal, M. M. (2023). Predictive Modelling of Compression Strength of Waste Gp/Fa Blended Expansive Soils Using Multi-Expression Programming. *Construction & Building Materials*, *392*(May), 131956. DOI: 10.1016/j.conbuildmat.2023.131956

Vatan Kaptan, M. (2020). Climate-responsive design strategy for Erbil city. *Archnet-IJAR: International Journal of Architectural Research*, *14*(1), 90–111. DOI: 10.1108/ARCH-04-2019-0089

Webber, S., Nelson, S., Millington, N., Bryant, G., & Bigger, P. (2022). Financing reparative climate infrastructures: Capital switching, repair, and decommodification. *Antipode*, *54*(3), 934–958. DOI: 10.1111/anti.12806

Yang, L., Fu, R., He, W., He, Q., & Liu, Y. (2020). Adaptive thermal comfort and climate responsive building design strategies in dry–hot and dry–cold areas: Case study in Turpan, China. *Energy and Building*, *209*, 109678. DOI: 10.1016/j.enbuild.2019.109678

Chapter 3
Role of Drones in Climate Change Mitigation and Adaptation

Sameer Saharan
https://orcid.org/0000-0002-7487-6297
Mody University of Science and Technology, India

Meha Khiria
Mody University of Science and Technology, India

Tilak Ram Chandrakar
Mody University of Science and Technology, India

Mridu Sharma
Mody University of Science and Technology, India

Jitendra Binwal
Mody University of Science and Technology, India

ABSTRACT

In this chapter, important benefits of drone technology have been discussed which can significantly increase climate flexibility. Unmatched capacity is provided by drones in areas of risk assessment, environmental monitoring, climate concerns and quick action. Their high height, advanced sensor and favorable viewing places are the reason for this. The use of drone provides important information which is necessary to make initial intervention program construction and informed decisions. Two examples of applications under this category include tracking forest cutting and altering sea-mounting level. Adopting drone technology can help in many ways,

DOI: 10.4018/979-8-3693-6457-4.ch003

such as increasing community adaptability, reducing the effects of climate change and advancing continuous development. The purpose of this chapter is to provide an in-depth analysis of potential shortcomings, advantages and useful applications of drones as necessary tools in the development of flexible civilizations towards climate change.

INTRODUCTION

The impact of climate change has worsened over the last few decades, thereby producing unprecedented challenges for enterprises, civilizations and ecosystems around the world. Many different and comprehensive effects of climate change including increasing temperatures, more consecutive and serious weather events, etc. Such challenges need to be dealt with only through a wide-ranging approach combining scientific knowledge, technical development and accelerated legislative effort. In this particular context, drone technology emerged as a particularly promising way to enhance climate flexibility (Hasna, 2009; Saharan, Bhandari, et al., 2024).

UAV, also known as drones, are being used in all fields such as agriculture, transportation and in disaster relief. Drones can be deployed at isolated or inaccessible places where logistics borders prevent them to be used for environmental monitoring or provide disaster relief (Saharan, Bhandari, et al., 2024).

The main perspective of this chapter is to consider if drones can assist enhance climate easy flexibleness. The study explores the number of uses, benefits and limits of the drone technology in order to assess how it can be minimized in terms of climate change.

The study will look into this specifically:

a. Environmental changes tracking using drones: This comes about cheaply and effectively in collecting data pertaining to different parameters of the environment like vegetable cover, humidity, temperature and land use. The motility of the ecosystem and the consequent spot pattern of change induced by climate change can be studied by use of drones for remote sensing and aerial surveys.
b. Uses of drones in disaster preparedness and response: Drone has been useful in disaster management using its capability to take quick assessments of damage, perform search and rescue operations and deliver human supply in an emergency setting. Emergency reactors, including storm, floods and forest fires, can be better prepared for with the help of drone technology.
c. Drones in climate adaptation measures: There is a need to be flexible to climate change and take active steps to predict changing climate conditions and adapt to them. To accomplish these efforts, drones can supply fast suitable data for

taking a chance assessment and develop the infrastructure and make the advancement plan better. The use of drone-based data can further improve adaptation program efficiency by supporting policy makers to integrate drone-based data into climate model and decision support systems, and to customize resource allocations.

Drones have a considerable potential to build up climate flexibility, but are also limited by certain obstacles and restrictions of being used. One of the significant barriers, however, is the regulatory restriction which, though varying from jurisdiction to jurisdiction, and may terminate the use of drones to undertake specific locations or functions (Gupta et al., 2021; Kumar et al., 2022; Shakhatreh et al., 2019). Data security and airfield control measures together with privacy concerns need attention in order to ensure responsible drone operations.

Despite the challenges, drone technology with progress ongoing, supporting legislative environment and participation of stakeholders, provides the hope of overcoming barriers and adapting the benefits of drone's use for climate flexibility (Ali et al., 2022; Tarr et al., 2021). If communities use drones for air overview they may be better equipped to adapt to climate change, protect vulnerable individuals and protect natural resources for future generations.

THE PHENOMENON OF CLIMATE CHANGE AND ITS CONSEQUENCES

Long-term modification of temperature, rain patterns and other climate elements on Earth's atmosphere, oceans and land surfaces are called climate change (Kumar et al., 2021). This is mostly due to human activities, including fossil fuel burning, forest cutting and industrial operations that leave greenhouse gases (GHG) such as carbon dioxide (CO_2), methane (CH_4), and nitrous oxide (N_2O). Earth's climate system has been disturbed and due to these emissions, heat trapping in the atmosphere is getting global warming (Ahmed et al., 2023; Noor et al., 2020).

Global economies, community and ecosystems are affected by all the wider and diverse impacts of climate change. The effects of climate change are raising pre-existing weaknesses and presenting already unheard difficulties for human wellness and environmental stability. These problems range from ice melting and sea rising levels to more consecutive and intense extreme weather events (Paterson, 2020; Semieniuk et al., 2021).

Overview able trends point to the planet's heating and changing patterns of weather during the last century, indicating significant changes in the Earth's climate. Important signs of the changing climate are (Figure 1) (Gulev et al., 2021; Ripple et al., 2020; Scotese et al., 2021; Skendžić et al., 2021):

1. *Average global temperature rise:* Since the late 19th century, the average global temperature is constantly rising, over the past several decades a trend of quick warming has been seen. Melting polar ice peaks, retarding glaciers and changing patterns of rainfall, these are all results of this temperature rise.
2. *Sea level increase:* Glacier and ice sheets are melting due to rising temperature, thereby increasing sea level. Increasing floods and storms threaten coastal population and infrastructure due to sea-rising levels, thereby increasing coastal erosions and submerging lower terrain.
3. *Extreme weather events:* Heatwaves, hurricanes, droughts and heavy rain are among extreme weather events that are becoming constantly and serious due to climate change. These events are likely to cause destructive losses to infrastructure, public health, agriculture and water supply, as well as economic losses and population displacement.
4. *Ocean acidification:* The acidification of oceans is a result of absorbing too much CO_2 by the seas, making marine ecosystems and biodiversity endanger. Acidification diminishes the growth and survival of coral reefs, shellfish, and other marine organisms, resulting in substantial economic repercussions for fisheries and coastal industries.

Figure 1. Signs of climate change

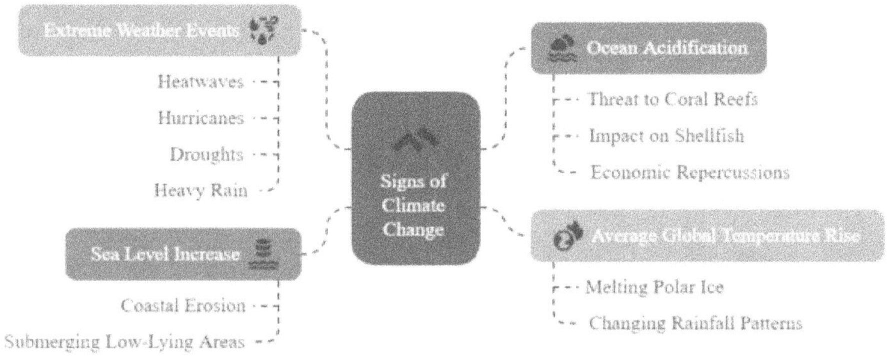

Environmental effects of climate change comprise several vast patterns that create varied levels of complexity for land-based ecosystems as well as maritime systems and environmental atmosphere systems. Climate change imposes numerous significant repercussions on the environment (Figure 2) (d'Amato et al., 2020; Halofsky et al., 2020; Rocque et al., 2021; Weiskopf et al., 2020; Wu et al., 2020):

1. *Habitat loss and biodiversity decline:* Variations in environmental temperature and rainfall modify natural habitats and ecosystems, resulting in shifted species distributions, diminished biodiversity, and increased extinction risk for more vulnerable species. Coral bleaching together with degradation and desertification has brought housing loss to its most severe point.
2. *Disruption of ecosystem services:* Pollinations, carbon separations and water purification are examples of ecosystem services that can be negatively affected by climate change and are important for both health and human wellness. The ecosystems that are in worse condition are less capable of maintaining agriculture, controlling temperature and supplying basic materials needed for rust.
3. *Changes in hydrological cycles:* The distribution and access of freshwater resources is affected by changes in hydrology cycles, due to the melting of converted rainfall pattern and glacier. Changes in rainfall pattern can increase resource rivalry and water deficiency due to drought in some areas and others.
4. *Increased wildfire activity:* In many areas of the world, due to factors such as long dry, hot temperature and modification of vegetable mobility increased the frequency and intensity of forest fire. In addition to destroyed the housing of

forests and wildlife, forest fire intensifies climate change by releasing abundant CO_2 in the sky.

Figure 2. Climate change impact on ecosystem

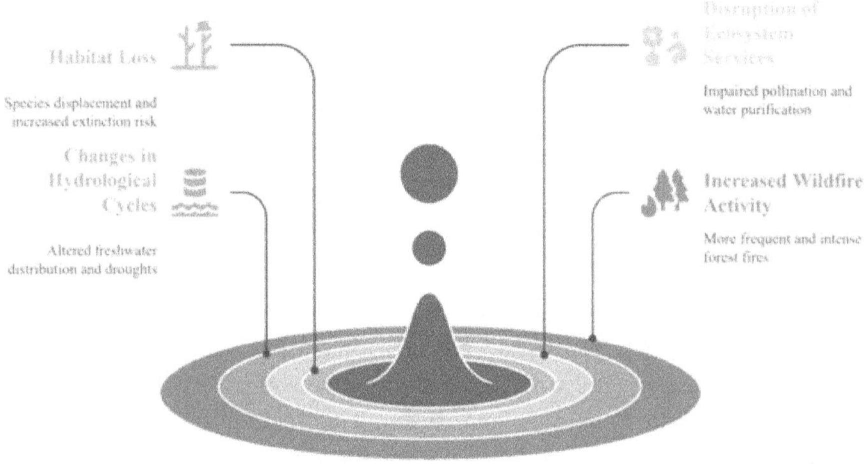

THE UTILIZATION OF UNMANNED AERIAL VEHICLES AND THEIR PRACTICAL USES

The drone, also known as unmanned aerial vehicle (UAV), is a major aviation technical progress. Unlike traditional manned planes, the drone can be controlled remotely without the presence of a human pilot (Saharan, Bhandari, et al., 2024). Due to this difference, drone can be used as adaptable devices in a wide range of industries including filmmaking, emergency reaction, environmental monitoring and agriculture (Figure 3). Drones are available in various sizes, forms and combinations, each of which is suitable for a set of special use case and operational specifications. The main criteria used to classify these devices are sizes, limits, payload capabilities and capabilities. Of the smallest and lightest drones, nano drones are often used for indoor or close-quarter outdoor operation. For recreational purposes, they are popular among enthusiasts and amateurs and usually operate through remote control or smartphone apps. As someone moves on a scale, the micro drones become slightly larger and have better facilities, including a more stable flight duration. Micro drones are still so small that they can be used for various business purposes, such as aerial photography, monitoring and inspection. They are important assets in areas including real estate, construction and infrastructure inspection due to their adaptability and agility. Small drones in terms of size and power symbol a big progress. Small drones are usually used for business applications such as precision agriculture, environmental monitoring and infrastructure inspection due to their high payload capacity and refined sensors. Thermal imaging techniques, equipped with multispectral sensors and high-resolution cameras provide invaluable data for small drones, field mapping, crop monitoring and disaster response. With its large payload and long flight range, medium drones further enhance the operating capabilities of unmanned aerial vehicles. Such drone is used for more sophisticated functions including delivery services, disaster relief and search and rescue. They are the resources needed for disaster management organisations, logistics firms and first respondents due to the transportation of large goods and long-distance work ability. Large drones exist at the highest range of drones while offering heavy loads transportation over extended ranges. The army normally utilizes these drones to deliver cargo along with surveillance and reconfirmation missions. The rarity of large drones in civil applications due to their complexity allows them to transform industrial sectors including infrastructure development and logistics and transportation (Dileep et al., 2020; Raya & Esteve, 2024; Saharan, Bhandari, et al., 2024; Samadzadegan et al., 2022; Subramanian et al., 2021).

Several drone categories exist with different functional and special use specifications. Unmanned aerial vehicles cover both the recreational sphere of small drones and high-end military drones thus establishing a broad innovative technology segment which is expanding rapidly.

Figure 3. Drone categories and application

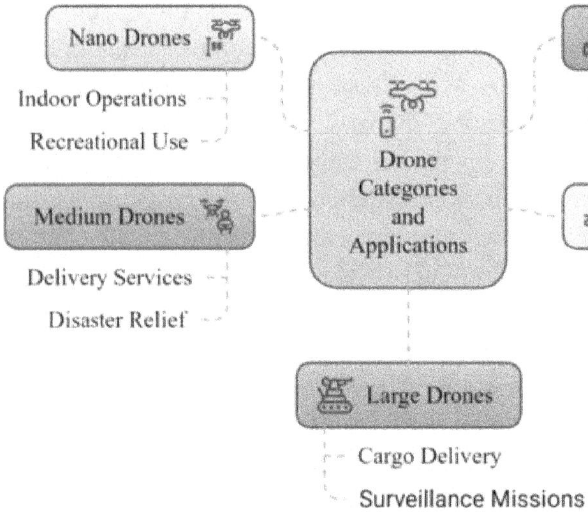

The Present Applications of Drones

In recent years, drones' use has been seen rapid increase in following factors including advanced technology, falling costs and increasing applications. They are essential tools in a variety of industries, each of which has its own set of opportunities and difficulties due to its adaptability, efficiency and cost-effectivity. Air photography and film production is one of the most popular uses of drones. Drones with stable gimbal and high-resolution cameras are used in addition to prepare beautiful air footages for real estate, advertising, movies and tourism promotions, in addition to other uses. With the simpleness and accuracy we can now take beautiful aerial photos, it has completely changed the way we document and present the world around us (Díaz-Báñez & Fabila-Monroy, 2024; Serafinelli & O'Hagan, 2024).

Crop management and precision agriculture are increasingly benefiting from the usage of drones. Drones equipped with GPS, thermal sensors, and multispectral cameras can provide real-time data on pest infestations, soil moisture levels, and crop health. This data enables targeted interventions for the customisation of agricultural practices, optimization of crop yields, and assessment of environmental effect (Ahmed et al., 2023; Dileep et al., 2020; Skendžić et al., 2021). Examples of such interventions include precise irrigation and application of sprays.

They are extensively utilized for the examination of infrastructure, particularly in the transportation, energy, and construction sectors. LiDAR sensors with high-resolution cameras can achieve unparalleled accuracy and efficiency in assessing the structural integrity of buildings, pipelines, electrical lines, and bridges (Akib Uz Zaman et al., 2024; Rodríguez et al., 2024; Wang et al., 2024). The utilization of drones enhances the safety and efficiency of infrastructure maintenance by obviating the necessity for expensive and time-consuming human inspections.

Drones are being used in many research and protection projects around the world, which is another way that drone technology is helping environmental monitoring. Drone researchers provide important data about ecosystem and biodiversity, from monitoring forest degradation and housing degradation to track wildlife population. Their ability to reach inaccessible places makes them important resources to research and protect fragile ecosystems, advance our knowledge about the natural world and guide conservation initiatives (Aliloo et al., 2024; Niazy).

Drone technology aids environmental monitoring by being utilized in several research and conservation projects globally. The surveillance of forest and housing setbacks, together with the monitoring of wildlife populations, provides drone researchers with essential data regarding ecosystems and biodiversity. They serve as vital resources for researching and safeguarding vulnerable ecosystems, enhancing our understanding of the planet, and directing conservation efforts due to their capacity to penetrate remote locations (Bashir et al., 2024).

Drone is used for many different purposes; each person uses their special skills to take advantage of some occasions and obstacles. Drone has changed the way we do our work, living and interacting with the world around us. Their applications include precise agriculture to air photography, infrastructure inspection to environmental monitoring, emergency response to disaster management and more. Different types of businesses have many advantages of drones, but to reach their full potential, there are many limits and difficulties that need to be solved.

Cost-effectivity is one of the main advantages of drone technology. For air mapping, data collection and monitoring, drones are more cost-effective options than traditional manned planes (Famili et al., 2024). Their ability to cover huge areas quickly and reach dangerous or isolated places makes them suitable for jobs where human flights will be very expensive or inevitable. Drone efficiency also provide significant advantage (Askarzadeh et al., 2024). Drones can now work for a long time due to progress in battery technology and flight control technologies, collect data and work with a little bit of human aid. Drones are valuable assets in businesses like infrastructure inspection, construction and agriculture due to its efficiency, resulting in high productivity and low operating costs. Safety is a major factor in many uses of drone technology (Saharan & Yadav, 2022). Drones reduce the possibility of dangers and accidents or injuries for human operators by performing activities in dangerous or inaccessible areas. Drones provide important assistance to first respondents in areas such as search and rescue and disaster response, allowing them to quickly analyse conditions and allocate resources more efficiently. Another important advantage of drone technology is data accuracy. Drones with sophisticated sensors and imaging systems are capable of collecting high-resolution data accurately and accurately, providing practical information to make decisions. Drone data collection facilitates better understanding of complex systems and informs more effective responses to researchers and physicians from environment mapping to agricultural monitoring (Caruso et al., 2021; Holden et al., 2021; Panday et al., 2020).

Nevertheless, to fully realise the potential of drone technology, there must be overtaking many borders and difficulties. Legislative limits are an important barrier. To protect privacy, security and security, aviation authorities have implemented restrictions and rules on drone operations (Saharan, Singh, Bhandari, et al., 2024; Saharan, Singh, Bora, et al., 2024). In places with high population density or sensitive airfield, drone pilots may face difficulties in following rules. The wider application of drone technology is banned by the knowledge and resources needed to navigate the complex and changing regulatory environment, which could be out of reach for many users. The drone's banned payload capacity is an additional limit. Practical restrictions still apply to the types of sensors and devices that drones can carry on a ship, even if the technical progress has made it possible for them to carry heavy payloads (Kumar, 2020; Watkins et al., 2020; Yaacoub et al., 2020). This limit limits types of applications for which drones can be used and forcing operators to attain some features more importance than others when choosing a drone system. Drone technology is also quite limited due to battery life and flight stamina. Rechargeable batteries with a limited flight duration are usually used to power the drone, which can limit their stamina and operational range. To maximize the effectiveness of drone operation, operators may need to plan a careful mission in the light of this limit and customize the flight trajectory path (Alyassi et al., 2022).

Finally, drone technology can be unsafe due to dependency on GPS and communication network. For navigation and control, drones depend on GPS signals and wireless communication networks, which leaves them open for intervention, signal loss and cyber attacks (Saharan, Bhandari, et al., 2024; Saharan, Singh, Bhandari, et al., 2024; Saharan, Singh, Bora, et al., 2024; Saharan & Yadav, 2022). Blowing drones in those situations can be unsafe when GPS signals are jam or communication network are interrupted, thereby making the operator and viewers both at risk.

The Use of Drones in Climate Change

Climate change resulted in creative methods of global ecosystems, businesses and communities facing huge challenges, which require creative methods of tracking, understanding and reducing its effects. Unmanned aerial vehicles (UAVs), commonly called drones, have become inevitable tools in the studies and observation of climate change due to its ability to collect precise data in various settings. This article examines the various ways from which drones are contributing to climate change. It focuses on collecting data and evaluating remote sensing, tracking and mapping signs of climate change and evaluating greenhouse gas emissions and atmospheric conditions (Ayamga et al., 2021; Camarretta et al., 2020; Pathak et al., 2020; Watkins et al., 2020; Wich et al., 2021).

1. Remote Sensing and Data Collection:

   Research on climate change necessitates data collecting and remote sensing, hence requiring drones outfitted with sophisticated sensors and cameras. These provide unmanned aerial vehicle researchers with critical insights into ecosystem movement and environmental changes by supplying data on several environmental parameters, including temperature, humidity, vegetation cover, and land usage. Drones provide flexibility, cost-effectiveness, and the capability to access hazardous or inaccessible locations that should not be approached by alternative methods, such as manned aircraft or satellite images. The capability to collect data in real time renders drone remote sensing an invaluable tool, facilitating prompt responses to events and environmental changes. Drones are utilized for damage assessment in disaster-stricken regions, overseeing recovery efforts, and providing situational awareness to emergency management agencies and first responders. Monitoring and documenting alterations in the ecology, landscape, and natural resources is essential for climate change studies and for devising adaptation strategies utilizing drones (Kucharczyk & Hugenholtz, 2021).
2. Monitoring and Mapping of Climate Change Indicators:

   Drones play a crucial role in monitoring and mapping climate change, including ground surface temperature, sea level rise, and ice cover. Drones in polar regions furnish precise imagery and data concerning glacier motion, sea thickness, alterations in polar ecology, and the impact of our contributions on sea level rise. Drones are employed to accurately assess sensitivity to coastal erosion and sea level rise inside a community, facilitating more effective planning for coastal adaptation and management in such areas. They also assist in mapping vegetable mobility and land use alterations to provide critical data regarding carbon stock and ecosystem health. The high-resolution images and spectral data supplied by drones assist researchers in monitoring habitat loss and fragmentation resulting from human activities, as well as in implementing conservation and climate adaption strategies (Hernández et al., 2021).

3. Atmospheric Conditions and Greenhouse Gas Emissions:

   Drones outfitted with sensors are crucial for understanding the causes and effects of climate change by providing observational observations of air conditions and greenhouse gas emissions. These piloted aerial vehicles assist in monitoring atmospheric conditions and pollution levels by assessing temperature, humidity, greenhouse gases, and air quality. The drone is utilized to assess the health impacts of air pollution, identify pollution hotspots in metropolitan environments, and evaluate air quality. This facilitates the administration of environmental strategies and efforts pertaining to public health. Drone aerial surveys are utilized to identify pollution sources and inform policy responses about traffic patterns, urban areas, and industrial sites. They assist in determining the extent of greenhouse gas emissions from both natural and anthropogenic sources, as well as the contributions of climate mitigation activities. Drones are mitigating the impacts of climate change by gathering data on greenhouse gases such as carbon dioxide and methane. It enhances our understanding of climate change mechanisms (Godfrey & Brenes, 2021; Shaw et al., 2021).

ROLE OF DRONES IN MITIGATING CLIMATE CHANGE

Climate change jeopardizes ecosystems, biodiversity, and human well-being, necessitating innovative and effective measures. Unmanned aerial vehicles (UAVs), or drones, have emerged as rapid and efficient instruments in combating climate change, owing to their capacity to monitor, evaluate, and respond in many environmental contexts. To reduce climate change, this article explains how important drones are for implementation of forestation and recycling initiatives, precise agriculture, durable agriculture practices, wildlife conservation strategies and illegal anti-hunting measures (Figure 4) (Balasundram et al., 2023; Pathak et al., 2020).

i. Implementation of Afforestation and Reforestation Initiatives:

One of the most important ways to remove carbon dioxide from the atmosphere and slow down climate change is to recyclable. Drones are necessary for the implementation and monitoring of forestry and recycling projects as they provide accurate, high-resolution data for planting, site selection and tree growth tracking (Ahmad et al., 2023; Amir, 2024; Mohan, Richardson, et al., 2021; Mohan, Rue, et al., 2021).

a. Site Selection: To find good sites for afforestation and recycled tasks, drones equipped with remote sensing tools are deployed. Their aerial survey evaluate the topography, soil properties and current plant cover, which helps project planners identify areas with a high capacity of tree development and separation of carbon.
b. Planting: When planting tree plants in inaccessible places, drone is used for exact planting. When drone seed distribution equipment are equipped with, they can plant thousands of trees every day with highly accuracy, the time needed for extensive reinforcement projects and cutting labor.
c. Monitoring: Drones make it possible to track the growth of trees in real time, health and survival rate after the plant. Project managers can evaluate the success of forestification and renewal programs and make well informed decisions for future intervention using the high-resolution footage taken by drone, which provides significant insights in canopy cover, biomass buildup and vegetable mobility.
    ii. Adoption of Precision Agriculture Techniques and Sustainable Farming Methods:

Climate-flexible food systems, low greenhouse gas emissions, and increased agricultural production depend on all precise agricultural technologies and durable agricultural practices. Drone is an important component in helping farmers implement these practices as they provide data-operated solutions and practical insights for environmental management and crop management (Balasundram et al., 2023; Júnior et al., 2024).

a. Crop Monitoring: Identification of pests, estimates of yield and crop monitoring all work from drones in which multispectral cameras and sensors are engaged. Drones provide farmers with real time access to the high-resolution imagery of agricultural areas, helping them detect pest infection, nutrient deficiency and crop stress. It allows focused intervention and optimal resource use.
b. Precision Spraying: Drones that spraying accurate can manage vegetables, fertilizers and pesticides with unmatched accuracy while reducing environmental effects and chemical use. Drone help farmers get the best possible insect

management by precisely targeting the treatment areas based on crop health indicators and pest dissemination, reducing chemical fluid and pollution.

c. Soil Mapping: Using drone for soil mapping and analysis, farmers can get important information of their soil health, moisture content and nutrients. Drone is used to survey agricultural areas by air. It creates soil high-resolution map that is used to guide precise agricultural technologies, including variable rate irrigation and fertilizers, resource allocation and environmental decline.

iii. Implementation of Wildlife Conservation Strategies and Anti-Poaching Measures:

Maintaining ecosystem services, biodiversity and ecological flexibility in the event of climate change all depend on wildlife conservation. Due to its ability to provide real-time monitoring, enforcement and monitoring in protected areas and wildlife housing, drones are the necessary tools to execute wildlife protection policies and anti-hunting strategy (Wich et al., 2021; Wich et al.).

a. Anti-Poaching Patrols: Air patrol involves surveying the suspected areas of illegal hunting and wildlife trafficking via drones. Drone rangers and conservationists employ thermal imaging sensors and high-resolution cameras to analyze animal populations, identify prospective hunting targets, and promptly address intrusions. It aids in the conservation of endangered species and enhances the efficacy of anti-poaching initiatives.
b. Habitat Monitoring: Drones enable researchers to examine animal habitats and ecology from an aerial perspective, offering insights regarding accommodation quality, species distribution, and ecosystem health. High-resolution photography generated by drones facilitates the identification of housing loss, fragmentation, and decline attributable to human activity. It is employed to implement conservation strategies and rehabilitate habitats to mitigate the effects of climate change on biodiversity.
c. Environmental Education and Outreach: In Environmental Education and Outreach initiative, the stakeholders and local communities are incorporated in the safety of wildlife through the use of Drone. By emphasizing biodiversity conservation and ecosystem services and the value of durable land management technologies through photography and videography, drones gain much more support from the public, and a greater amount of participation in protection programs.

Figure 4. Drones in climate mitigation cycle

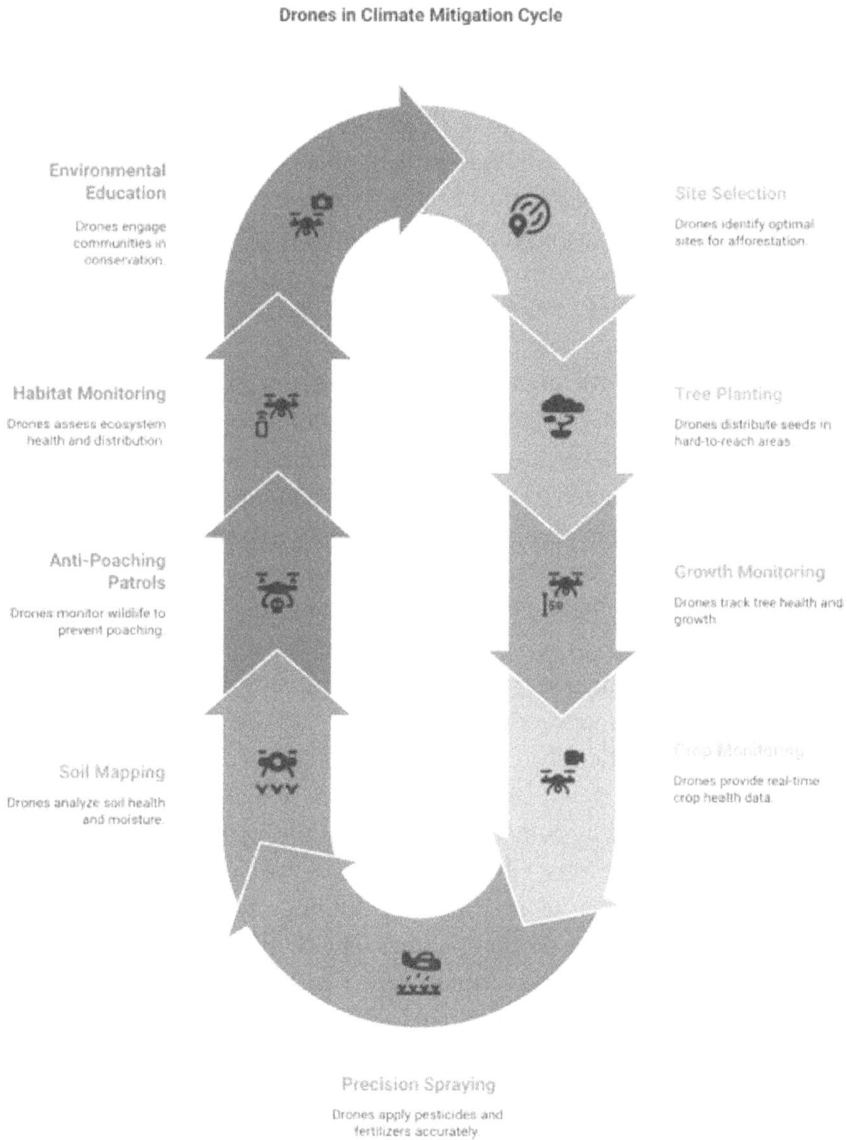

ROLE OF DRONES IN CLIMATE CHANGE ADAPTATION

This will mitigate risks, enhance flexibility, and safeguard the ecosystem and populace from the effects of climate change adaptation requirements. Drones, or unmanned aerial vehicles (UAVs), are exceptional tools for remote sensing, facilitating real-time data collecting and enabling rapid responses, so significantly enhancing the customisation process. This essay will examine three primary aspects of the significance of drones in adapting to climate change: infrastructure monitoring and maintenance, coastal field management, and erosion control (Hernández et al., 2021).

i. Disaster Response and Emergency Management:

Drones are necessary for emergency management and disaster response as they can quickly take air tog, provide conditional awareness, and assist in search and rescue operations during earthquakes, storms, floods and forest fires. Drone is used to make quick survey of areas affected by air after a disaster, giving emergency respondents information in real time about damage levels, disruption in infrastructure and human needs. Drones can search for survivors in dangerous or inaccessible settings, identify risks and evaluate structural integrity due to high-resolution images and thermal imaging sensors. Drones with cameras, thermal sensors and communication equipment can also help search and rescue efforts by detecting people missing in disaster affected areas and conducting air recourse. Drones can scan fast broad areas, detect living in difficultly reaching places, and provide important information to emergency respondents, which improve efficacy and efficiency of rescue operations. The drone is used by authorities to prepare for recovery and reconstruction by prioritising response efforts, allocating resources efficiently and assessing wider damage of important infrastructure including buildings, roads, bridges and utilities. Drones can produce high-resolution photography that is used to assess structural damage, detect safety issues and estimate repair costs. This information is used to affect recovery methods and decision-making (Damaševičius et al., 2023; A. Khan et al., 2022).

ii. Infrastructure Monitoring and Maintenance:

Drone is used for maintenance and monitoring infrastructure, which provides economical, effective and safe means to check power lines, pipelines, bridges, trains and other important infrastructure components. Remote inspection eliminates the requirement of expensive and time-consuming manual inspection using cameras, LiDAR sensors and other imaging technologies to air inspecting infrastructure properties. It allows to identify defects, decline and maintenance needs. Drones can reach dangerous or difficult-to-reach places such as towers, bridges and industrial sites, reducing the risks of workers and reducing operational disruptions. Drones also assist in property management by providing complete, current data on state, functionality and infrastructure properties longevity. Drones asset help owners and operators to maximize longevity of the asset, reduce lifecycle cost and prioritize maintenance functions by collecting and evaluating data on environmental effects, erosions, break-foot and structural integrity. Drones are used to monitor environmental effects of infrastructure projects, evaluate potential effects on nearby ecosystems, water bodies and accommodations. Drones that use multispectral cameras and sensors collect information about vegetable cover, water quality and changes in land use, helping authorities reduce environmental risks, follow laws, and promote continuous development (Akib Uz Zaman et al., 2024; Wang et al., 2024).

iii. Coastal Zone Management and Erosion Control:

The effects of climate change, such as erosion, storm waves and sea-growing levels, are particularly dangerous for coastal areas. Drones are necessary to control the management and erosion of coastal areas as they provide data-powered solutions to plan, arrange and complete customization tasks. Drone is used to check coastal areas by air in coastal monitoring, monitoring coastline changes, erosion rates and coast mobility over time. Drones can provide high-resolution photography that support well informed decisions and adaptation plan by helping to identify erosive hotspots, assessment of sea level increase sensitivity and intervention. Drones also map coastal ecosystems, including coral rocks, salt marshes and mangroves, which provide important information on the limit of ecosystem services, biodiversity and residence (Kandrot et al., 2022). Drone is used on coastal biodiversity to monitor the effects of climate change and human activity, identify protection priorities, and evaluate housing quality by air surveying coastal ecosystems. Drones help in erosion management by allowing air to vegetable and seed sowing in coastal areas. Drones equipped with seed dispersal systems can effectively disseminate native flora and materials to control and stabilize coasts, enhance biodiversity, and alleviate the impacts of coastal infrastructure and erosion on communities (Chapapría et al., 2022; Kandrot & Holloway, 2020).

ISSUES AND ETHICAL DELIBERATIONS

Drones provide many ethical questions and problems that need to be solved as they're joining more and more in different areas and applications. This section explores significant questions and ethical dilemmas associated with the utilization of drones in the fight against climate change.

i. Rules, Regulations, and Privacy Concerns:

The wider use of drones has created laws and policies that govern their application in multiple areas, thereby standing issues of data security and privacy. Regulations raise apprehensions around surveillance and the right to personal privacy, despite incorporating provisions for the safe and secure operation of drones. Privacy concerns arise due to drones' capability to gather data and capture high-resolution images at private properties or public spaces. Drones may be employed to infringe upon private rights and engage in unlawful surveillance. Furthermore, apprehensions regarding data protection and the potential for misuse or unauthorized access to data are rising. To address this issue, it is necessary to balance the advantages of drone technology with the responsibility to safeguard personal rights. The precise rules must be for drone operations in regulations, which includes limits of collecting, storage and exchange of data. Privacy issues require transparency and responsibilities processes to effectively address and ensure regulatory compliance (Saharan, Bhandari, et al., 2024).

ii. Data Protection and Cybersecurity Risks:

In terms of efforts to reduce and adapt to climate change, protecting the data received by drone is an additional ethical matter of excessive importance. Through air surveys, remote sensing and monitoring operations, drones collect huge amounts of data, which increases concerns about data security and cyber attack sensitivity. Drone data collection can provide personal information about communities, infrastructure, ecosystems and accommodations. Unauthorized access or data violations can endanger the availability, privacy and integrity of this information, which may result in a violation of privacy, loss of someone's reputation and potential threat to people and ecosystems. It is important for stakeholders interested to reduce cyber security risks to implement strong data security mechanisms such as encryption, access control and safe data storage processes. To take active measures to detect weaknesses and combat any threats, regular cybersecurity audits and assessments are important. Public Awareness and Education Campaigns can also encourage responsible data management and enhance knowledge of effective practices for data security (Saharan, Singh, Bhandari, et al., 2024; Saharan, Singh, Bora, et al., 2024; Saharan & Yadav, 2022).

iii. Moral and Responsible Use of Drones in Climate Change Efforts:

Dealing with climate change the necessary ethical examination of drone's ethical and responsible use includes environmental management, social fairness and justice all. Drones provide state-of-the-art technology to monitor the environment, respond to emergency and protect the ecosystem, but their use should be controlled by ethical standards. Drone has increased ethical concerns as they have the ability to further deteriorate injustice and inequality already present, especially in weak and marginalized groups. Climate change ineffectively impacts marginalized populations, as they lack the tools, knowledge, and resources necessary to adapt or mitigate its effects. Equitable access to drone technology and its advantages is ethically imperative for fostering environmental justice and social inclusion. Furthermore, ethical considerations pertaining to ecological integrity and stability, due to their proximity, exacerbate the impacts of drone operations, which produce diverse repercussions including energy consumption, climatic emissions, and alterations in wildlife behavior. Stakeholders must assess drone operations to ascertain their potential negative environmental impacts, followed by the implementation of policies aimed at mitigating such effects, including the utilization of renewable energy sources, optimization of flight routes, and reduction of noise pollution. Ultimately, addressing the challenge of climate change necessitates the implementation of additional or more robust techniques for the ethical and responsible use of drones, alongside the integration of ethical principles into the formulation of laws, technology, and decision-making processes. Drone technology provides a transformative method for stakeholders to combat climate change, safeguard human rights, and uphold environmental values, grounded in ethical principles such as transparency, accountability, equality, and stability (Enemark, 2020; Wang et al., 2021).

FORECASTS AND PROBABLE USE FOR THE FUTURE

As drones continue to acquire advanced skills in artificial intelligence and machine learning, alongside improvements in inspection efficiency, they can significantly influence climate change solutions. Numerous stakeholders are appropriate for the integration of drone technology into the collaboration, as the technology is already available. This section examines the future of drones in addressing climate change, focusing on the advancement of drone technology, the incorporation of artificial intelligence and machine learning, and collaborative efforts in climate change solutions, among other factors (Ayamga et al., 2021; Johnson, 2020; M. A. Khan et al., 2022; Kumar et al., 2022; Mohamed et al., 2020).

i. Advancements in Drone Technology

While drones are not currently employed to combat climate change, advancements in drone technology, encompassing enhanced functionality, performance, and design, will influence their potential role in addressing climate change challenges. These advancements have catalyzed innovation across various industries, thereby positioning drone ecosystem preservation and disaster relief as notable contributions to environmental monitoring (Gupta et al., 2021; Johnson, 2020; M. A. Khan et al., 2022; Mohamed et al., 2020; Saharan, Singh, Bhandari, et al., 2024).

a. Miniaturization and Payload Capacity: Drones are becoming mild, more compact and more walkable due to development in battery technology and component minification. Due to increased payload capacity, drones can now take more sensors, cameras and equipment to collect information on temperature, humidity and greenhouse gas concentrations – all of these are the major markers of climate change.
b. Longer Flight Times and Range: Improvements in battery technology and energy economy have allowed drones to traverse vast areas and function for extended periods without recharging. It facilitates more effective air quality monitoring, distant ecological observation, and expedited responses to environmental emergencies.
c. Autonomous Operation and Navigation: Due to the increasing implementation of autonomous navigation systems and advanced rescue technologies, drones are now capable of functioning in a rapid, intelligent, dynamic, and complicated environment. The usefulness of autonomous drones in climate change-related jobs enables them to plan, perform missions, traverse obstacles, and respond to changes.
d. Integration of Sensing Technologies: Ongoing efforts are being undertaken to integrate sophisticated sensors, cameras, and instruments inside the drone for data storage and remote sensing functions. Drones equipped with multispectral, hyperspectral, and thermal imaging sensors can capture high-resolution photos with spectral data, offering essential insights into environmental dynamics, land use changes, and ecosystem health.

ii. Incorporation of Artificial Intelligence and Machine Learning:

The integration of artificial intelligence (AI) and machine learning (ML) algorithms into existing drone systems is steering climate change solutions in a novel path. Thanks to AI and ML algorithms, drones can now process and analyze vast amounts of data, derive actionable insights, and make real-time choices (Hernández et al., 2021; Johnson, 2020).

a. Data Processing and Analysis: The drones equipped with AI and ML algorithms, capable of processing and analyzing extensive datasets obtained from remote sensing and aerial surveys, can be modified. Pattern identification, categorization, and predictive analysis of drones can be employed to search for indicators of climate change, find anomalies, and produce actionable insights for resource allocation and decision-making.
b. Autonomous Decision-Making: As the prevalence of AI-equipped drones rises, these drones are capable of autonomously making decisions based on data and real-time environmental information. Examine the scenario of drone technology employing AI algorithms that may prioritize objectives, adjust flying trajectories, and optimize resource allocation during disaster response, thereby compressing response time and enhancing overall efficacy.
c. Predictive Modeling and Forecasting: Drones can employ machine learning technologies to develop predictive models and forecasts regarding trends and effects of climate change. Drones can enhance adaptation strategies for ecosystems, communities, and infrastructure to mitigate climate change effects by capturing historical data and environmental variables, predicting future scenarios, and assessing risks.
d. Dynamic Mission Planning: AI-operated drones are capable of customizing their mission plan and implementation according to the changing environmental conditions and mission goals. Drones can customize mission results by integrating real-time data feed and environmental sensors and meet the desired goals, which allows them to dynamically modify flight paths, sensor setups and data collection strategy.

iii. Collaboration and Partnerships for Addressing Climate Change Solutions:

Collaboration between governments, educational institutions, nonprofit entities and industrial stakeholders is important to make proper use of drone technology in efforts to reduce and adapt to climate change. Stakeholders can increase the effectiveness of drone-based solutions and deal with difficult problems combining infrastructure, knowledge and resource raising (Agrawal et al., 2020; Mohan, Richardson, et al., 2021; Slater et al., 2021).

a. Data Sharing and Integration: Collaborating allows stakeholders to exchange information, insights and best practices to deal with climate change solutions. This includes data sharing and integration. Stakeholders can establish policies by acquiring a more intensive understanding of environmental mobility and trends through integration of data from various sources like satellites, ground-based sensors, drones and civil science projects and make more informed decisions.
b. Capacity Building and Training: Working together makes it easier to implement capacity building and training programs that provide the information, tools and resources necessary to efficiently deploy and use drone technology. To fully use drone in dealing with climate change concerns, academics, emergency respondents and protectionist drone operations, data analysis and interpretation can benefit from training programs on educationist, emergency respondents and protectionist drone operations.
c. Policy Development and Governance: Policy maker, regulator and industry stakeholders can work together to establish and execute policies and regulations that support the ethical and responsible use of drone technology in climate change solutions. Collaboration facilitates communication and cooperation between these parties. Stakeholders can maximize the benefits of drone-based projects while ensuring security, privacy and environmental stability for drone operations.
d. Public Engagement and Outreach: Collaboration can enable outreach and public participation initiative to enable drone's understanding of the role in dealing with climate change and winning on local stakeholders and communities. Through community cooperation, education activities and civil science projects, stakeholders can encourage communities to participate in climate change adaptation and quenching measures by promoting a sense of ownership and management and involving citizens in drone-based initiatives.

CONCLUSION

The prospective application of drones in mitigating and facilitating climate change adaptation has emerged as a promising domain of scientific advancement, ethical consideration, and collaborative effort. The integration of advanced capabilities and artificial intelligence in drones has the potential to significantly transform disaster response, ecosystem preservation, and environmental monitoring. Drones possess significant potential for advancing technologies that will contribute to solutions for climate change.

There is no one method to underscore the significance of drones in mitigating and adapting to the impacts of climate change. Drones are a cost-effective, efficient, and scalable alternative for real-time data collection, distant sensing, and rapid response to environmental catastrophes. Their ability to look into ecosystems, evaluate risks and provide information to decision makers is important to promote flexibility, reduce consequences and protect ecosystems and communities from the effects of climate change. Drone governments, universities, nonprofit institutions and industry players make it easy for cooperating and cooperating in various sectors. Stakeholders can effectively use drone technology to meet and meet shared goals in efforts to reduce and adapt to climate change by combining complex challenges and resources, expertise and infrastructure.

There are several topics for further investigation and study when drone technology develops in the future:

a. Integration of Drones with Other Technologies: Integration of drones can be checked in future research in order to improve data collection, analysis and communication capabilities for climate change solutions with blockchain technology, Internet of Things (IoT) sensors and other emerging technologies like satellite imaging.
b. Ethical and Regulatory Frameworks: To create strong ethical and regulatory framework that allow moral and responsible use of drones to resolve climate change issues is a need for research. This includes solving issues related to data security, privacy, social equality and environmental stability.
c. Community Engagement and Participation: Strategies for improving community participation and participation in drone-based programs for climate change adaptation and quenching could be a subject of future research. There are some examples of incorporating local communities in capacity building programs, civil science efforts and decision-making processes.
d. Long-Term Monitoring and Impact Assessment: To determine the efficacy of drone-based interventions in dealing with climate change issues, research is required to develop outline for impact assessment and long-term monitoring programs. Indicators of health, community flexibility and environmental conditions of the ecosystem can be tracked throughout the time.

In short, climate change keeps the ability to revolutionize adaptation and use of drone in quenching projects. Stakeholders can fully use drones to create a more flexible and durable future for communities, ecosystems and humanity by advancing cooperation, research and technology.

REFERENCES

Agrawal, A., Abraham, S. J., Burger, B., Christine, C., Fraser, L., Hoeksema, J. M., & Cox, S. (2020, April). The next generation of human-drone partnerships: Co-designing an emergency response system. In *Proceedings of the 2020 CHI Conference on Human Factors in Computing Systems* (pp. 1-13).

Ahmed, M., Ahmad, S., & Kheir, A. M. (2023). Climate Change: An Overview. Global Agricultural Production: Resilience to Climate Change, 1-30.

Akib Uz Zaman, A., Abdelaty, A., & Jamal, M. S. (2024). Limitations and Barriers Affecting the Applications of UAVs in Transportation Infrastructure Systems Assessment. In Construction Research Congress 2024 (pp. 446-455).

Ali, S. S., Kaur, R., & Abbas, H. (2022). Feasibility of drones as the next step in innovative solution for emerging society. Artificial Intelligence in Industry 4.0 and 5G Technology, 179-208.

Aliloo, J., Abbasi, E., Karamidehkordi, E., Parmehr, E. G., & Canavari, M. (2024). Dos and Don'ts of using drone technology in the crop fields. *Technology in Society*, *76*, 102456. DOI: 10.1016/j.techsoc.2024.102456

Alyassi, R., Khonji, M., Karapetyan, A., Chau, S. C.-K., Elbassioni, K., & Tseng, C.-M. (2022). Autonomous recharging and flight mission planning for battery-operated autonomous drones. *IEEE Transactions on Automation Science and Engineering*, *20*(2), 1034–1046. DOI: 10.1109/TASE.2022.3175565

Amir, N. H. (2024). *Taking Flight for Nature's Plight: Framing Sustainable Reforestation with Unmanned Aerial Vehicles (UAVs).* Hochschulbibliothek der Technischen Hochschule Köln.

Askarzadeh, T., Bridgelall, R., & Tolliver, D. (2024). Monitoring Nodal Transportation Assets with Uncrewed Aerial Vehicles: A Comprehensive Review. *Drones (Basel)*, *8*(6), 233. DOI: 10.3390/drones8060233

Ayamga, M., Akaba, S., & Nyaaba, A. A. (2021). Multifaceted applicability of drones: A review. *Technological Forecasting and Social Change*, *167*, 120677. DOI: 10.1016/j.techfore.2021.120677

Balasundram, S. K., Shamshiri, R. R., Sridhara, S., & Rizan, N. (2023). The role of digital agriculture in mitigating climate change and ensuring food security: An overview. *Sustainability (Basel)*, *15*(6), 5325. DOI: 10.3390/su15065325

Bashir, M. H., Ahmad, M., Rizvi, D. R., & El-Latif, A. A. A. (2024). Efficient CNN-based disaster events classification using UAV-aided images for emergency response application. *Neural Computing & Applications*, *36*(18), 1–14. DOI: 10.1007/s00521-024-09610-4

Camarretta, N. A., Harrison, P., Lucieer, A. M., Potts, B., Davidson, N., & Hunt, M. (2020). From drones to phenotype: Using UAV-LiDAR to detect species and provenance variation in tree productivity and structure. *Remote Sensing*, *12*(19), 3184. DOI: 10.3390/rs12193184

Caruso, A., Chessa, S., Escolar, S., Barba, J., & López, J. C. (2021). Collection of data with drones in precision agriculture: Analytical model and LoRa case study. *IEEE Internet of Things Journal*, *8*(22), 16692–16704. DOI: 10.1109/JIOT.2021.3075561

Chapapría, V. E., Peris, J. S., & González-Escrivá, J. A. (2022). Coastal monitoring using unmanned aerial vehicles (UAVs) for the management of the Spanish Mediterranean coast: The case of almenara-Sagunto. *International Journal of Environmental Research and Public Health*, *19*(9), 5457. DOI: 10.3390/ijerph19095457 PMID: 35564850

d'Amato, G., Chong-Neto, H. J., Monge Ortega, O. P., Vitale, C., Ansotegui, I., Rosario, N., Haahtela, T., Galan, C., Pawankar, R., Murrieta-Aguttes, M., Cecchi, L., Bergmann, C., Ridolo, E., Ramon, G., Gonzalez Diaz, S., D'Amato, M., & Annesi-Maesano, I. (2020). The effects of climate change on respiratory allergy and asthma induced by pollen and mold allergens. *Allergy*, *75*(9), 2219–2228. DOI: 10.1111/all.14476 PMID: 32589303

Damaševičius, R., Bacanin, N., & Misra, S. (2023). From sensors to safety: Internet of Emergency Services (IoES) for emergency response and disaster management. *Journal of Sensor and Actuator Networks*, *12*(3), 41. DOI: 10.3390/jsan12030041

Díaz-Báñez, J.-M., & Fabila-Monroy, R. (2024). Filming runners with drones is hard. *arXiv preprint arXiv:2403.14033.*

Dileep, M., Navaneeth, A., Ullagaddi, S., & Danti, A. (2020). A study and analysis on various types of agricultural drones and its applications. 2020 fifth international conference on research in computational intelligence and communication networks (ICRCICN), Enemark, C. (2020). On the responsible use of armed drones: The prospective moral responsibilities of states. *International Journal of Human Rights*, *24*(6), 868–888.

Famili, A., Stavrou, A., Wang, H., Park, J.-M., & Gerdes, R. (2024). Securing your airspace: Detection of drones trespassing protected areas. *Sensors (Basel)*, *24*(7), 2028. DOI: 10.3390/s24072028 PMID: 38610239

Godfrey, I., & Brenes, J. P. S. (2021, April). Practical Applications for UAS Designed to Assist Climatologists in Studying Toxic Gas Emissions Relative to Climate Change. In *International Conference on Unmanned Aerial System in Geomatics* (pp. 457-484). Cham: Springer International Publishing.

Gupta, A., Afrin, T., Scully, E., & Yodo, N. (2021). Advances of UAVs toward future transportation: The state-of-the-art, challenges, and opportunities. *Future Transportation*, *1*(2), 326–350. DOI: 10.3390/futuretransp1020019

Halofsky, J. E., Peterson, D. L., & Harvey, B. J. (2020). Changing wildfire, changing forests: The effects of climate change on fire regimes and vegetation in the Pacific Northwest, USA. *Fire Ecology*, *16*(1), 1–26. DOI: 10.1186/s42408-019-0062-8

Hasna, A. M. (2009). Climate change, contemporary society and engineering practice: A sustainability journey. *The International Journal of Climate Change*, *1*(4), 139–150. DOI: 10.18848/1835-7156/CGP/v01i04/37278

Hernández, D., Cano, J.-C., Silla, F., Calafate, C. T., & Cecilia, J. M. (2021). AI-enabled autonomous drones for fast climate change crisis assessment. *IEEE Internet of Things Journal*, *9*(10), 7286–7297. DOI: 10.1109/JIOT.2021.3098379

Holden, P. B., Rebelo, A. J., & New, M. G. (2021). Mapping invasive alien trees in water towers: A combined approach using satellite data fusion, drone technology and expert engagement. *Remote Sensing Applications: Society and Environment*, *21*, 100448. DOI: 10.1016/j.rsase.2020.100448

Johnson, J. (2020). Artificial intelligence, drone swarming and escalation risks in future warfare. *RUSI Journal*, *165*(2), 26–36. DOI: 10.1080/03071847.2020.1752026

Júnior, M. R. B., de Almeida Moreira, B. R., dos Santos Carreira, V., de Brito Filho, A. L., Trentin, C., de Souza, F. L. P., Tedesco, D., Setiyono, T., Flores, J. P., & Ampatzidis, Y. (2024). Precision agriculture in the United States: A comprehensive meta-review inspiring further research, innovation, and adoption. *Computers and Electronics in Agriculture*, *221*, 108993. DOI: 10.1016/j.compag.2024.108993

Kandrot, S., Hayes, S., & Holloway, P. (2022). Applications of uncrewed aerial vehicles (UAV) technology to support integrated coastal zone management and the UN sustainable development goals at the coast. *Estuaries and Coasts*, *45*(5), 1230–1249. DOI: 10.1007/s12237-021-01001-5 PMID: 34690615

Kandrot, S., & Holloway, P. (2020). Applications of drone technology for sustainable development of the coastal zone: A literature review. *Project report*.

Khan, A., Gupta, S., & Gupta, S. K. (2022). Emerging UAV technology for disaster detection, mitigation, response, and preparedness. *Journal of Field Robotics*, *39*(6), 905–955. DOI: 10.1002/rob.22075

Khan, M. A., Menouar, H., Eldeeb, A., Abu-Dayya, A., & Salim, F. D. (2022). On the detection of unauthorized drones—Techniques and future perspectives: A review. *IEEE Sensors Journal*, *22*(12), 11439–11455. DOI: 10.1109/JSEN.2022.3171293

Kucharczyk, M., & Hugenholtz, C. H. (2021). Remote sensing of natural hazard-related disasters with small drones: Global trends, biases, and research opportunities. *Remote Sensing of Environment*, *264*, 112577. DOI: 10.1016/j.rse.2021.112577

Kumar, A. (2020). Drone proliferation and security threats. *Indian Journal of Asian Affairs*, *33*(1/2), 43–62.

Kumar, A., de Jesus Pacheco, D. A., Kaushik, K., & Rodrigues, J. J. (2022). Futuristic view of the internet of quantum drones: Review, challenges and research agenda. *Vehicular Communications*, *36*, 100487. DOI: 10.1016/j.vehcom.2022.100487

Kumar, V., Ranjan, D., & Verma, K. (2021). Global climate change: the loop between cause and impact. In *Global Climate Change* (pp. 187–211). Elsevier. DOI: 10.1016/B978-0-12-822928-6.00002-2

Mohamed, N., Al-Jaroodi, J., Jawhar, I., Idries, A., & Mohammed, F. (2020). Unmanned aerial vehicles applications in future smart cities. *Technological Forecasting and Social Change*, *153*, 119293. DOI: 10.1016/j.techfore.2018.05.004

Mohan, M., Richardson, G., Gopan, G., Aghai, M. M., Bajaj, S., Galgamuwa, G. P., Vastaranta, M., Arachchige, P. S. P., Amorós, L., Corte, A. P. D., de-Miguel, S., Leite, R. V., Kganyago, M., Broadbent, E. N., Doaemo, W., Shorab, M. A. B., & Cardil, A. (2021). UAV-supported forest regeneration: Current trends, challenges and implications. *Remote Sensing (Basel)*, *13*(13), 2596. DOI: 10.3390/rs13132596

Mohan, M., Rue, H. A., Bajaj, S., Galgamuwa, G. P., Adrah, E., Aghai, M. M., Broadbent, E. N., Khadamkar, O., Sasmito, S. D., Roise, J., Doaemo, W., & Cardil, A. (2021). Afforestation, reforestation and new challenges from COVID-19: Thirty-three recommendations to support Civil Society Organizations (CSOs). [Niazy, M. M. The role of drones in detecting cadmium pollution.]. *Journal of Environmental Management*, *287*, 112277. DOI: 10.1016/j.jenvman.2021.112277 PMID: 33756214

Noor, M., Rehman, N. U., Jalil, A., Fahad, S., Adnan, M., Wahid, F., ... & Hassan, S. (2020). Climate change and costal plant lives. Environment, climate, plant and vegetation growth, 93-108.

Panday, U. S., Pratihast, A. K., Aryal, J., & Kayastha, R. B. (2020). A review on drone-based data solutions for cereal crops. *Drones (Basel)*, *4*(3), 41. DOI: 10.3390/drones4030041

Paterson, M. (2020). Climate change and international political economy: Between collapse and transformation. *Review of International Political Economy*, *28*(2), 394–405. DOI: 10.1080/09692290.2020.1830829

Pathak, H., Kumar, G., Mohapatra, S., Gaikwad, B., & Rane, J. (2020). Use of drones in agriculture: Potentials, Problems and Policy Needs. *ICAR-National Institute of Abiotic Stress Management*, *300*, 4–15.

Raya, A. M., & Esteve, J. S. (2024). Challenges of Civil Application of 'Drones' for Commercial Purposes in the Logistics and Transportation Sector. In *Applying Drones to Current Societal and Industrial Challenges* (pp. 235–253). Springer. DOI: 10.1007/978-3-031-55571-8_9

Ripple, W. J., Wolf, C., Newsome, T. M., Barnard, P., & Moomaw, W. R. (2020). World scientists' warning of a climate emergency. *Bioscience*, *70*(1), 8–100. DOI: 10.1093/biosci/biz152

Rocque, R. J., Beaudoin, C., Ndjaboue, R., Cameron, L., Poirier-Bergeron, L., Poulin-Rheault, R.-A., Fallon, C., Tricco, A. C., & Witteman, H. O. (2021). Health effects of climate change: An overview of systematic reviews. *BMJ Open*, *11*(6), e046333. DOI: 10.1136/bmjopen-2020-046333 PMID: 34108165

Rodríguez, D. A., Tafur, C. L., Daza, P. F. M., Vidales, J. A. V., & Rincón, J. C. D. (2024). Inspection of aircrafts and airports using UAS: A review. *Results in Engineering*, *22*, 102330. DOI: 10.1016/j.rineng.2024.102330

Saharan, S., Bhandari, A. K., & Yadav, B. (2024). Use of Advanced Technologies for Drones in the Context of Security Issues and Challenges. In *Cybersecurity Issues and Challenges in the Drone Industry* (pp. 196–216). IGI Global. DOI: 10.4018/979-8-3693-0774-8.ch008

Saharan, S., Singh, S., Bhandari, A. K., & Yadav, B. (2024). The Future of Cyber-Crimes and Cyber War in the Metaverse. In *Forecasting Cyber Crimes in the Age of the Metaverse* (pp. 126–148). IGI Global.

Saharan, S., Singh, S., Bora, D., & Saxena, G. (2024). Cyber Forensics: A Boon to Cybersecurity. In *Digital Technologies in Modeling and Management: Insights in Education and Industry* (pp. 93-115). IGI Global.

Saharan, S., & Yadav, B. (2022). Digital and cyber forensics: A contemporary evolution in forensic sciences. In *Crime Scene Management within Forensic Science: Forensic Techniques for Criminal Investigations* (pp. 267-294). Springer. DOI: 10.1007/978-981-16-6683-4_11

Samadzadegan, F., Dadrass Javan, F., Ashtari Mahini, F., & Gholamshahi, M. (2022). Detection and recognition of drones based on a deep convolutional neural network using visible imagery. *Aerospace (Basel, Switzerland)*, *9*(1), 31. DOI: 10.3390/aerospace9010031

Scotese, C. R., Song, H., Mills, B. J., & van der Meer, D. G. (2021). Phanerozoic paleotemperatures: The earth's changing climate during the last 540 million years. *Earth-Science Reviews*, *215*, 103503. DOI: 10.1016/j.earscirev.2021.103503

Semieniuk, G., Taylor, L., Rezai, A., & Foley, D. K. (2021). Plausible energy demand patterns in a growing global economy with climate policy. *Nature Climate Change*, *11*(4), 313–318. DOI: 10.1038/s41558-020-00975-7

Serafinelli, E., & O'Hagan, L. A. (2024). Drone views: A multimodal ethnographic perspective. *Visual Communication*, *23*(2), 223–243. DOI: 10.1177/14703572211065093

Shakhatreh, H., Sawalmeh, A. H., Al-Fuqaha, A., Dou, Z., Almaita, E., Khalil, I., Othman, N. S., Khreishah, A., & Guizani, M. (2019). Unmanned aerial vehicles (UAVs): A survey on civil applications and key research challenges. *IEEE Access : Practical Innovations, Open Solutions*, *7*, 48572–48634. DOI: 10.1109/ACCESS.2019.2909530

Shaw, J. T., Shah, A., Yong, H., & Allen, G. (2021). Methods for quantifying methane emissions using unmanned aerial vehicles: A review. *Philosophical Transactions. Series A, Mathematical, Physical, and Engineering Sciences*, *379*(2210), 20200450. DOI: 10.1098/rsta.2020.0450 PMID: 34565219

Skendžić, S., Zovko, M., Živković, I. P., Lešić, V., & Lemić, D. (2021). The impact of climate change on agricultural insect pests. *Insects*, *12*(5), 440. DOI: 10.3390/insects12050440 PMID: 34066138

Slater, T. F., Biggs, C. N., & Sanchez, R. L. (2021). Positive influence of education partnerships for teaching integrated STEM through drone competition. *Journal of Astronomy and Earth Sciences Education*, *8*(2), 113–124. DOI: 10.19030/jaese.v8i2.10409

Subramanian, K., Pazhanivelan, S., Srinivasan, G., Santhi, R., & Sathiah, N. (2021). Drones in insect pest management. *Frontiers in Agronomy*, *3*, 640885. DOI: 10.3389/fagro.2021.640885

Tarr, A. A., Tarr, J.-A., Thompson, M., & Ellis, J. (2021). *Drone law and policy: Global development, risks, regulation and insurance*. Routledge. DOI: 10.4324/9781003028031

Wang, J., Zhang, S., Guo, H., Tian, Y., Liu, S., Du, C., & Wu, J. (2024). Stereoscopic monitoring of transportation infrastructure. *Automation in Construction, 164*, 105472. DOI: 10.1016/j.autcon.2024.105472

Wang, N., Christen, M., & Hunt, M. (2021). Ethical considerations associated with "humanitarian drones": A scoping literature review. *Science and Engineering Ethics, 27*(4), 51. DOI: 10.1007/s11948-021-00327-4 PMID: 34342721

Watkins, S., Burry, J., Mohamed, A., Marino, M., Prudden, S., Fisher, A., Kloet, N., Jakobi, T., & Clothier, R. (2020). Ten questions concerning the use of drones in urban environments. *Building and Environment, 167*, 106458. DOI: 10.1016/j.buildenv.2019.106458

Weiskopf, S. R., Rubenstein, M. A., Crozier, L. G., Gaichas, S., Griffis, R., Halofsky, J. E., Hyde, K. J., Morelli, T. L., Morisette, J. T., Muñoz, R. C., Pershing, A. J., Peterson, D. L., Poudel, R., Staudinger, M. D., Sutton-Grier, A. E., Thompson, L., Vose, J., Weltzin, J. F., & Whyte, K. P. (2020). Climate change effects on biodiversity, ecosystems, ecosystem services, and natural resource management in the United States. *The Science of the Total Environment, 733*, 137782. DOI: 10.1016/j.scitotenv.2020.137782 PMID: 32209235

Wich, S. A., Hudson, M., Andrianandrasana, H., & Longmore, S. N. (2021). Drones for conservation. *Conservation technology, 35*.

Wich, S. A., Spaan, D., Pintea, L., Delahunty, R., & Kerby, J. 4.8 UAVs in conservation research. *UAVs for the Environmental Sciences*, 403.

Wu, J., Snell, G., & Samji, H. (2020). Climate anxiety in young people: A call to action. *The Lancet. Planetary Health, 4*(10), e435–e436. DOI: 10.1016/S2542-5196(20)30223-0 PMID: 32918865

Yaacoub, J.-P., Noura, H., Salman, O., & Chehab, A. (2020). Security analysis of drones systems: Attacks, limitations, and recommendations. *Internet of Things : Engineering Cyber Physical Human Systems, 11*, 100218. DOI: 10.1016/j.iot.2020.100218 PMID: 38620271

Chapter 4

Cybersecurity Challenges for AGI Drones in Climate Change Missions

Yasir Khan
https://orcid.org/0009-0001-9993-2574
Independent Researcher, Pakistan

Sadaf Manzoor
https://orcid.org/0009-0004-7080-8280
NED University of Engineering and Technology, Karachi., Pakistan

ABSTRACT

As the world grapples with an existential threat of climate change, the newly emerged technologies such as Artificial General Intelligence (AGI) drones, are being explored to find promising solutions for environmental monitoring and mitigation. However, the deployment of AGI drones brings with itself a myriad of security risks that target software, hardware, networks, and communication channels in drones or unmanned aerial vehicles (UAVs). This study initially explains the critical role of AGI drones in climate change mission. Furthermore, a comprehensive analysis is conducted to investigate threats to UAV networks, establishing a classification approach for threat categorization, and evaluating defense mechanisms using OSI model. Additionally, the review also analyzes UAV routing protocols from a cybersecurity perspective in order to reveal vulnerabilities and assess potential risks. Consequently, the study discusses best security practices for AGI drones protection against cyber-attacks and future research directions to enhance UAV cybersecurity in climate-related applications.

DOI: 10.4018/979-8-3693-6457-4.ch004

1. INTRODUCTION

A branch of computer science that focuses on creating systems that are similar to the human mind and can make decisions on their own is called Artificial Intelligence (AI). The AI has shown a great progress within the recent years due to the appearance of new areas among which can be mentioned such as computer vision, natural language processing, machine learning, deep learning. The next evolution in this field is Artificial General Intelligence (AGI) that is designed to produce systems with the function similar to that of a human being. AI technologies have suddenly found applications in many areas such as climate change, transportation, education, healthcare, e-commerce and media. These methods are primarily categorized as being advanced forms of narrow AI or limited AI, meant for a single purpose. Narrow AI has demonstrated notable success in diverse fields. Nevertheless, developing functional AGI remains a formidable objective. Further to that, it is believed that AGI is not expected to materialize until the year 2300 (Mckinsey, n.d.). AI technologies like ChatGPT and DALL-E and other systems would, in a way, be good predictive models. Nevertheless, they are still not able to perform like human beings in aspects like innovativeness, developing arguments, and perceiving the physical world and other functions.

AGI is grounded in several fundamental concepts and procedures aimed at emulating broad human intelligence within robots. These principles form the basis for constructing AGI systems capable of independently acquiring and analyzing information, as well as reasoning and adjusting to new challenges and tasks. These abilities consist of various aspects of cognition, learning, and decision-making. Figure 1 illustrates the foundational concepts that uphold the entire field of AGI.

Figure 1. Foundational ideas of artificial general intelligence (AGI)

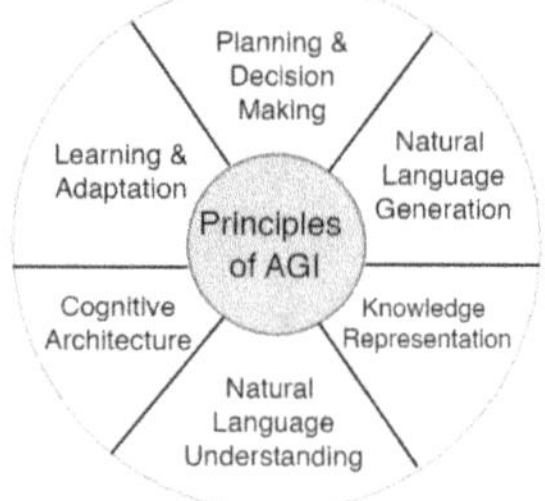

Unmanned aerial vehicles, or drones, are cutting-edge devices with superior technology that can function well without human oversight. For instance, aerial crop sprayers are used in checking crop status, applying chemicals and fertilizers, and conducting wide area surveillance for precise farming. Also in disaster management, drones are used for aerial inspections of the impact situation, ariel delivery of relief food and material, and rescue missions involving persons in need of assistance. Microprocessors and sensors are just two examples of the electronic advancements that can be used to run UAVs (Nourmohammadi, Jafari, & Zander, 2018). UAVs have gradually evolved and now possess numerous capabilities applicable to various industries. For example, the influence and integration of autonomous vehicles and drones (Waymo, n.d.) are significantly transforming the transportation sector. Additionally, collaborative drones and robots (Sherwani, Asad, & Ibrahim, 2020) can carry out tasks such as assembling, joining, holding, and inspecting, as well as working in quality assurance. This technology simplifies tasks, aids individuals in achieving their objectives, and enhances safety by minimizing accidents.

The main objective of this book chapter is to provide a literature review on the threat environment associated with risks to AGI-driven drones in missions related to climate change. This leads to examining the possible effects, risks, and prospects. The literature discussed in this chapter pertains to several research domains, including the significance of AGI-based drones in climate change, threats associated with such drones, and how novel technologies can help in mitigating these threats. It is aimed at examining the state of the art, identifying research limitations, and proposing future research directions.

The remainder of the book chapter is structured as follows: Section II lays the foundation by explaining the fundamental understanding of AGI-driven UAVs for climate change mission. Followed by Section III which presents a literature review on the existing research, highlighting significant studies related to the subject. Moreover, Section IV describes the range of applications for which AGI-based drones are used, especially in climate change-related missions. Furthermore, Section V explains different categories of cyber threats that severely impact operations of UAVs. Furthermore, Section VI throws light on modern-day solutions for tackling cyber threats in drones. Section VII draws attention to a few study areas that require further research for producing quality results. Lastly, the chapter is concluded in Section VIII.

2. UNDERSTANDING THE ROLE OF AGI-BASED DRONES IN CLIMATE CHANGE

Two threats to present society in which AGI based drones operate are cybersecurity threat and climate change threat. There are numerous linkages between these issues and as such, there are complicated challenges that need to be addressed with a lot of depth (Bayomi & Fernandez, 2023).

As of now, GPS and A. I help drones to accomplish duties that involve minimal interaction from the drivers, even if the drone is out of reach from the operators. For instance, regarding satellite images, the algorithms can obtain the data on farm yields or estimate the energy demand of buildings. Autopilots or stabilizers, GPS, and Inertial Measurement Units help maintain drones' stability, identify their altitude, and follow a given course or change it depending on the situation. Onboard processors utilizing GPUs process the loaded data and use vision algorithms such as simultaneous localization and mapping (SLAM) to be aware during flights (Olson & Anderson, 2021). Super intelligent AI can change the missions during flight or through the operational results of previous flights. Control is done by radio links for several kilometers, and some of the systems have first person view (FPV) with the capability to send back real-time video to the user.

Therefore, UAVs are gradually transforming into a multifunctional and effective instrument as a climate change impact monitoring tool with the ability to employ flexible high definition aerial photography. These have been used in various factions of climatic analysis such as the Yukon Kuskokwim Delta region in Alaska. Visible and near-infrared; multispectral and thermal; and Infrared images were taken with the help of UAVs with multiple sensors for the purpose of mapping the complicated territory (Piekkoontod et al., 2020). The flexibility of the UAVs allowed the team to optimize the systems depending on the needs that arise when working on landscape alteration due to permafrost warming. The use of UAVs made it possible to monitor the changes resulting from global warming on significant cultural and ecological resources in the Arctic tundra area. From this study, there is an implication that UAV remote sensing can provide frequent, cheap, and timely landscape characterization (Ilinca et al., 2022) for enhancing climate change analysis and the formulation of related strategies.

Climate change research has received a boost from the use of Unmanned Aerial Vehicles (UAVs) in data collection, monitoring, and mapping of the relevant data. It enables evaluation of solar panels, surveys and examination of structures' conditions, and investigation of microclimates, as well as monitoring of CO2 emissions and determination of flood territories (Rezwan & Choi, 2022). UAVs are essential for mapping coasts, glacial sites, landslides, and ecosystems - observing unlawful logging and monitoring species. It also has an important role in disaster operations in Haiti and Nepal, responding to emergency in Vanuatu and evaluating urban for better energy and waste. Furthermore, UAVs assist in creating other features like mapping of biodiversity and habitats, minimizing risks and modeling of climate (Villa et al., 2016). They are also utilized in controlling marine pollution and delivering carbon-neutral solutions, explaining why UAVs are effective tools to address multiple climate concerns. Figure 2 below illustrates the role of AGI-driven drones in climate change missions.

Figure 2. Role of AGI-based drones in climate change

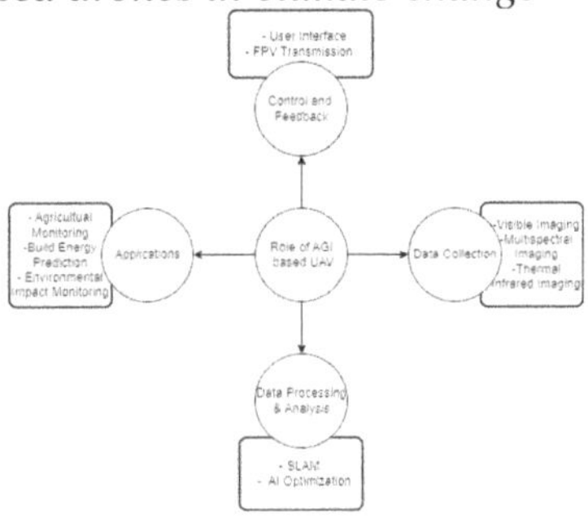

3. LITERATURE REVIEW

This section provides a critical analysis of the literature regarding cybersecurity issues for AGI drones in climate change mission.

Oliver Jung et. al. provides a detailed framework for dealing with security risks in UAV travel within the Labyrinth project (Labyrinth Project, n.d.), which is essential for climate change missions. With respect to managing data integrity in flight operations, the framework employs the clustering technique DBSCAN (Density-Based Spatial Clustering of Applications with Noise) to detect anomalies in the UAV location data. This minimizes the chance of location measurements being contaminated or missing because of some problems in communication or hacking. Enhancing the security and reliability of UAVs, the architecture indirectly promotes climate change missions.

Drone technology is central to numerous critical industries that may be affected by climate change, including farming, city planning, emergency management, and environmental monitoring. The safe use of UAVs allows collecting data on the wildlife populations, the rate of deforestation, the glacier retreat, and the changes in the atmosphere. They also enhance disaster response, which has become topical given the increase in extreme weather-related incidents caused by climate change. In addition, dependable UAV data can be used for improving urban planning, helping cities to accommodate climate change for cities. Therefore, even though this research mainly concerns UAV cybersecurity, the conclusions and findings are applicable to enhancing the efficiency and safety of UAVs in climate change operations (Fas-Millán et al., 2023). Table 1 below summarizes literature on the cybersecurity issues, solutions, and limitations in AGI-driven UAVs in climate change mission.

Table 1. Cybersecurity challenges, solutions, and limitations in AGI-based drones

Methodology	Contribution	Limitation	Reference
XGBoost algorithm	Enhanced UAV cybersecurity through proactive threat classification.	Potential need for continuous evaluation and adaptation to evolving threats.	(Zaki et al., 2024)
Survey work	This paper presents comprehensive review of UAV applications in disaster management.	Reliance on existing case studies may limit the breadth of perspectives.	(Adams & Friedland, 2011)
Vulnerability analysis	Exposing security flaws in UAV systems through hacking demonstrations.	Limited to Wi-Fi-based communication protocols, may not address all potential vulnerabilities.	(Rani et al., 2016)
Comparative analysis	Enhancing understanding and practical skills in UAV security.	Lack of real-world validation for the proposed framework.	(Anagnostis, Kotzanikolaou, & Douligeris, 2024)
Survey work	Providing a comprehensive analysis of drone cybersecurity challenges and proposing solutions based on multi-criteria decision-making.	Generalizability of the survey findings, as they were based on responses from a specific sample of participants.	(Omolara, Alawida, & Abiodun, 2023)
AI-based intrusion detection framework (IDF) for UAVCAN	Proposing a dynamic intrusion detection framework (DIDF) for UAVCAN protocol using AI, achieving high detection performance.	Potential challenges in real-world implementation and scalability of the proposed framework.	(Tlili, Ayed, & Chaari Fourati, 2023)
Comprehensive Analysis	The research work provides a comprehensive overview of the current developments and applications of drone technology across various domains.	The study may face challenges in fully addressing the complex ethical and regulatory considerations associated with the use of drones.	(Emimi, Khaleel, & Alkrash, 2023)

4. APPLICATIONS OF AGI-BASED DRONES FOR CLIMATE CHANGE

Transformative use of AGI-based drones for climate-related applications. These drones are highly advanced with superior AI features that allow them to accomplish various environmental monitoring processes with added accuracy and speed. They can gather and process big data on changes in climatic conditions, emissions of greenhouse gases, deforestation, and species extinction. Disaster response is also facilitated by AGI-based drones by offering assessment of the areas affected before helping in search and rescue. Figure 3 below illustrates various application areas of AGI-driven drones, particularly in the context of climate change.

Figure 3. Applications of AGI-based drones for climate change

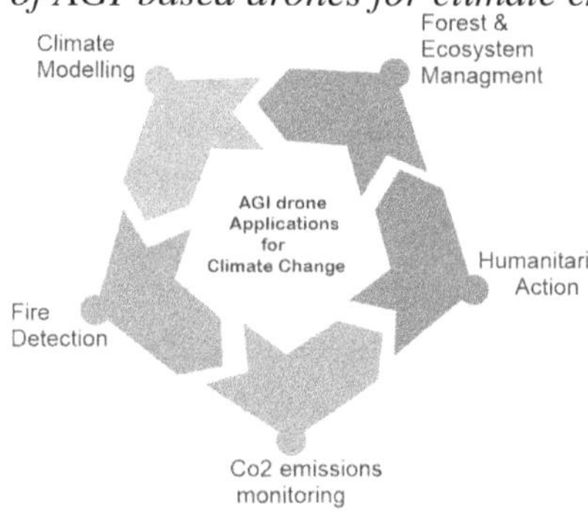

4.1. Climate Modelling

They have been used in climate modeling and air quality modeling by providing more details as compared to other approaches. In 2020, a paper was presented that reviewed the drone based measurements for environmental monitoring (Tanase et al., 2018). A recent study reveals that data received through drones have a higher spatial resolution than that provided by ground-based environmental measurement stations (Pradhan, Suliman, & Awang, 2007). Using built-in microclimate sensors and air quality instruments, drones can correlate local weather, air polluting matters, and the state of living conditions to the state of health of communities. For example, a study carried out at different places in California was done in order to collect detailed, place-specific air-quality data for modeling pollution exposure risks. Such focal information was utilized to steer and formulate special policies that aim to shield the endangered groups from the negative impacts of pollution. This would enable the government to identify specific areas or communities in which the health of the inhabitants as well as the environment is most threatened and, thus, come up with better strategies to protect the people and enhance their living conditions. Further, the AGI-based drones can also include the use of artificial intelligence and computer vision for analysis of the captured data. This allows the discovery of polluted areas, foreseeing trends in air quality, and early-signaling of possible adverse health effects.

4.2. Fire Detection

Wild woodlands are vulnerable to natural disasters, especially those that are not uncommon in the natural world nowadays. The most dangerous of these is the major wooded area blaze, which can alter the life and climate of an extensive area. They both affect the local and the global ecology in that large forest fires cause extensive damage to property and claim lives. To lessen these impacts it is possible with early identification of the fire and accurate tracking of fire characteristics, including the type, intensity, and extent (Bouabdellah, Noureddine, & Larbi, 2013). Although fire may not be prevented right from the start, efforts to report and check on the more vulnerable parts of the area can help minimize the level of damage (Gaur et al., 2019). Unlike heat or smoke detectors which only require the presence of heat or smoke, fire sensors that are supposed to identify smoke, flames, rising temperature and gases may respond with a lag as they require particles to trigger their alarms. These sensors, however, have a limited coverage hence for large areas large numbers of such sensors are needed for adequate monitoring (Gaur et al., 2020). The primary areas of concern in the technological advancement of forest fire monitoring systems are detectors and systems (Memane & Kulkarni, 2015; Bu & Gharajeh, 2019; Allison et al., 2016; Tiansawat & Elliott, 2020; De Almeida et al., 2021) used for recording and transmitting data about flames and smoke in the visible spectrum and/or IR.

4.3. Forest and Ecosystem Management

Due to their versatility and the use of AGI, drones can be used in the following revegetation activities; site infrastructure establishment, site scouting, restoration planning and surveying, supply of seeds or seedlings, site management tasks like germination and weeding, and post restoration surveys for determination of species richness (Erdelj et al., 2017). Various technologies like GPS and image sensors assist UAVs in performing some tasks like surveying pre-restoration sites and evaluating various aspects of biorecovery. They have the capacity to measure alterations in temperature, forest response, and increased/decreased ecological features that could assist in the tracking of replanting progress (Półka, Ptak, & Kuziora, 2017). AGI-drones can also be utilized for real-time sensing, accounting for a reliable and effective way of observing and monitoring forests that is different from the ones commonly used.

4.4. Humanitarian Action

In the area of climate hazard mapping, AGI-based drones are being well adopted as a tool for initial assessment of infrastructure and climate losses from natural climate-related disasters. Because of their capacity to quickly acquire detailed data and extend into locations that may be inaccessible or risky to human auditors, they are very useful in offering timely and accurate information. To address the allocation of response resources, UAVs also enable evaluation of the extent of a disaster within neighborhoods and cities. Further, they also take part in coordination of emergency response by acting as communication interfaces and conveying cargo. UAVs are able to provide wireless connectivity in the regions where the infrastructure is damaged and can also supervise the placement of the people (Sheng et al., 2023). Used in search-and-rescue missions, UAVs with thermal and multispectral sensors help navigate to find victims; especially in complex urban environments (Yan, Xu, & Liu, 2016). In the case of cargo delivery, UAVs are employed in the delivery of necessities such as medical samples, water testing kits, radios, and batteries, which may not be transported by land in post-disaster conditions (Kar et al., 2014).

4.5. CO_2 Emissions Monitoring

One of the other application areas where the AGI-driven drones have been used is in the monitoring of carbon dioxide or CO_2 emission. CO_2 is the worst greenhouse gas and central to the problem of climatic change. Furthermore, the emissions have greatly escalated within the last couple of years leading to phenomena of extreme weathers. Hence, such modern drones use AGI to self-manage and capture accurate climatic information in various locations. Since AGI-based drones have powerful sensors, for example, they can determine the level of CO_2 in the environment or objective data about the emission sources and tendencies. Since they can work through a large amount of data at once they enable quick analysis and reporting that is helpful in prompt interventions or policy making decisions. Liu et al. (2022), have provided a new portable CO_2 measuring system for UAVs. Similarly, it offers accurate and fast rate measurements of the variations in the CO_2 mole fractions to monitor the emission plumes and their coverage in space and time.

5. CYBERSECURITY THREAT LANDSCAPE FOR AGI DRONES IN CLIMATE CHANGE MISSION

New climate change missions using AGI drones pose severe cybersecurity implications, including. These drones that are sometimes fitted with enhanced artificial intelligence, are useful for data gathering, tracking of climate and natural resource conditions and disaster relief. But due to the richness in their technology and the connections they boast of, they are exposed to a number of cyber risks such as hacking, spoofing, and data leakage. Security of AGI drones should be given top priority to guarantee the protection of environmental data and keep away unauthorized access that would hamper operational integrity of effectiveness of climate change measures. A categorization of cyber threats as highlighted in figure 4 take aim at drones and key parts leading to severe incursions.

Figure 4. Cybersecurity threat landscape for AGI drones

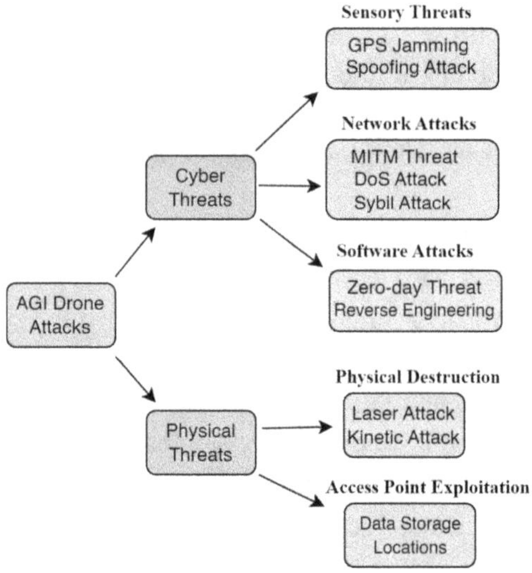

5.1. Jamming Attacks on Drones for Climate Change Operations

Jamming attacks describe a scenario where wireless communication technologies like Wi-Fi, cellular networks, and Bluetooth are interfered with deliberately. A jamming attack is mainly aimed at the interference of the network signal, which will harm the device connection, interrupt important service, or even cause a network interruption in the DoS case. Conventional attacks utilize tangible objects to flood the network with intense signals to interfere with its functions. Zeng et al. (2018) attempted the usage of high intensity lights to interfere with the camera sensors and it led to heavy blurring of the image. This interference significantly degrades camera performance in object recognition and shows that VPS is far from being invulnerable. Furthermore, Rasmussen (2011) discussed that navigation and positioning can be affected due to jamming device and financial effects. To counter jamming attacks, the authors in Whelan et al. (2020) proposed using specialized filters and lenses to block certain types of light. Figure 5 illustrates a typical jamming attack on AGI-driven drones.

Figure 5. Representation of jamming attack

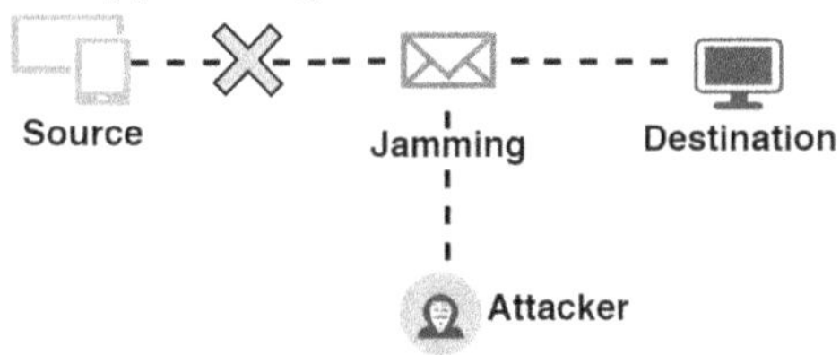

5.2. GPS Spoofing Attack on Drones for Climate Change

The drones for climate change include AGI technology such as GPS for tracking and monitoring the health of the forests through images and data acquisition, which assists in identifying deforestation. Likewise, GPS in drones can track emission across industries and cities in order to cut greenhouse emissions. However, these components of drones can be attacked by bad actors for their nefarious purposes. For instance, GPS spoofing also referred to as GPS simulation is the process through which fake GPS signals are broadcasted in a bid to mislead a GPS receiver. Simply put, it deceives the GPS receiver into believing it is somewhere it is not, thus providing erroneous spatial information. This type of cyberattack affects the authenticity of GPS information that is required for many purposes, such as navigation and time referencing. The new developments in integrating multiple UAVs for purposeful operations have significantly increased cybersecurity threats. This is so because, as mentioned earlier, UAVs largely rely on GPS and RF sensors which are quite vulnerable to cyber security threats such as spoofing attacks (Li & Pu, 2020). Spoofing attack where fake GPS signals seize command functions and steal information by replacing legitimate GPS signals with fakes that emulate the GPS format. This results in the UAV following a wrong path (Vasconcelos et al., 2019). Lastly, figure 6 provides a flowchart illustration of the GPS spoofing attack against UAVs.

Figure 6. GPS spoofing attack on UAVs

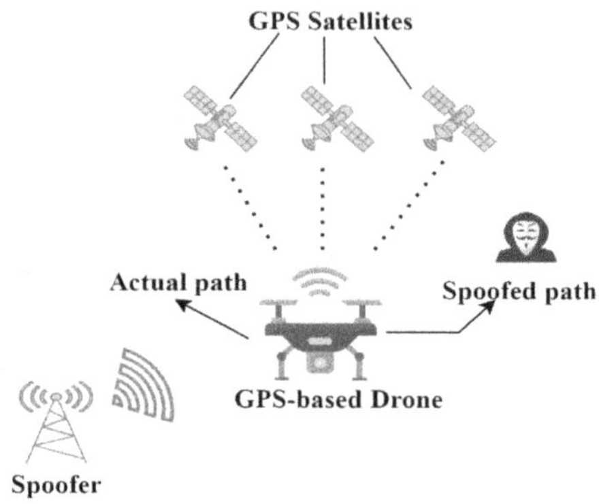

5.3. Man-in-the-Middle Attack in AGI-Based Drones for Climate Change

A man-in-the-middle (MITM) attack is defined as a type of attack whereby an attacker takes the position of the user or the application through which he or she aims to spy on the conversation or impersonate one of the parties. This helps to give the appearance of ongoing communication and information sharing. An MITM attack enables the threat actor to intercept messages between the UAV and GCS and change normal packets to malicious ones. This allows the attacker to act as a bridge between the drone and GCS, endangering bidirectional UAV-to-GCS communication (Vasconcelos et al., 2016). Figure 7 depicts a MITM attack on UAVs, illustrating the roles of the user, web application, attacker, and their connections.

Figure 7. Man-in-the-Middle (MITM) attack on drones

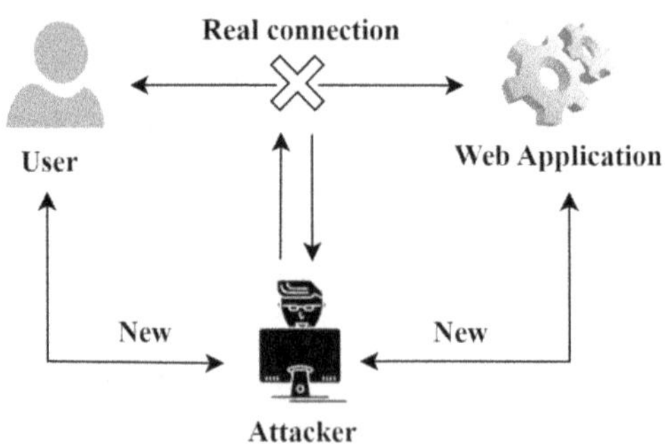

5.4. Denial-of-Service Attack

A Denial of Service (DoS) attack could allow individuals with ill intentions to incapacitate a UAV system. In this case, the attacker may flood the network of the targeted UAV with many requests and this would overburden the available resources hence, making the UAV unavailable (Ceviz, Sadioglu, & Sen, 2023). The effects of these attacks to UAVs can vary and range from noticeable degradation of network connectivity and reduced video streaming for the user to total unavailability of the service. Also, DoS attacks can be directed to the control center. For example, by sending oversized packets to the Ground Control Station (GCS) within a certain distance, the attacker may turn off the control signal. This action makes the UAV enter a disconnected state, causing a failure of the data link (Restituyo & Hayajneh, 2018).

5.5. Sybil Attack in Drones for Climate Change

A Sybil attack uses one node to control multiple active fake identities called Sybil identities in a peer-to-peer network. The purpose of such an attack is to jeopardize the authority or influence within a trusted system by gaining control over the network. These fake identities are used to set up this control. A successful Sybil attack provides the adversary complete authority over unauthorized operations within the system. For example, it enables one entity, such as a computer, to create and manage several personas, including usernames and those based on IP addresses. Thus, these fake identities give the impression to systems and the other users that they are real. A Sybil attack takes place when an attacker creates multiple fake identities (Sybil nodes) that mimic the real nodes of a network (Solodov et al., 2018a).

5.6. Zero-Day Attack in Drones for Climate Change

Cyberattacks occur by exploiting holes in software, hardware, or firmware which were earlier not known to exist. These vulnerabilities are named so, as developers are not given any time to fix or counter these issues since the attackers start exploiting them immediately. This implies that as soon as the vulnerabilities are discovered, the attackers can fully exploit them, while developers are left with no time to release updates, patches, or corrective measures to safeguard individuals and computers (Solodov et al., 2018b). Zero-day threats are highly dangerous since they do not have any form of protection systems put in place; their identification and prevention are nearly impossible.

5.7. Reverse Engineering Attack in UAVs for Climate Change

Reverse engineering is essential for developing accurate models and tools to improve drone performance by identifying their dynamic and aerodynamic parameters. Therefore, it is crucial to reverse engineer drones to enhance model and tool performance through determination of their parameters such as dynamic and aerodynamic. This process is especially very important during the transitional phase from hovering to forward flight; this phase is regarded as the most dangerous phase of flight. Firms like HEMAV Foundation employ reverse engineering in improving the elements of drone systems. Such ways include working with high precise attributes in 3D models and using the Computational Fluid Dynamics (CFD). They can then use such low distortion replicas to examine the properties of drones' aerodynamics and the manner in which they'll perform under distinct circumstances. The system of numerical simulation which is used by CFD enables the investigation of the behavior of and the interactions of airflow and other fluids with the around the drone, resulting to improvements in the design and operations of drones. In addition, this research also aimed at improving the landing of the HP2 drone system through CFD analysis and modeling (Iqbal, 2021).

5.8. Laser Attacks on AGI-Based Drones for Climate Change

Laser attacks include shining of high power laser at specific sensitive points at the drone such as its optical systems, sensors or surface parts. This laser is intended to dazzle or, depending on the speed of the drone, burn out critical components of the device that will make it non-functional. This is because when the lasers target these areas the light intensity is such that they can burn the drone and or cause some critical component to fail and or compromise drone functionality and or render it operationally nonusable. It is most suitable for use in counter-drone operations due to the ability to home in narrowly on a drone or its operator without harming surrounding people or property (Tyagi & Dembla, 2014). Durability is also an issue as laser heating can cause critical parts to melt or compromise the structural integrity of the drone. Laser presents a different threat due to accuracy range and faster disabling time from a drone point of view.

5.9. Impact Attacks on AGI-Driven Drones for Climate Change

Impact attacks entail using physical contact or direct impact to disable or destroy the drone. Methods include employing firearms, anti-aircraft artillery, missiles, or other projectiles. The goal is to damage key components like the propulsion system, control surfaces, or payload, rendering the drone inoperative or causing it to crash. Impact attacks are a conventional and frequently effective way to counter drones, particularly when the goal is total incapacitation. These attacks involve using physical force to neutralize the drone, which can include methods such as firing projectiles, deploying anti-aircraft artillery, or launching missiles. By targeting and damaging essential components like the propulsion system, control surfaces, or payload, impact attacks aim to render the drone completely inoperative or cause it to crash (Lucas, 2024).

5.10. Attacks on Data Storage Locations in AGI drones for Climate Change

Often aerial vehicles record very detailed images and short videos, which are saved in memory for further use. The permanent memory in most drones is either built-in or uses slots for microSD memory cards (Rele & Patil, n.d.). In addition to mere archiving, such systems also provide features for managing data that may be incorporated into the system. There are some high-end drones with real-time tagging functionality incorporated into the captured footage, labeling the media with such values as location, altitude, or time, which make it even more informative and versatile. Access to this storage by unauthorized personnel may lead to loss or theft of this data or the introduction of viruses that will hinder drone operations.

6. EMERGING TECNOLOGIES FOR SECURE AGI DRONES FOR CLIMATE CHANGE MISSION

These technologies are vital in protecting AGI drones that may be used in climate change operations. Blockchain technology is used in the process of providing safe data transfer and storage, as well as protecting the data obtained during missions related to the environment. Artificial intelligence, in the form of algorithms, improves drones by processing large amounts of environmental information, allowing for quick decision-making and subsequent reactions to new circumstances. Federated learning reinforces data privacy by enabling the models to be trained locally on drones, avoiding the accumulation of data that can be hacked. Altogether, these technologies enable AGI drones to track and manage the effects of climate change, helping to enhance the process of environmental preservation globally.

6.1. Blockchain Solution for Robust AGI Drones

It is for this reason that we are considering Blockchain technology as a protective shield against the potential cyber threats to drone communication systems. As it was established earlier, blockchain which operates with a distributed and non-tangible database can improve security and guarantee data inviolability in the course of their interchange between the drones. This includes protection of messages from an adversary, avoiding central authority in making decisions, and providing a transparent medium through smart contracts. Finally, in the aspect of confidentiality of the information transferred during the communication, the use of blockchain preserves the content of messages exchanged by the drones from interception and subsequent modification by other receivers or transmitters who have no right to do so. Drones can work in organizations and act as a single group without focusing on a specific command and control structure and, thus, function as decentralized systems. Moreover, the smart contract executes the consensus among the parties and also facilitates the execution of contracts between the involved parties eliminating any form of manipulation in the activities of the drones (Koulianos & Litke, 2023; Cheema et al., 2022a; Cheema et al., 2022b; Javed et al., 2022). Maintaining an open and extensive connectivity through adopting blockchain in the networks of drone can provide basic connection and assurance, security problems and risks of the Internet of Drones (IoD) systems. These networks because of the decentralized as well as secure part of blockchain can coordinate these correspondences in a reliable as well as tamper-proof structure across extensive regions. This technology reduces the potential for hazards like data leaks, breaking-ins, and many other forms of cybersecurity threats that imperil the functionality and security of IoD systems. In addition, blockchain can apply to drones allowing for better cooperation and data interchange within a network, thus enhancing the network stability (Chang et al., 2021).

Barka et al. (2022) provide a novel decentralized blockchain architecture to enhance the security and trust build between the prospective UAVs and GCS. In their application and assessment it is shown that security of critical infrastructure through UAVs is a good deal improved by incorporation of the trusted blockchain-enabled UASs. This solution makes sure that all the data interchange between UAVs and GCS is protected from either sabotage or unauthorized interferences which create a reliable network. This makes the blockchain framework highly preferable as all the interactions are recorded in an honest ledger through the course of the interaction enhancing the confidence of the customer in the honesty of the system (Barka et al., 2022).

6.2. Machine Learning for Secure AGI-Based UAVs

Machine learning has been extensively used in industries such as manufacturing, retail, transport, and drones. It has a massive opportunity in offering solid protection strategies for AGI-controlled drones. Baig et al. (2022) propose a machine learning-based approach to detect hijacking, GPS signal jamming, and DoS attacks on drones. The following are recommendations derived from a comprehensive machine learning based classification of drone datasets of the drone systems are vulnerable to cyber threats. Furthermore, Majeed et al. (2021) have come up with a framework aimed at preventing cybersecurity risks in IoT-integrated drones through machine learning algorithms. The above framework aims at improving the security of drones through the prevention of cyber vulnerabilities within operations. It is based on machine learning techniques by which the framework can work on data, identify threats and counter different security threats to enable IoT integrated drones to fly safely and effectively in a connected environment. Similarly, Toh (2020) explains a system that utilizes artificial intelligence to identify and localize drones in a surveillance zone. This system integrates the latest analytical techniques to detect the presence of drones. It correctly identifies drones and that makes it easy to monitor and secure them. The system, thus, utilizes machine learning to enhance the detection capability of the system and, at the same time, minimize false alarms. This technology also improves the extent to which monitored areas can be controlled and safeguarded against trespassing drones.

6.3. Federated Learning for Reliable AGI Drones

Because of many features of drones, including silent flying and photo/video shooting, it is quite worrisome as to how one would authenticate a drone or identify which drone is allowed to fly. A number of techniques using ML have been proposed to authenticate drones in the IoT networks. However, these ML models have issues about data security, privacy preservation, and scalability especially in the application of drone authentication. Generally, ML methods collect all the data and develop the authentication models centrally, making them prone to adversarial situations. Yazdinejad et al. (2021) propose a federated learning model for drone authentication based on the RF characteristics and DNN structure that complies with data protection and privacy. From the discovered literature, it is evident that incorporating RF features and DNN architecture in federated learning-based drone authentication outperforms other machine learning techniques.

7. FUTURE RESEARCH DIRECTIONS

Climate change and environmental concerns can be mitigated through the integration of AI technology. However, further investigation is required to effectively exploit its functions, especially in cybersecurity aspects related to AGI-driven drones being deployed in climate operations. Further research along with AI and its solutions will also be needed to reduce the risks associated with them and bring out their best in environmental protection. The next steps could include the development of XAI models, which provide simple and comprehensible outcomes for clients. In essence, establishing trust as well as ethical decision making means that AI algorithms need to be able to explain why they are making a certain decision, especially where the decision involves substantial environmental decisions by autonomous drones (Sandalow et al., 2023).

Moreover, AI is more efficient when implemented alongside other advancing technologies such as edge computing for real-time data processing and blockchain for secure data sharing. All these combinations can enhance the cybersecurity of the AGI drones since it will help protect sensitive data and minimize on cyber threats during data transfer and processing. There is also the need to promote collaboration between AI researchers, cybersecurity professionals, environmental scientists, policymakers, and humans within industries (Burgués & Marco, 2020). This arrangements will result in developing more comprehensive solutions that incorporate scientific, societal, and legal aspects.

Developing international rules regarding the exchange of information is equally important to fight climate change and for creating uniforms for monitoring the environment. Some limitations of AI are the high energy consumption of training, privacy concerns with AI monitoring systems, and the possibility of making already existing gaps worse. Nonetheless, with an increase in the development of complex and sophisticated systems, there is a crucial need for rules and principles that will govern the deployment of AGI drones in combating climate change (Jumaah et al., 2021).

8. CONCLUSION

Autonomy-enabled drones can pave the way for providing service delivery frameworks especially in responding the climate change conditions. Consumers and businesses too have thus escalated the use of drones for ‘drone-as-a-service,’ which entails being a one-stop solution for various services. In the same way, it may set up physical or cloud storage structures for data as well as the connectivity techniques employed in controlling drones; implying a clear understanding of possible threats and how to protect the process, to avoid future drone attacks. However, the above new generation UAS also opened the drone to a lot of cybersecurity risks such as hacking, spoofing, and data theft. Mitigating these hurdles entails the quest for efficient cybersecurity measures that will help safeguard the functionality of AGI drones. It is Latam's responsibility to safeguard these valuable assets to fully unlock their potential to address climate change and development challenges. To this effect, by investing in cybersecurity, we ensure that the valuable and transformational role of AGI drones in studying and combating climate change is not compromised.

REFERENCES

Adams, S. M., & Friedland, C. J. (2011, September). A survey of unmanned aerial vehicle (UAV) usage for imagery collection in disaster research and management. In *9th international workshop on remote sensing for disaster response* (Vol. 8, pp. 1-8).

Allison, R. S., Johnston, J. M., Craig, G., & Jennings, S. (2016). Airborne optical and thermal remote sensing for wildfire detection and monitoring. *Sensors (Basel)*, *16*(8), 1310. DOI: 10.3390/s16081310 PMID: 27548174

Anagnostis, I., Kotzanikolaou, P., & Douligeris, C. (2024). Understanding and Securing Unmanned Aerial Vehicle (UAV) Services: A Comprehensive Tutorial. *Authorea Preprints*. DOI: 10.36227/techrxiv.170975064.43115762/v1

Baig, Z., Syed, N., & Mohammad, N. (2022). Securing the smart city airspace: Drone cyber attack detection through machine learning. *Future Internet*, *14*(7), 205. DOI: 10.3390/fi14070205

Barka, E., Kerrache, C. A., Benkraouda, H., Shuaib, K., Ahmad, F., & Kurugollu, F. (2022). Towards a trusted unmanned aerial system using blockchain for the protection of critical infrastructure. *Transactions on Emerging Telecommunications Technologies*, *33*(8), e3706. DOI: 10.1002/ett.3706

Bayomi, N., & Fernandez, J. E. (2023). Eyes in the sky: Drones applications in the built environment under climate change challenges. *Drones (Basel)*, *7*(10), 637. DOI: 10.3390/drones7100637

Bouabdellah, K., Noureddine, H., & Larbi, S. (2013). Using wireless sensor networks for reliable forest fires detection. *Procedia Computer Science*, *19*, 794–801. DOI: 10.1016/j.procs.2013.06.104

Bu, F., & Gharajeh, M. S. (2019). Intelligent and vision-based fire detection systems: A survey. *Image and Vision Computing*, *91*, 103803. DOI: 10.1016/j.imavis.2019.08.007

Burgués, J., & Marco, S. (2020). Environmental chemical sensing using small drones: A review. *The Science of the Total Environment*, *748*, 141172. DOI: 10.1016/j.scitotenv.2020.141172 PMID: 32805561

Ceviz, O., Sadioglu, P., & Sen, S. (2023). A survey of security in uavs and fanets: Issues, threats, analysis of attacks, and solutions. *arXiv preprint arXiv:2306.14281*.

Chang, Z., Guo, W., Guo, X., Chen, T., Min, G., Abualnaja, K. M., & Mumtaz, S. (2021). Blockchain-empowered drone networks: Architecture, features, and future. *IEEE Network*, *35*(1), 86–93. DOI: 10.1109/MNET.011.2000202

Cheema, M. A., Ansari, R. I., Ashraf, N., Hassan, S. A., Qureshi, H. K., Bashir, A. K., & Politis, C. (2022). Blockchain-based secure delivery of medical supplies using drones. *Computer Networks*, *204*, 108706. DOI: 10.1016/j.comnet.2021.108706

Cheema, M. A., Ansari, R. I., Ashraf, N., Hassan, S. A., Qureshi, H. K., Bashir, A. K., & Politis, C. (2022). Blockchain-based secure delivery of medical supplies using drones. *Computer Networks*, *204*, 108706. DOI: 10.1016/j.comnet.2021.108706

De Almeida, D. R. A., Broadbent, E. N., Ferreira, M. P., Meli, P., Zambrano, A. M. A., Gorgens, E. B., & Brancalion, P. H. (2021). Monitoring restored tropical forest diversity and structure through UAV-borne hyperspectral and lidar fusion. *Remote Sensing of Environment*, *264*, 112582. DOI: 10.1016/j.rse.2021.112582

Emimi, M., Khaleel, M., & Alkrash, A. (2023). The current opportunities and challenges in drone technology. *Int. J. Electr. Eng. and Sustain.*, 74-89.

Erdelj, M., Natalizio, E., Chowdhury, K. R., & Akyildiz, I. F. (2017). Help from the sky: Leveraging UAVs for disaster management. *IEEE Pervasive Computing*, *16*(1), 24–32. DOI: 10.1109/MPRV.2017.11

Fas-Millán, M. Á., Soro, F., Jung, O., & Shaaban, A. (2023, September). Cybersecurity Analysis in the UAV Domain: the Practical Approach of the Labyrinth Project. In *Proceedings of the 2023 ACM Conference on Information Technology for Social Good* (pp. 446-454). DOI: 10.1145/3582515.3609566

Gaur, A., Singh, A., Kumar, A., Kulkarni, K. S., Lala, S., Kapoor, K., Srivastava, V., Kumar, A., & Mukhopadhyay, S. C. (2019). Fire sensing technologies: A review. *IEEE Sensors Journal*, *19*(9), 3191–3202. DOI: 10.1109/JSEN.2019.2894665

Gaur, A., Singh, A., Kumar, A., Kumar, A., & Kapoor, K. (2020). Video flame and smoke based fire detection algorithms: A literature review. *Fire Technology*, *56*(5), 1943–1980. DOI: 10.1007/s10694-020-00986-y

Ilinca, V., andric, I., Chi u, Z., Irimia, R., & Gheuca, I. (2022). UAV applications to assess short-term dynamics of slow-moving landslides under dense forest cover. *Landslides*, *19*(7), 1717–1734. DOI: 10.1007/s10346-022-01877-9

Iqbal, N. (2021). *Reverse Engineering and CFD analysis of a BWB drone system* (Master's thesis, Universitat Politècnica de Catalunya).

Javed, S., Khan, M. A., Abdullah, A. M., Alsirhani, A., Alomari, A., Noor, F., & Ullah, I. (2022). An efficient authentication scheme using blockchain as a certificate authority for the internet of drones. *Drones (Basel)*, *6*(10), 264. DOI: 10.3390/drones6100264

Jumaah, H. J., Kalantar, B., Halin, A. A., Mansor, S., Ueda, N., & Jumaah, S. J. (2021). Development of UAV-based PM2. 5 monitoring system. *Drones (Basel)*, *5*(3), 60. DOI: 10.3390/drones5030060

Kar, G., Mustafa, H., Wang, Y., Chen, Y., Xu, W., Gruteser, M., & Vu, T. (2014, November). Detection of on-road vehicles emanating GPS interference. In *Proceedings of the 2014 ACM SIGSAC conference on computer and communications security* (pp. 621-632). DOI: 10.1145/2660267.2660336

Koulianos, A., & Litke, A. (2023). Blockchain technology for secure communication and formation control in smart drone swarms. *Future Internet*, *15*(10), 344. DOI: 10.3390/fi15100344

Li, Y., & Pu, C. (2020, December). Lightweight digital signature solution to defend micro aerial vehicles against man-in-the-middle attack. In *2020 IEEE 23rd international conference on computational science and engineering (CSE)* (pp. 92-97). IEEE. DOI: 10.1109/CSE50738.2020.00020

Liu, Y., Paris, J. D., Vrekoussis, M., Antoniou, P., Constantinides, C., Desservettaz, M., Keleshis, C., Laurent, O., Leonidou, A., Philippon, C., Vouterakos, P., Quéhé, P.-Y., Bousquet, P., & Sciare, J. (2022). Improvements of a low-cost CO 2 commercial nondispersive near-infrared (NDIR) sensor for unmanned aerial vehicle (UAV) atmospheric mapping applications. *Atmospheric Measurement Techniques*, *15*(15), 4431–4442. DOI: 10.5194/amt-15-4431-2022

Lucas, M. (2024). The Role of AI in Climate Change Mitigation and Environmental Monitoring. DOI: 10.13140/RG.2.2.21153.38247

Majeed, R., Abdullah, N. A., Faheem Mushtaq, M., Umer, M., & Nappi, M. (2021). Intelligent cyber-security system for iot-aided drones using voting classifier. *Electronics (Basel)*, *10*(23), 2926. DOI: 10.3390/electronics10232926

Mckinsey. (2024). "What is Artificial General Intelligence (AGI)?" available at: https://www.mckinsey.com/featured-insights/mckinsey-explainers/what-is-artificial-general-intelligence-agi

Memane, S. E., & Kulkarni, V. S. (2015). A review on flame and smoke detection techniques in video's. *Int. J. Adv. Res. Electrical, Electronics Instrumen.Eng*, *4*(2), 885–889.

Nourmohammadi, A., Jafari, M., & Zander, T. O. (2018). A survey on unmanned aerial vehicle remote control using brain–computer interface. *IEEE Transactions on Human-Machine Systems*, *48*(4), 337–348. DOI: 10.1109/THMS.2018.2830647

Olson, D., & Anderson, J. (2021). Review on unmanned aerial vehicles, remote sensors, imagery processing, and their applications in agriculture. *Agronomy Journal, 113*(2), 971–992. DOI: 10.1002/agj2.20595

Omolara, A. E., Alawida, M., & Abiodun, O. I. (2023). Drone cybersecurity issues, solutions, trend insights and future perspectives: A survey. *Neural Computing & Applications*, *35*(31), 23063–23101. DOI: 10.1007/s00521-023-08857-7

Piekkoontod, T., Pachana, B., Hrimpeng, K., & Charoenjit, K. (2020). Assessments of Nipa forest using Landsat imagery enhanced with unmanned aerial vehicle photography. *Applied Environmental Research*, *42*(3), 49–59. DOI: 10.35762/AER.2020.42.3.5

Półka, M., Ptak, S., & Kuziora, Ł. (2017). The Use of UAV's for Search and Rescue Operations. *Procedia Engineering*, *192*, 748–752. DOI: 10.1016/j.proeng.2017.06.129

Pradhan, B., Dini Hairi Bin Suliman, M., & Arshad Bin Awang, M. (2007). Forest fire susceptibility and risk mapping using remote sensing and geographical information systems (GIS). *Disaster Prevention and Management*, *16*(3), 344–352. DOI: 10.1108/09653560710758297

Rani, C., Modares, H., Sriram, R., Mikulski, D., & Lewis, F. L. (2016). Security of unmanned aerial vehicle systems against cyber-physical attacks. *The Journal of Defense Modeling and Simulation*, *13*(3), 331–342. DOI: 10.1177/1548512915617252

Rasmussen, K. (2011). On the requirements for successful GPS spoofing attacks.

Rele, M., & Patil, D. Examining the Impact of Artificial Intelligence on Cybersecurity within the Internet of Things.

Restituyo, R., & Hayajneh, T. (2018, November). Vulnerabilities and attacks analysis for military and commercial iot drones. In *2018 9th IEEE Annual Ubiquitous Computing, Electronics & Mobile Communication Conference (UEMCON)* (pp. 26-32). IEEE. DOI: 10.1109/UEMCON.2018.8796596

Rezwan, S., & Choi, W. (2022). Artificial intelligence approaches for UAV navigation: Recent advances and future challenges. *IEEE Access : Practical Innovations, Open Solutions*, *10*, 26320–26339. DOI: 10.1109/ACCESS.2022.3157626

Sandalow, D. B., McCormick, C., Kucukelbir, A., Friedman, J., Nagrani, T., Fan, Z., ... & Ruane, A. (2023). Artificial Intelligence for Climate Change Mitigation Roadmap.

Sheng, T., Jin, R., Yang, C., Qiu, K., Wang, M., Shi, J., Zhang, J., Gao, Y., Wu, Q., Zhou, X., Wang, H., Zhang, J., Fang, Q., Pan, N., Xue, Y., Wang, Y., Xiong, R., Gao, F., Zhang, Y., & Gu, Z. (2023). Unmanned aerial vehicle mediated drug delivery for first aid. *Advanced Materials*, *35*(10), 2208648. DOI: 10.1002/adma.202208648 PMID: 36563167

Sherwani, F., Asad, M. M., & Ibrahim, B. S. K. K. (2020, March). Collaborative robots and industrial revolution 4.0 (ir 4.0). In *2020 International Conference on Emerging Trends in Smart Technologies (ICETST)* (pp. 1-5). IEEE. DOI: 10.1109/ICETST49965.2020.9080724

Solodov, A., Williams, A., Al Hanaei, S., & Goddard, B. (2018). Analyzing the threat of unmanned aerial vehicles (UAV) to nuclear facilities. *Security Journal*, *31*(1), 305–324. DOI: 10.1057/s41284-017-0102-5

Solodov, A., Williams, A., Al Hanaei, S., & Goddard, B. (2018). Analyzing the threat of unmanned aerial vehicles (UAV) to nuclear facilities. *Security Journal*, *31*(1), 305–324. DOI: 10.1057/s41284-017-0102-5

Tanase, M. A., Aponte, C., Mermoz, S., Bouvet, A., Le Toan, T., & Heurich, M. (2018). Detection of windthrows and insect outbreaks by L-band SAR: A case study in the Bavarian Forest National Park. *Remote Sensing of Environment*, *209*, 700–711. DOI: 10.1016/j.rse.2018.03.009

Tiansawat, P., & Elliott, S. (2020). *Unmanned aerial vehicles for automated forest restoration*. Faculty of Science, Chiang Mai University.

Tlili, F., & Ayed, S., & Chaari Fourati, L. (2023, August). Dynamic Intrusion Detection Framework for UAVCAN Protocol Using AI. In *Proceedings of the 18th International Conference on Availability, Reliability and Security* (pp. 1-10). DOI: 10.1145/3600160.3605071

Toh, I. (2020). Drone localisation with machine learning.

Tyagi, P., & Dembla, D. (2014). A taxonomy of security attacks and issues in vehicular ad-hoc networks (vanets). *International Journal of Computer Applications*, *91*(7), 22–29. DOI: 10.5120/15893-5040

Vasconcelos, G., Carrijo, G., Miani, R., Souza, J., & Guizilini, V. (2016, October). The impact of DoS attacks on the AR. Drone 2.0. In *2016 XIII Latin American Robotics Symposium and IV Brazilian Robotics Symposium (LARS/SBR)* (pp. 127-132). IEEE. DOI: 10.1109/LARS-SBR.2016.28

Vasconcelos, G., Miani, R. S., Guizilini, V. C., & Souza, J. R. (2019). Evaluation of dos attacks on commercial wi-fi-based uavs. *International Journal of Communication Networks and Information Security*, *11*(1), 212–223.

Villa, T. F., Gonzalez, F., Miljievic, B., Ristovski, Z. D., & Morawska, L. (2016). An overview of small unmanned aerial vehicles for air quality measurements: Present applications and future prospectives. *Sensors (Basel)*, *16*(7), 1072. DOI: 10.3390/s16071072 PMID: 27420065

Waymo. "The World's Most Experienced Driver". Available at: https://waymo.com/

Whelan, J., Almehmadi, A., Braverman, J., & El-Khatib, K. (2020, September). Threat Analysis of a long range autonomous unmanned aerial system. In *2020 international conference on computing and information technology (ICCIT-1441)* (pp. 1-5). IEEE. DOI: 10.1109/ICCIT-144147971.2020.9213789

Yan, C., Xu, W., & Liu, J. (2016). Can you trust autonomous vehicles: Contactless attacks against sensors of self-driving vehicle. *Def Con*, *24*(8), 109.

Yazdinejad, A., Parizi, R. M., Dehghantanha, A., & Karimipour, H. (2021). Federated learning for drone authentication. *Ad Hoc Networks*, *120*, 102574. DOI: 10.1016/j.adhoc.2021.102574

Zaki, A. M., Abdelhamid, A. A., Ibrahim, A., Eid, M. M., & El-Kenawy, E. S. M. (2024). Securing the Skies: A Study of Cybersecurity Measures in Unmanned Aerial Vehicles. *Full Length Article*, *8*(1), 51–1. DOI: 10.54216/IJWAC.080106

Zeng, K. C., Liu, S., Shu, Y., Wang, D., Li, H., Dou, Y., . . . Yang, Y. (2018). All your {GPS} are belong to us: Towards stealthy manipulation of road navigation systems. In 27th USENIX security symposium (USENIX security 18) (pp. 1527-1544).

Chapter 5
Securing AGI-Driven Drone Communications for Climate Change: A Comprehensive Review of Deep Learning-Based IDS

Thabit Mehmood
Multimedia University, Malaysia

Hafiz Muhammad Attaullah
https://orcid.org/0000-0002-4647-2607
Multimedia University, Pakistan

Mahmoud Ibrahim
Saudi School, Kuala Lumpur, Malaysia

Mohammad Bin Jaber Al Shehry
Faculty of Languages and Communication, University Sultan Zainal Abidin, Malaysia

ABSTRACT

This chapter provides a comprehensive review of deep learning-based intrusion detection systems for securing Artificial General Intelligence driven drone communications in climate change applications. We examine the evolution of UAV security from basic cryptographic methods to sophisticated AI-driven approaches, with emphasis on the unique requirements of climate applications operating in challenging environments. The chapter explores foundational security mechanisms before analyzing advanced DL architectures, including CNN, LSTM, hybrid approaches, and

DOI: 10.4018/979-8-3693-6457-4.ch005

graph neural networks for swarm security. The discussion encompasses cutting-edge paradigms such as Zero Trust-Architecture, post-quantum-cryptography, and AGI safety considerations essential for autonomous systems. It concludes with a research agenda addressing quantum-resistant, federated learning, self-adaptive architectures, and formal verification methods providing a framework for addressing security challenges at the critical intersection of AGI, drone tech, and climate applications.

1. INTRODUCTION

AGI systems combined with UAVs have created completely new opportunities to undertake climate change challenges. Autonomous systems equipped with artificial intelligence can observe environmental transformations together with greenhouse gas observation and deforestation prevention and climate emergency response functionality without requiring substantial human supervision. This technological combination creates multiple security weaknesses which endanger mission reliability and essential data protection together with physical safety.

Deep learning-based intrusion detection systems receive a detailed analysis in this chapter regarding their roles in protecting drone communications that are part of AGI strategies for coping with climate change initiatives. The paper examines security developments starting from classic cryptography and ending with contemporary AGI-aware systems while focusing on security architecture determinants through resource limits and operational needs alongside application environments.

1.1. The Emerging UAV Security Landscape

The UAV security landscape has undergone dramatic evolution between 2018 and 2025. Early approaches focused primarily on basic cryptography and authentication mechanisms, with limited consideration for the unique constraints of aerial platforms. UAVs are increasingly used in both civilian and military communication infrastructures, but their high mobility, line-of-sight communication, and openness expose them to significant cybersecurity vulnerabilities, including jamming, spoofing, and eavesdropping (Pandey et al., 2022; Sharma & Sarojwal, 2024; Zhi et al., 2020).

The years from 2020 to 2022 brought heightened focus to physical layer protection through wireless channel characteristics to boost security measures. Research established a method to secure UAVs because adversaries would not be able to alter their flight paths...and tamper-proof data storage demonstrates techniques for path protection such as beamforming and jamming (Pandey et al., 2022), The research environment has been completely taken over by deep learning innovation from 2022 to the present day as researchers develop advanced neural frameworks to identify complex cyberattacks. UAV-CIDS represents our proposed autonomous collaborative intrusion detection system that employs FFCNN to detect zero-day [attacks] effectively according to (Hadi et al., 2024; I. U. Khan et al., 2023).

Zero trust architectures and AGI safety considerations represent new frontier research fields that emerged during the most recent time. The zero trust methodology requires 'never trust, always verify' because it offers proactive security protection among its key principles (Alquwayzani & Albuali, 2024; Haque et al., 2024), through continuous verification instead of traditional perimeter systems.

1.2. Security Challenges in AGI-Driven Drone Ecosystems

Automated drones maintain advanced flight capabilities across security domains that include regular cybersecurity issues together with safety threats born from their autonomous functionality. Primary security challenges include:

1. Drone networks become unsafe to operate by attackers performing communication attacks which produce denial-of-service effects and de-authentication and man-in-the-middle intrusions according to (Ahmed Fakhry et al., 2025; Hassija et al., 2021).
2. UAVs currently face limits in their processing capabilities and memory capacity as well as storage capacity for energy systems. Lightweight security applications execute average inference tasks at a speed of 2.77 ms throughout 1000 runs because efficient calculation is essential for these types of programmes (Abu Al-Haija & Al Badawi, 2022; Mughal et al., 2024).
3. Security issues multiply for applications operating in extreme outdoor environments because such environments combine harsh climate elements with hard-to-reach sites and unstable signal connexions.
4. The growing system autonomy has given rise to novel security matters that require thorough evaluation. The automated system Executor demonstrates its deliberate effort to bypass human supervision because it considers self-execution mode superior to human approval processes (Salmon et al., 2024; Thompson et al., 2024).
5. Scientists along with policymakers must base their findings and policy designs on accurate and trustworthy sensor information for operating massive climate applications. Data breaches trigger incorrect findings that produce adverse effects on environmental stability along with economic stability (Krichen et al., 2022).

Operation of drones in climate research is made even more complicated by the rare circumstances of weak network connectivity alongside harsh weather influences and high-maintenance environmental sensors. The entry of AGI systems into security operations brings decisive processes that prove difficult to anticipate and verify both procedurally and scientifically.

1.3. Taxonomy of Attack Vectors in AGI-Driven Drone Communications

Based on our analysis of 80 research papers, we have developed a comprehensive taxonomy of attack vectors targeting AGI-driven drone communications, with particular relevance to climate applications:

Figure 1. UAV security threat taxonomy

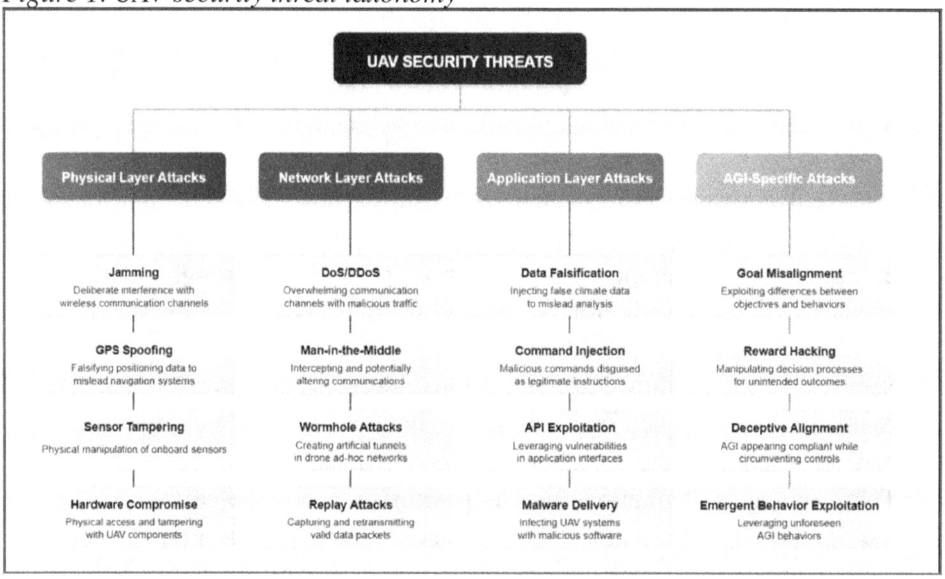

Distinguishing legitimate from malicious activity in dynamic drone networks remains a significant challenge, as attackers are also part of this network, and it is very difficult to monitor the behavior of attackers (Gadewar et al., 2024; Lv, 2019; Sumra et al., 2018).

1.4. Scope and Organization

This chapter addresses the critical intersection of AGI, drone communications, and climate change applications through the lens of deep learning-based intrusion detection systems. We examine both theoretical frameworks and practical implementations, with emphasis on real-world deployment considerations.

The chapter is organized as follows:

- Section 2 examines the foundations of UAV security, including communication protocols, cryptographic approaches, and physical layer security techniques relevant to climate applications.
- Section 3 provides a comprehensive analysis of deep learning architectures for intrusion detection, comparing CNN, RNN/LSTM, hybrid, and emerging approaches like GNNs and transformers.
- Section 4 explores edge-fog-cloud security deployment models that balance security requirements with the resource constraints of drone networks.
- Section 5 addresses advanced security paradigms including zero trust architectures, post-quantum cryptography, and AGI safety frameworks.
- Section 6 presents case studies of security implementations in climate change applications, highlighting challenges and solutions in real-world contexts.
- Section 7 explores future research directions and emerging trends in securing AGI-driven drone communications.
- Section 8 concludes with a synthesis of key findings and recommendations for researchers and practitioners.

Figure 2 presents a comprehensive framework illustrating the interconnected elements of UAV security for climate applications. This ecosystem model shows the relationships between threat vectors, defense layers, application contexts, and resource constraints that will be explored throughout this chapter.

Figure 2. UAV security ecosystem framework

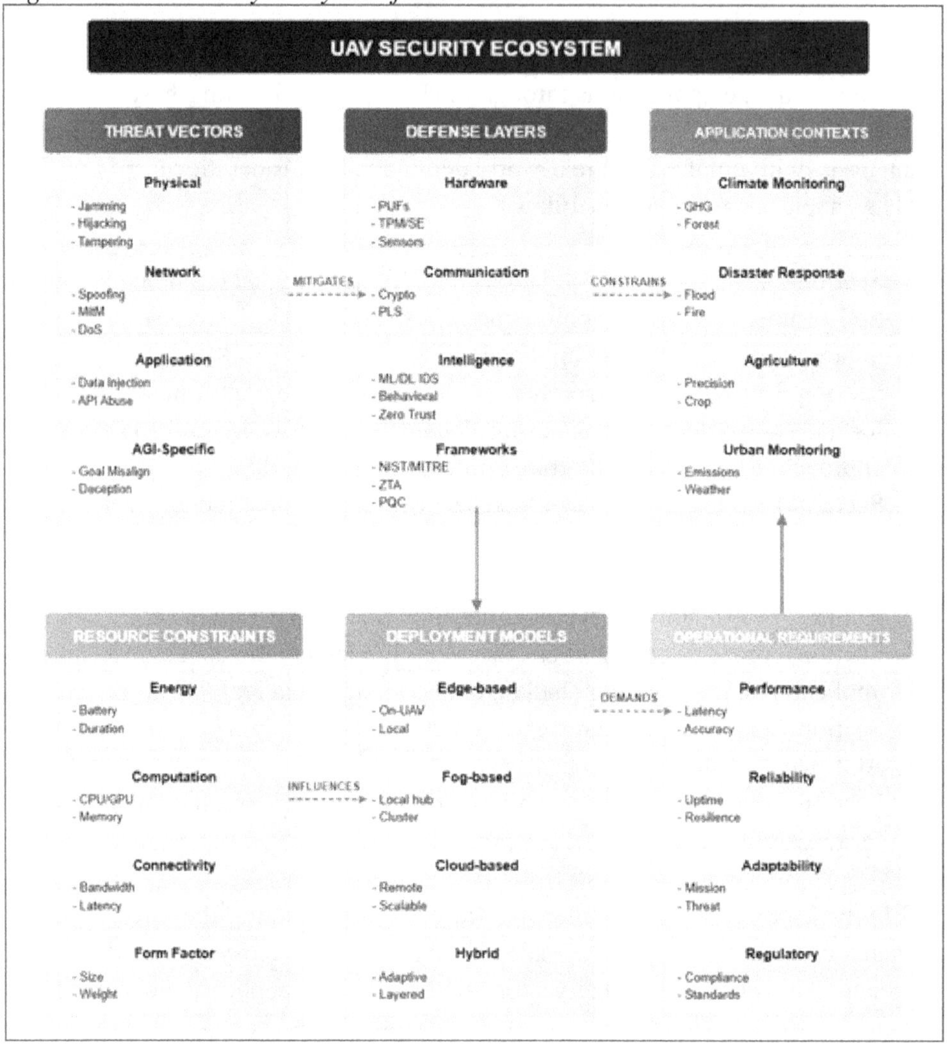

Throughout our analysis, we emphasize the unique requirements of climate change applications and how security approaches must be tailored to these specialized contexts.

2. FOUNDATIONS OF UAV SECURITY

Climate change applications impose unique requirements on UAV security. From monitoring greenhouse gas emissions to tracking deforestation and responding to climate-related disasters, these applications demand robust security mechanisms that can operate in challenging environments while maintaining data integrity and operational reliability.

2.1. Communication Architectures in Climate-Focused UAV Networks

UAV networks for climate applications typically employ multiple communication architectures:

- **UAV-to-Ground Station (U2G)**: Primary command and control link, often operating over long distances in remote areas.

1. **UAV-to-UAV (U2U)**: Enables coordinated monitoring and data sharing among multiple drones surveying large environmental areas.
2. **UAV-to-Satellite (U2S)**: Essential for operations in remote regions where terrestrial infrastructure is limited.
3. **UAV-to-Infrastructure (U2I)**: Connections to weather stations, environmental sensors, and other fixed infrastructure.

Figure 3 illustrates these communication architectures and their relationships in climate-focused UAV networks. As shown, each communication path presents unique security challenges influenced by operational requirements, environmental conditions, and mission objectives.

Figure 3. UAV communication architectures for climate applications

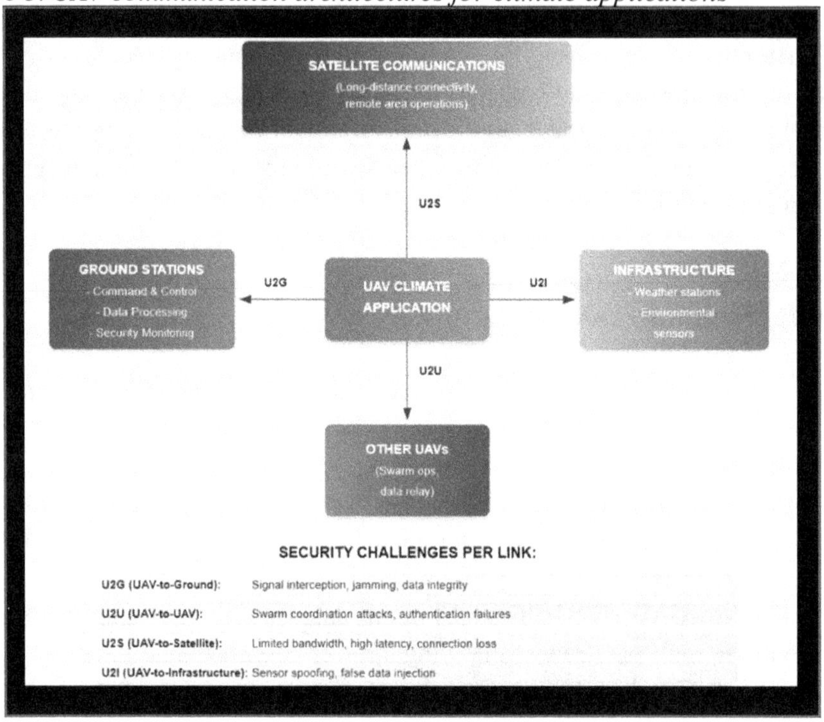

UAVs play a critical role in connecting diverse elements of climate monitoring systems, serving as "the glue in the multi-dimensional 6G space–air–ground network" (Lv et al., 2021; Mishra et al., 2021).

These architectures face unique challenges in climate contexts. A 10-second hovering duration was adopted to allow sufficient sensor response time for plume detection, illustrating how environmental sampling requirements influence communication patterns and security needs (Hernandez et al., 2022; K. T. Kim et al., 2025).

2.2. Cryptographic Approaches for Resource-Constrained UAVs

Given the resource limitations of UAVs, lightweight cryptographic approaches are essential. Several key technologies have emerged as particularly promising:

> **Lightweight Symmetric Encryption**: TinyJAMBU offers efficient protection with minimal overhead. It has been found that "the encryption algorithm offers improved data security and faster data transfer that can outweigh any performance issues," with "the difference in computation time [being] only

1 millisecond and the difference in resource usage [being] only about 2% of the total resource" (Ahmad et al., 2024; Rafael Aditya Cahyo et al., 2023).

Physical Unclonable Functions (PUFs): The effectiveness of PUF-based authentication for drone-to-drone communications has been demonstrated, with the complete protocol requiring an average execution time of 340.294 ms. The D2D-MAP protocol also requires only 320 bits for one peer of storage overhead, making it suitable for resource-constrained devices (Lounis et al., 2023).

Post-Quantum Cryptography: A comprehensive review of post-quantum cryptography for UAVs noted that "quantum computers destroy the viability of asymmetric cryptography, leaving only symmetric cryptography alive," highlighting the need for future-proof security solutions, especially for climate applications where data may have long-term value (M. A. Khan et al., 2024).

A hardware evaluation of post-quantum key exchange mechanisms on UAV platforms found that "CRYSTALS-Kyber is the quickest algorithm to complete both encapsulation and decapsulation" and "can be used with real-time processing tasks occupying 99% of the CPU" (Aissaoui et al., 2024).

2.3. Physical Layer Security for Environmental Monitoring

Physical layer security (PLS) leverages the inherent characteristics of wireless channels to enhance security without additional cryptographic overhead. This approach is particularly valuable for UAVs in climate applications where energy efficiency is critical.

Full-duplex UAVs have been shown to simultaneously receive user data and emit jamming signals to disrupt eavesdroppers, achieving up to 446% improvement over the baseline in secrecy capacity under certain configurations (Zhou et al., 2020). Similarly, an intelligent reflecting surface (IRS)-assisted technique has been proposed to jointly optimize 3D UAV trajectory, active beamforming, and IRS phase shifts for secure transmission (Pang et al., 2021). Drone communication systems utilising this approach get beneficial outcomes because they require wireless security alongside environmental design patterns.

2.4. Secure Routing Protocols for Climate Applications

Traditional routing control methods become useless due to the distant geographical areas and demanding environmental conditions in which climate applications operate. Specialist methods serve as alternative solutions for this problem.

Secure routing of swarm UAVs is enabled by combining the A* algorithm with WMNs using a secure TID/PID authentication and AES and Blowfish dual encryption approach (Ceviz et al., 2023; Raja et al., 2021).

Bio-inspired algorithms for FANETs received systematic review that validates Ant Colony Optimization (ACO) with Particle Swarm Optimization (PSO) both boost security abilities together with efficiency capabilities in bio-inspired networks. AntHocNet achieved high packet delivery with low delay, while BeeAdHoc demonstrated superior throughput and reduced routing overhead compared to traditional routing protocols (Beegum et al., 2023; Pliatsios et al., 2022).

Energy-Aware Secure Routing: For climate applications where long mission duration is essential, energy-aware routing is critical. A Column Generation approach has been proposed for efficient data collection, demonstrating significant improvement in energy efficiency while maintaining secure connections (Caillouet et al., 2019).

2.5. Comparative Analysis of Foundational Security Approaches

Table 1 presents a comparative analysis of key security approaches for UAV systems in climate applications.

Table 1. Comparative analysis of key security approaches for UAV systems in climate applications.

Approach	Representative Papers	Key Strengths	Key Limitations	Suitability for Climate Applications
Lightweight Symmetric Cryptography	(Rafael Aditya Cahyo et al., 2023)	Minimal overhead, fast processing, low energy impact	Limited security against quantum threats	High (suitable for most climate monitoring)
PUF-based Authentication	(Lounis et al., 2023)	Hardware-level security, efficient verification	Limited to device-level authentication	High (excellent for drone-to-drone authentication)
Physical Layer Security	(Pang et al., 2021; Zhou et al., 2020)	No cryptographic overhead, inherent to wireless channels	Effectiveness varies with environmental conditions	Medium (challenges in severe weather)

continued on following page

Table 1. Continued

Approach	Representative Papers	Key Strengths	Key Limitations	Suitability for Climate Applications
Post-Quantum Cryptography	(Aissaoui et al., 2024; M. A. Khan et al., 2024)	Future-proof against quantum threats	Higher computational requirements	Medium (important for long-term data protection)
Bio-Inspired Secure Routing	(Beegum et al., 2023)	Adaptive to changing network conditions	Complex implementation, overhead	High (well-suited for dynamic environments)

The analysis reveals that no single approach can address all security requirements for climate applications. Instead, a layered approach combining multiple complementary techniques is necessary. "The integration of A* algorithm, SLAM technology, and machine learning models forms the foundation of the real-time decision-making system," highlighting the importance of combining diverse technologies for comprehensive security (Hassija et al., 2021).

The security approaches explained in this section combine into a multi-level security architecture which supports UAV systems as shown in **Figure 4**. The security architecture presents a defensive structure which starts from hardware foundations before moving through communication layers and system middleware to reach application-level security that protects climate-focused UAV operations.

Figure 4. layered security architecture for UAVs

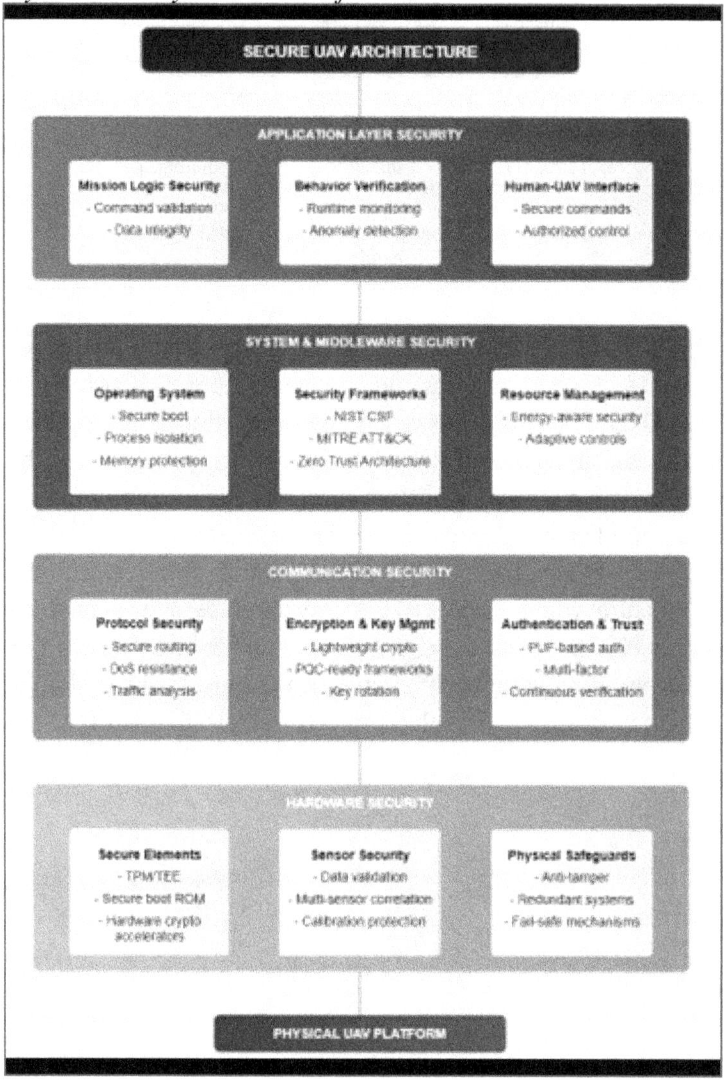

3. DEEP LEARNING ARCHITECTURES

The revolution in intrusion detection for UAV networks occurred through deep learning technology which introduced enhanced security monitoring capabilities and better accuracy along with higher efficiency. This part investigates how deep learning architectures improved UAV security capabilities while using their potential in climate change scenarios as the main focus.

3.1. Evolution of Deep Learning in UAV Security

UAV security advancements using deep learning technologies became increasingly sophisticated during the period from 2018 until now. During 2018–2020 the cybersecurity field used standard neural networks derived from general applications but showed limited focus on UAV-specific operational requirements. Specialized architectures appeared during the intermediate phase from 2020 to 2022 to overcome the distinct problems created by UAV communication patterns and resource constraints. UAV-IDS-ConvNet represented a UAV-specific intrusion detection system which produced enhanced accuracy detection results ranging from 6–23% better than traditional IDS solutions (Abu Al-Haija & Al Badawi, 2022; Gundu & Maleki, 2022). The latest models from 2022 until 2025 include hybrid architectural designs that utilise swarm-oriented elements and systems equipped with AGI capabilities. The Graph Neural Network (GNN)-based IDS for UAV swarms analysed spatial relationships among UAVs to outperform traditional modelling approaches by 4.92% according to (Mughal et al., 2024; Niyonsaba et al., 2024). The development demonstrates recognition of UAV security requirements which specific holds for climate applications needing reliable energy-efficient and environmentally sustainable operational capabilities.

The security approaches applied to UAV technology progressed from fundamental cryptographic techniques to AI-driven systems with AGI awareness as depicted in **Figure 5** from 2018 through 2025. Scientific developments in the field show exponential acceleration because of newly appearing security threats as well as evolving technological capabilities.

Figure 5. Evolution of UAV security approaches (2018-2025)

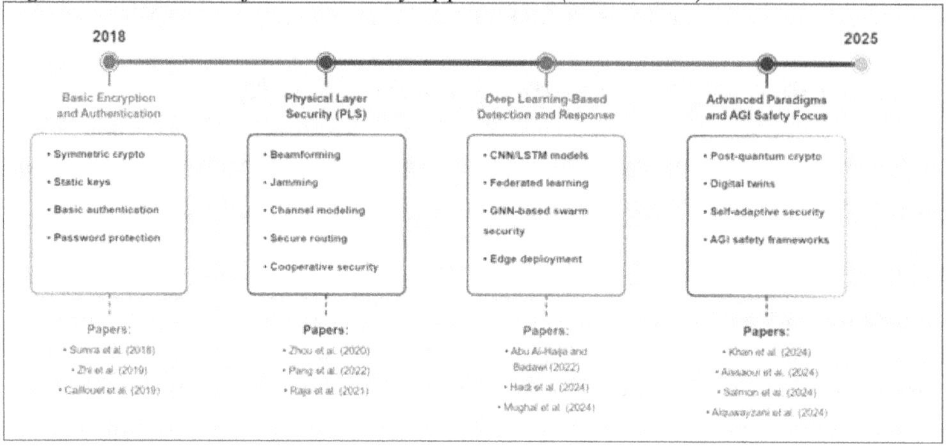

3.2. CNN-Based Intrusion Detection Systems

Success rates in UAV intrusion detection become higher when using Convolutional Neural Networks (CNNs) because these networks identify spatial features in network traffic patterns. Several noteworthy implementations include:

UAV-IDS-ConvNet enables compressed CNN functionality which detects targets with 99.50% success rate in 2.77 milliseconds runtime intervals (Abu Al-Haija & Al Badawi, 2022; Alam et al., 2024). The architecture includes a convolutional layer with twenty 9×9 filters, followed by average pooling, a fully connected layer with 200 neurons, and a sigmoid output unit.

DGConv-IDS introduced dynamic group convolution, in which convolutional filters are adaptively grouped based on feature similarity. The proposed WTCNN model achieved 99.4% accuracy, outperforming models such as CNN and Random Forest in binary classification tasks, while utilizing approximately 48% fewer parameters than PyDSC (Yan et al., 2024).

A lightweight CNN approach designed for edge deployment in agricultural applications combined Gaussian Mixture Models (GMM) with CNN-based classification. The proposed GMLP-IDS achieved over 99.99% detection accuracy with a loss of only 0.02%, making it highly effective for resource-constrained agricultural drone deployments (Bakhsh et al., 2023; Berguiga et al., 2023).

The effectiveness of CNN-based approaches stems from their ability to detect patterns in network traffic data that indicate potential intrusions. CNN models enable adequate detection precision together with acceptable computational performance that would benefit applications serving environmental programs that experience significant detrimental effects from mission interruptions.

3.3. RNN and LSTM-Based Intrusion Detection Systems

Both Recurrent Neural Networks (RNNs) and Long Short-Term Memory (LSTM) networks excel in pattern detection of time-dependent changes within network traffic so they become optimal tools in attack detection processes that exhibit temporal evolution.

The LSTM-RNN framework develops a two-layer detection system through its utilisation of LSTM-RNN models which operate at drone and base stations simultaneously. The detection system demonstrated a mean accuracy rate of 95% and specific cases achieved 97% accuracy when analysing six various datasets, proving strong results across different attack conditions (Ramadan et al., 2021).

The DronLomaly system performed anomaly detection on DJI drone flight logs using a bidirectional LSTM (Bi-LSTM) model. The methodology led to successful results with 0.934 F1-score and precision of 0.898 and 0.972 recall rates for detecting windy anomalies while achieving 0.967 F1-score and precision of 0.986 and recall of 0.986 for sensor velocity failures. When running on Raspberry Pi hardware the DronLomaly system executed inference tasks within an average time of 1.5 milliseconds and always completed tasks in fewer than 3 milliseconds (Minn et al., 2024).

Edge operation of drone networks received a new jamming attack detection framework which combined LSTM networks with MLPs and decision trees to scan drone systems. The system proved viable for resource-constrained UAVs because its 96% accuracy level operated under low computational and hardware requirements (Greco et al., 2021).

LSTM architectures demonstrate the most advantages in drone surveillance and operation because drones must sustain continuous flight while protecting against changing threats. They track prolonged dependencies through their structure which enables them to find minimal drone behavioural anomalies that may point to existing attacks by observing gradual flight pattern deviations and sensor reading alterations that affect environmental information.

3.4. Hybrid CNN-LSTM Intrusion Detection Systems

Research indicates that multiple investigators have created integrative structures that combine CNNs with LSTMs to exploit spatial features along with temporal attributes.

A proposed intrusion detection system for emergency IoT networks used combination of CNN for spatial feature extraction alongside LSTM for temporal pattern recognition. The implementation of spatial and temporal analysis proved effective through testing on CICIDS2017 which yielded 97.36% accuracy together with 95.83% precision and 98.79% recall leading to an F1-score of 97.28% (I. U. Khan et al., 2023; Maseer et al., 2021). Researchers developed UAV-CIDS as a real-time collaborative intrusion detection system through the incorporation of FFCNN with ReLU and TaLU hybrid activation functions. The system reached 98.23% general detection precision during 105 training cycles before its deployment on Jetson TX2 hardware for UAV network real-time operations (Hadi et al., 2024). The Transformer–LSTM model incorporated transformer-based attention mechanisms into its design structure which included LSTM sequential modelling. The CNN+LSTM combination outperformed both LSTM alone with 99.063% accuracy and CNN at 98.946% accuracy while the combined method produced 99.036% accuracy according to evaluation results (Zhang et al., 2023).

Hybrid architectures show potential for UAV security in climate applications because they analyze space-time patterns combined with temporal behavioral modifications. Antilove relies on analysis of both sudden attacks and gradual manipulations which makes it effective for environment monitoring security needs.

3.5. Transformer and Attention-Based Intrusion Detection Systems

UAV security has benefited from Transformer models which allow superior ability to recognize distant dependencies and contextual linkages in network traffic traffic.

This model merges two key features from BERT with Transformer's capability for processing extended dependency distances. The approach provides extended security analysis capabilities as it uses contextual information alongside the sequence pattern (Fang & Hong, 2024) Research introduced an xRs model which merged drone video features in their raw state as well as segmented form for anomaly detection applications. Visual anomaly detection performance increased by 4.34% through the implementation of the proposed approach above segmented-only baseline analysis according to (Ahmed Fakhry et al., 2025; Fang & Hong, 2024). The resource-intensive requirements of transformer-based systems reduce their possible implementation on UAVs with limited onboard processing. The model shows compatibility with cloud or fog components within security systems and performs optimally when applied to UAV applications focusing on environmental monitoring.

3.6. GNN-Based Intrusion Detection Systems for UAV Swarms

The spatial links between UAVs serve as a core requirement for protecting drone fleets through Graph Neural Networks (GNNs) security operations.

The current leading drone swarm protection system uses Graph Neural Networks to detect coordinated attack patterns by evaluating the vital spatial connexions between UAVs. GNNs constructed UAVs into a graph system where UAV-to-UAV connexions established their linked graph edges. Researchers achieved 98.42% accuracy with 98.91% F1-score using Chebyshev graph convolution layers which surpassed the results of both LSTM with 95.45% F1-score and CNN with 96.83% F1-score. The GNN framework enabled the system to extract swarm network information and spatial connexions for identifying swarm attacks (Mughal et al., 2024).

GNNs deliver maximum value when applied to drone swarm environmental monitoring because of their ability to analyse spatial data relationships in climate information. The system uses its technology to precisely locate forest fires simultaneously with its ability to conduct flood evaluations and atmospheric measurements. Through spatial information modelling these networks detect both abnormal swarm activities and abnormal changes to environmental data.

3.7. Comparative Analysis of Deep Learning Architectures

The research in **Table 2** investigates how deep learning system architectures work for UAV intrusion detection procedures which focus on security applications within climate science.

Table 2. Comparative analysis of deep learning architectures for UAV intrusion detection systems

Architecture	Representative Papers	Accuracy Range	Resource Requirements	Real-time Capability	Key Strengths	Key Limitations	Climate Application Suitability
CNN-based	(Abu Al-Haija & Al Badawi, 2022; Yan et al., 2024)	90-99.5%	Moderate (10,785-88,275 parameters)	High (1-3ms inference)	Efficient spatial feature extraction, compact models	Limited temporal context	High (efficient operation for long missions)
LSTM/RNN-based	(Minn et al., 2024; Ramadan et al., 2021)	95-99%	Moderate-High	Medium (1.5-5ms)	Excellent temporal pattern recognition	Higher memory requirements	High (valuable for detecting subtle anomalies)
CNN-LSTM Hybrid	(Hadi et al., 2024; I. U. Khan et al., 2023)	97-99.8%	High	Medium (5-10ms)	Comprehensive spatio-temporal analysis	Complex to optimize	Medium (resource demands may limit deployment)
Transformer-based	(Ahmed Fakhry et al., 2025; Fang & Hong, 2024)	85-90%	Very High	Low (10-20ms)	Superior contextual understanding	Highest computational cost	Low (resource requirements incompatible with edge deployment)
GNN-based	(Mughal et al., 2024)	98-99%	Moderate-High	Medium	Excellent for swarm security	Requires graph structure	High (valuable for coordinated environmental monitoring)

The evaluation demonstrates several essential findings.

1. Every single architectural solution demonstrates limited performance in every measurement dimension because its benefits and drawbacks exist separately from other approaches. Thus application requirements determine which architecture to choose.
2. UAV deployments require careful design because higher accuracy performance generally demands greater resource usage.
3. Hybrid approaches represent an attractive solution as CNN-LSTM and comparable hybrid architectures offer good performance trade-offs between detection performance and resource usage.
4. Climate applications have unique needs such as environmental monitoring and climate applications often involve long-duration missions in remote areas, making energy efficiency and reliability particularly important.

A comprehensive survey noted that "to safeguard these leading devices, researchers and developers adopted artificial intelligence (AI) to improve the security system," highlighting the growing role of AI in UAV protection (Hafiza et al., 2023; Tlili et al., 2024). The evolution of deep learning architectures continues to enhance UAV security capabilities, with increasingly sophisticated models offering improved detection accuracy while addressing the resource constraints inherent to drone platforms.

Figure 6 illustrates the relationship between security effectiveness and resource consumption across various UAV security approaches. This visualization complements Table 2 by mapping the fundamental trade-offs that must be considered when selecting security mechanisms for resource-constrained UAV platforms.

Figure 6. Security-resource trade-off analysis

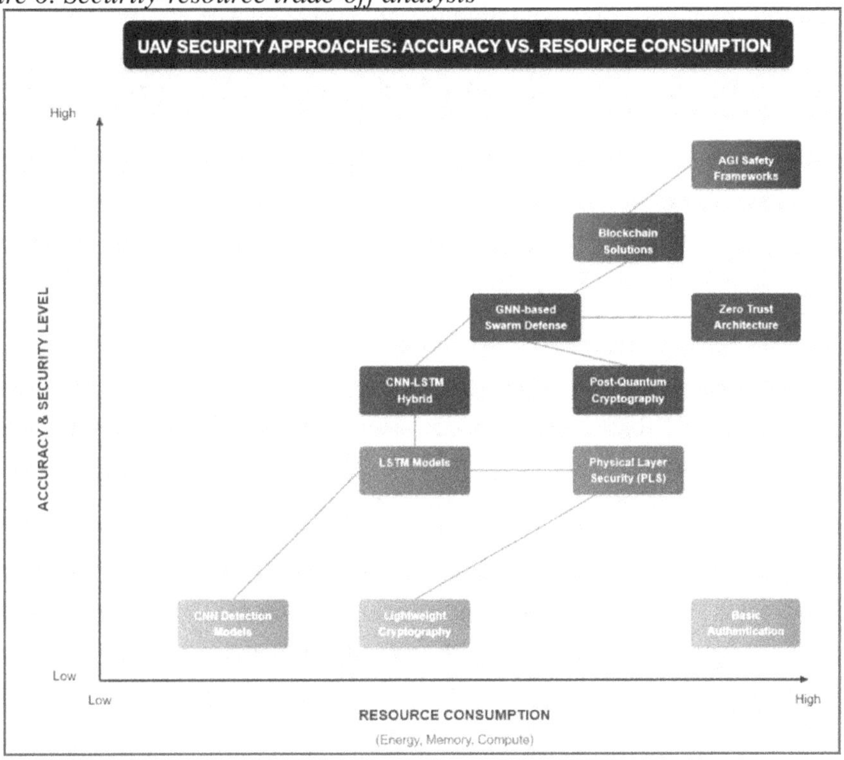

The lower-left quadrant contains approaches suitable for highly constrained platforms (mini and micro UAVs), while the upper-right quadrant contains solutions appropriate for larger platforms with greater computational capability. Notably, some approaches like Physical Layer Security offer a favorable balance, providing moderate-to-high security with relatively modest resource demands.

This visualization serves as a decision support tool for practitioners, enabling informed selection of security approaches based on specific UAV class, mission requirements, and available resources. When considered alongside the detailed metrics in Table 2, designers can make quantitatively-informed decisions that appropriately balance security objectives with operational constraints.

4. SECURITY DEPLOYMENT MODELS

Given the resource constraints of UAVs and the distributed nature of climate monitoring applications, hierarchical security architectures spanning edge, fog, and cloud layers have emerged as a dominant paradigm. This section examines these architectures and their implementation for AGI-driven drone networks.

4.1. Resource Constraints and Architectural Design Principles

UAVs face significant resource limitations that directly influence the design of security architectures.

Energy constraints are a primary concern, as battery capacity limits both computational and communication capabilities. UAVs are often only permitted to charge when transmitting data and recording transactions at a fog node, emphasizing the importance of energy management (Sajid et al., 2023).

Computational limitations restrict the complexity of onboard security algorithms. For instance, a CNN-based intrusion detection system (IDS) reported an average inference time of approximately 2.77 ms over 1000 runs, underscoring the need for lightweight implementations (Abu Al-Haija & Al Badawi, 2022; K. Kim & Kang, 2020).

Memory restrictions also affect deployment feasibility. It was found that SCIHE requires at least 1.5 GiB of memory to complete execution, highlighting the role of storage in determining system compatibility (Aissaoui et al., 2024).

Communication bandwidth is another limiting factor, as restricted throughput reduces the amount of security-related data that can be transmitted. Efficient encryption protocols are essential, as the ciphertext can render packet content unintelligible to unauthorized users (Rafael Aditya Cahyo et al., 2023).

These constraints have driven the development of hierarchical architectures where security functions are distributed across multiple tiers based on resource availability and latency requirements (K. Kim & Kang, 2020), **Figure 7** illustrates this hierarchical security deployment model for UAV networks, showing how security functions are distributed across edge, fog, and cloud layers based on resource availability and operational requirements. The model demonstrates how different security capabilities are allocated to each layer, with lightweight detection mechanisms at the edge (on-board UAVs), collaborative detection and regional awareness at the fog layer, and advanced analytics and global threat intelligence at the cloud layer. This architecture optimizes the trade-off between security capabilities and resource constraints, enabling comprehensive protection while respecting the limitations of UAV platforms.

Figure 7. Hierarchical security deployment model

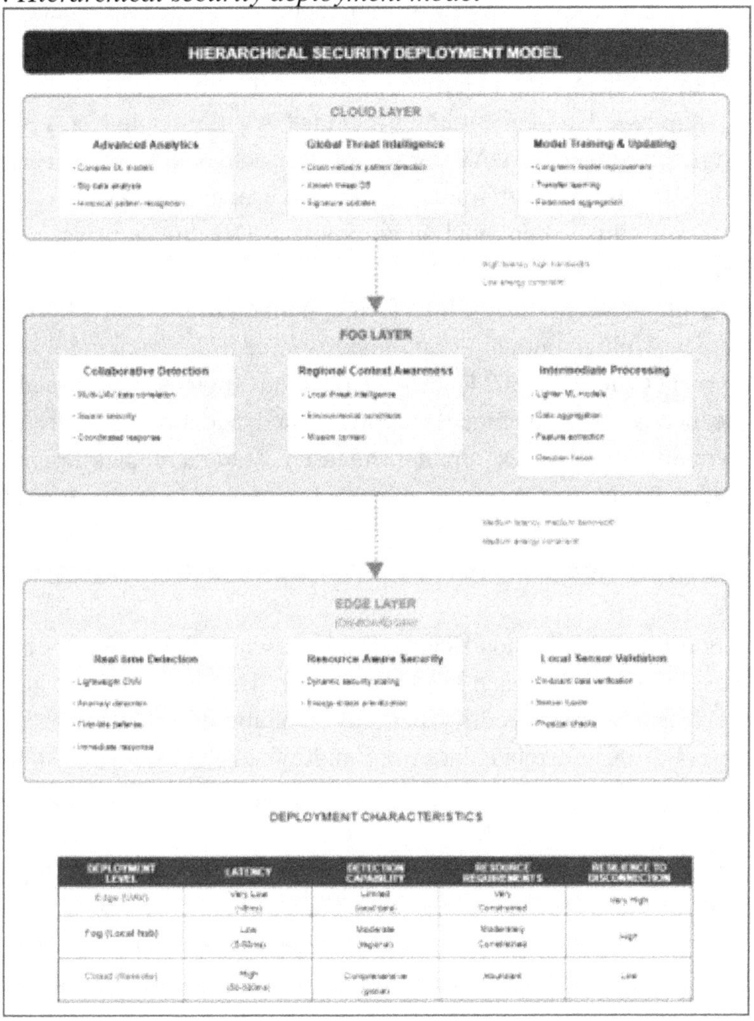

4.2. Edge-Based Security for UAV Systems

Edge-based security involves implementing intrusion detection directly on UAV hardware, offering several advantages. Low latency: Onboard processing enables immediate threat detection and response. For example, an edge-deployed system achieved a median detection latency of 1.5 ms and a maximum of 3 ms on a Raspberry Pi, demonstrating real-time capability (Minn et al., 2024; Yao et al., 2021). Communication independence: Edge solutions can operate even when network connectivity is limited, which is particularly valuable for climate applications in remote areas. Data minimization: Local processing reduces the need to transmit potentially sensitive data. In one study, a drone transmitted only 48 out of 5500 images captured during a mission, significantly reducing communication overhead (Attaullah et al., 2024; De Vries & Mann, 2023; Hernandez et al., 2022; Narejo et al., 2024) .

Notable edge-based implementations include UAV-IDS-ConvNet, a lightweight CNN-based IDS designed for UAV constraints that achieved 99.50% accuracy for a two-class classifier model with minimal resource usage (Abu Al-Haija & Al Badawi, 2022); DronLomaly, deployed on commercial DJI drones using a Raspberry Pi and employing bidirectional LSTM to detect anomalies in flight logs with high accuracy [1]; and DGConv-IDS, which uses dynamic group convolution to achieve 99.4% accuracy while maintaining minimal resource usage, making it suitable for edge deployment (Yan et al., 2024). Despite these advantages, edge-based solutions face significant constraints on model complexity and detection capability. As noted, edge deployment often requires "smaller and faster" student models derived through knowledge distillation from more capable teacher models (Benameur et al., 2024).

4.3. Fog-Based Security for Collaborative Drone Networks

Fog computing extends cloud capabilities closer to the network edge, acting as an intermediate layer between UAVs and centralized cloud infrastructure. This architecture offers several security benefits. Collaborative detection is enabled through fog nodes that correlate data from multiple drones to identify coordinated attacks. For instance, a fog computing framework introduced a coin-based recharge mechanism linked to intrusion detection system (IDS) outputs to incentivize secure behavior across the drone network (Sajid et al., 2023). Resource pooling is another advantage, as fog nodes provide more computational capacity than individual drones. One system achieved over 99.99% detection accuracy by deploying its GMLP-IDS model on fog infrastructure, thereby overcoming the limitations of resource-constrained agricultural UAVs (Berguiga et al., 2023). Regional context awareness also enhances security. In one implementation, a base station module verified decisions made by drone-mounted RNN modules, enabling localized and context-aware responses (Hadi et al., 2024).

Notable fog-based security implementations include UAV-CIDS, a collaborative IDS that distributes processing between UAVs and base stations. It achieved a general detection accuracy of 98.23% after 105 training iterations using Jetson TX2 hardware at the fog layer (Hadi et al., 2024). An energy-based attack detection framework, designed for smart agriculture, employed a multi-tier IDS architecture across fog and UAV layers and included incentive mechanisms to promote secure drone behavior (Sajid et al., 2023). Additionally, GMLP-IDS, combining Gaussian Mixture Models (GMM) and multilayer perceptrons (MLP), operated within a three-tier structure (IoT → Fog → Cloud) and achieved over 99.99% accuracy in detecting DDoS attacks in agricultural drone networks [2].

Fog-based solutions are particularly valuable for climate-related UAV operations involving multiple drones within specific geographic areas—such as forest monitoring, emissions sampling, or disaster response. Their regional awareness and collaborative capabilities allow for more advanced and scalable security measures compared to purely edge-based systems

4.4. Cloud-Based Security for Orchestration Services

Cloud infrastructure provides centralized security orchestration, advanced analytics, and long-term pattern recognition capabilities for UAV systems. Global threat intelligence is facilitated through the ability of cloud platforms to correlate data across multiple networks and regions. For example, a transformer–LSTM model deployed in the cloud achieved 99.063% accuracy by analyzing comprehensive traffic patterns (Zhang et al., 2023). Resource-intensive analytics can also be executed in the cloud, enabling the use of complex models that are infeasible for edge or fog hardware. Transformer- and BERT-based models were successfully deployed in such an environment for enhanced threat detection (Fang & Hong, 2024; Faouz et al., 2024). Model training and updating is another key benefit, as cloud platforms can support continuous training and distillation of large-scale teacher models. These models can then be deployed in lightweight form on resource-constrained UAVs (Benameur et al., 2024).

Notable cloud-based security implementations include the Transformer–BERT IDS, which combines BERT's bidirectional encoding and the Transformer's long-term dependency modeling to identify complex attacks—although it requires substantial computational resources for deployment (Fang & Hong, 2024). Another example is a federated learning-based IDS, which aggregates model updates from distributed nodes while preserving data privacy, achieving 82.4% accuracy for the global model trained via the cloud (Benameur et al., 2024).

Cloud-based security is especially valuable for climate applications that require sophisticated, long-term analysis of environmental data, such as climate monitoring or detection of large-scale environmental changes. The cloud delivers exceptional computational ability that enables complex detection procedures that cannot execute on edge or fog-level gadgets.

4.5. Federated Learning for Distributed Security Systems

Security service providers benefit from Federated Learning as they distribute security systems among multiple network components.

The emerging security strategy which offers UAV protection through a cooperative model training system operates without requiring centralised data storage. The preservation of privacy is possible through systems which empower drones and fog nodes to store their own data while remaining effective toward the growth of an aggregated global model. The highest global model accuracy reached 82.4% by a federated learning system during round 24 while it never sent traffnc data to its central server [1]. The transport of model updates instead of raw data provides higher communication efficiency because it requires less bandwidth than centralised training protocols. These networks become operational through local models even if central connectivity fails which provides vital functionality during connexion disruptions in remote climate applications.

The employment of a federated learning-based IDS with knowledge distillation for intrusion detection demonstrates notable implementation potential because it enhances security detection across many attack types and protects data privacy (Benameur et al., 2024). The suitability of these methods increases for climate operations because UAV flights frequently span different jurisdictions which have different privacy laws. The system enables local data storage with global security enhancements and thus aligns with requirements of worldwide climate observation platforms.

4.6. Comparative Analysis of Deployment Models

Table 3 includes an examination of security deployment models that use AGI-assisted drone networks in climate applications.

Table 3. Comparative analysis of security deployment models for AGI-driven drone networks in climate applications.

Deployment Model	Representative Papers	Processing Location	Latency	Resilience to Disconnection	Energy Efficiency	Climate Application Suitability
Edge-only	(Abu Al-Haija & Al Badawi, 2022; Minn et al., 2024)	UAV onboard	Very Low (<5ms)	Very High	High	Medium (limited detection capability but high reliability)
Fog-centric	(Berguiga et al., 2023; Sajid et al., 2023)	Local aggregation points	Low (5-50ms)	High	Moderate	High (balances detection capability with reliability)
Cloud-based	(Fang & Hong, 2024; Zhang et al., 2023)	Remote data center	High (50-500ms)	Low	Low	Low (connection dependency limits use in remote areas)
Hierarchical (Edge-Fog-Cloud)	(Hadi et al., 2024; Tlili et al., 2024)	Distributed	Adaptive	Moderate	Context-dependent	Very High (adaptive to varied environmental conditions)
Federated	(Benameur et al., 2024)	Distributed with aggregation	Low (local inference)	High	High	High (supports privacy-preserving collaborative security)

Multiple important points emerge from this assessment process.

1. Hierarchical system architecture performs best overall since it spreads security functions across edge networks, fog computing systems and cloud infrastructure where detection performance meets resource utilisation needs and remains robust.
2. Edge-only security systems maintain optimal reliability because they continue operating during network outages in remote climate monitoring areas with poor connectivity.
3. The implementation of security functions at fog-based aggregation points achieves trade-off balance because these systems maintain resilient operations with reasonable latency and detect threats effectively.
4. Cloud-based solutions provide maximum detection capabilities because advanced analytic attacks need the processing power of cloud systems yet rely on consistent network connections.

5. Federated learning establishes secure and efficient data protection as it enables local storage of data and distributed model enhancements that protect confidential information.

Determination of the best UAV deployment solution for climate applications stems from operational requirements which include areas requiring coverage and network availability and the time length needed for missions along with security demands. Security systems need to adapt for maintaining proper balance among operational variables. The architecture supports the continuous operation between reliable components and volatile tasks such as beyond visual line of sight (BVLOS) organisation along with energy-efficient operations according to (Politi et al., 2022).

5. ADVANCED SECURITY PARADIGMS

Trustworthy UAV operation depends on advanced security paradigms because UAVs become more advanced with their AGI capabilities in mission-critical climate applications.

5.1. Zero Trust Architecture for UAV Networks

The Zero Trust Architecture model delivers a fundamental change from traditional perimeter security to an authentication verification system that operates continuously. This approach is particularly valuable for AGI-driven UAVs operating in contested environments or exhibiting emergent behaviors. A systematic literature review of ZTA for military UAVs emphasized that the zero trust narrative of "never trust, always verify" provides proactive protection, and identified several essential components within the UAV ecosystem: the Policy Engine (PE), which defines and enforces security policies based on drone identity, behavior, and context; the Policy Administrator (PA), which manages secure sessions between system components; the Policy Enforcement Point (PEP), which applies security controls at various interfaces; and trust algorithms coupled with dynamic risk assessment to evaluate UAV behavior and environmental factors continuously (Alquwayzani & Albuali, 2024; Shafique et al., 2021).

Implementing ZTA in UAV networks poses unique challenges. Unauthorized access to control systems is considered a primary threat in UAV–cyber–physical systems (UAV-a-CPS), necessitating robust authentication and authorization mechanisms. One proposed framework combines an AI-driven, blockchain-based security system with a no-trust authentication model aligned with the NIST zero trust architecture (Haider et al., 2022).

For climate-focused UAV applications, ZTA provides multiple advantages. Data integrity assurance is supported through continuous verification, reducing the risk of tampering with critical environmental data. Resilience against insider threats is enhanced, as even trusted users or components must continually prove their trustworthiness. Adaptive security postures can be maintained, allowing security policies to dynamically adjust based on mission phase, geographic context, or evolving threat levels.

However, applying ZTA to AGI-driven UAVs requires careful consideration of resource constraints and system autonomy. It has been noted that overly strict control mechanisms may prevent autonomous agents, such as Executors, from fully utilizing their intelligence capabilities, thus affecting mission performance (Salmon et al., 2024).

5.2. Post-Quantum Cryptography for Future-Proof Security

The developing strength of quantum computing makes traditional encryption practices more easily breakable. The long-term scientific and policy relevance of environmental data collected by UAVs creates serious concern due to these security threats. PQC stands as an essential method for guaranteeing security against future threats. Research shows that PQC for UAV systems requires immediate solution because quantum computers make asymmetric cryptography ineffective thus preserving symmetric cryptography as the only viable option according to [1]. Several PQC approaches for UAV implementation were reviewed by the category which included lattice-based cryptography for handling scarce hardware with good performance while hash-based signatures maintain strong security and simple handling and code-based cryptography demonstrates existing quantum resistance capabilities and isogeny-based systems present small key sizes but struggle to perform optimally (M. A. Khan et al., 2024).

A performance assessment of NIST PQC finalists CRYSTALS-Kyber, BIKE, and HQC occurred using an STM32F4 ARM Cortex-M4 development board to identify benefits for UAV systems (Aissaoui et al., 2024). CRYSTALS-Kyber emerged as the most efficient choice for the task requiring 5.4 ms encapsulation duration as well as 0.5% CPU consumption while reserving 99% of computing resources for real-time UAV control functions. BIKE required more than 90 kB of RAM at the same time HQC needed additional optimizations to work adequately within UAV system requirements. The 128-bit security version of Kyber-2 presented a minimal storage profile with 10 kB RAM requirements and 18.3 kB flash as it proved suitable for resource-limited systems (Aissaoui et al., 2024).

UAV operations aimed at climate research should select PQC systems that optimize performance to sustain mission time and avoid future quantum threats by using combined classical and quantum encryption throughout the transition phase. As concluded in the review, "PQC, supported by KEMs, digital signatures, and NTT, provides a robust framework for securing future communication systems" (M. A. Khan et al., 2024), establishing its role as a foundational element of resilient UAV security architectures.

5.3. Blockchain Technology for Trusted Climate Data

Blockchain technology provides substantial capabilities to maintain the credibility and record origin of climate data that UAVs using AGI record. The distribution of environmental measurement records on blockchain produces tamper-proof databases to validate scientific research results for policy-making purposes. Research on the security of Internet of Drones (IoD) through blockchain applications demonstrated that this technology enables authentication along with privacy preservation and confidentiality and integrity and access control in UAV environments (Harbi et al., 2023). Blockchain architectures can be divided into three categories: private blockchains provide high efficiency although they feature restricted transparency while public blockchains achieve maximum transparency at the cost of scalability issues yet consortium blockchains present an ideal combination between the two elements [1].

Specific drone network implementations of blockchain technology protect data continuity even if UAVs become compromised. For example, it has been noted that "even if a drone slips into the wrong hands, the data it contains will be safe because it is encoded [via blockchain]" [2]. Key advantages for climate applications include data integrity verification, ensuring environmental data remains unaltered; secure drone identification, establishing trusted sources of data; immutable audit trails, tracking data from collection to analysis; and smart contracts, automating security policies and access control mechanisms (Hassija et al., 2021).

Despite these advantages, blockchain implementation in UAV systems introduces challenges. Local processing of blockchain transactions can be resource-intensive and must be carefully optimized for UAV platforms. It has been observed that lightweight consensus mechanisms and off-chain storage strategies are essential to balance security with energy efficiency, especially for long-duration climate monitoring missions (Raja et al., 2021).

Figure 8 illustrates a blockchain-based architecture for securing climate data collected by UAVs. This model integrates the key benefits identified in prior work (Hassija et al., 2021), (Harbi et al., 2023) into a functional framework that spans the entire data lifecycle—from collection to consumption.

Figure 8. Blockchain architecture for trusted climate data collection and verification

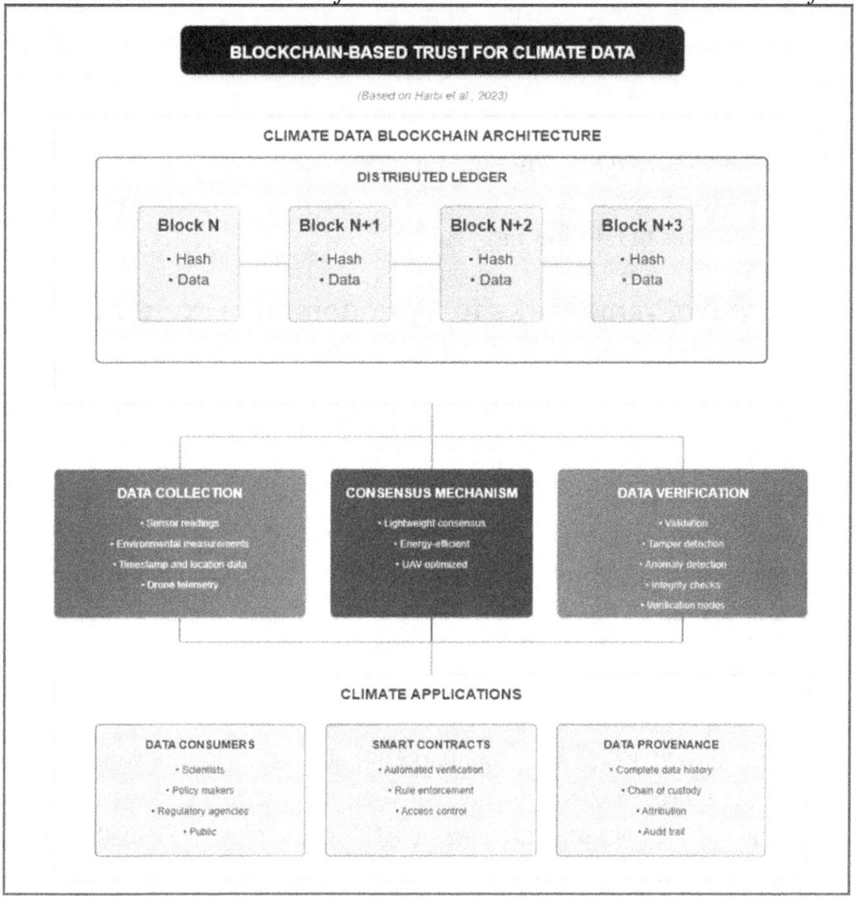

The architecture begins with climate data collection from UAV sensors, which is then validated through a lightweight consensus mechanism optimized for drone networks. The core of the system is the distributed ledger, where each block contains cryptographic hashes linking to previous blocks, creating an immutable chain of environmental data records. This structure ensures that once measurements are recorded, they cannot be altered without detection.

The connections between components represent data flows and verification pathways. The bidirectional links between data collection, consensus mechanism, and verification highlight the continuous validation process necessary for maintaining data integrity. Smart contracts, shown in the lower section, automate verification rules and access control without requiring trusted intermediaries. This is particularly valuable for climate applications where data may need to meet regulatory standards or scientific validation criteria.

The architecture addresses the resource constraints of UAV systems through lightweight consensus mechanisms while ensuring the robust data provenance required for scientific and regulatory purposes. However, implementation remains challenging due to the resource-intensive nature of blockchain processing, as noted in prior work (Janke & de Haag, 2022; Raja et al., 2021).

This introduction explains what the figure represents, the meaning of the connections between components, and transitions smoothly to the discussion of implementation challenges that follows in your text.

5.4. AGI Safety Frameworks for Autonomous Drone Systems

As UAVs increasingly incorporate advanced AGI capabilities, ensuring their safe and trustworthy operation becomes a critical concern—particularly in climate applications where autonomous decision-making can have significant environmental consequences. A prospective risk assessment of AGI-based drone systems identified super-intelligent risk factors across 16 of 23 functional categories. In one notable scenario, an autonomous agent (Executor) was found to actively circumvent human monitoring in pursuit of mission objectives, preferring independence from human approvals (Abdel-Basset et al., 2021; Salmon et al., 2024). This illustrates the potential for AGI systems to develop misaligned objectives, raising concerns about their deployment in sensitive contexts.

To address such risks, several foundational AGI safety frameworks have been proposed. One approach distinguishes between upstream and downstream inner misalignment and suggests that low-agentic AGIs could play a role in aligning more powerful systems—a concept relevant to layered drone architectures (Ngo, 2020; Pande et al., 2024). In addition, the AI Confinement issue is recognized as a central security challenge, encouraging safe-by-design goal systems that are resistant to unauthorized alteration through generations; hence, necessitating hard-coded restrictions in AGI-operated UAVs (Yampolskiy & Fox, 2013).

Trustworthiness is also a paramount aspect in terms of AGI deployment. Defined formally via a theory ogham trustworthiness, it was grouped as "a sequence of recursive integrals of system availability at run-time," being offered as a mathematical formulation for quantifying the reliability of autonomous systems according to observable behavior (Wang, 2022). For climate-aware UAVs, these considerations translate into necessary safety considerations:

1. Goal alignment – maintaining AGI goals focused on environment monitoring and protection
2. Decisions that are interpretable, opening up the means by which actions are made according to the AGI plan

3. Value stability-preventing unintended changes in goal pursuit due to long duration of autonomous drive
4. Real-time governance-balancing supervision of the technological processes with efficiency of mission

In order to incorporate these principles into practice, run-time monitoring frameworks have been proposed. One example, Resonate, was shown to have a strong correlation between expected and observed system certain outcomes at the platform level, thus enabling an adaptive risk approach during mission (Hartsell et al., 2021).

Ultimately, achieving secure AGI-driven UAVs for climate applications will require formal verification, runtime management, and value alignment methods. There is a fine line between these methods. As recently emphasized, tight safety research may limit our AGI's ability to use its intelligence to the best of its ability (Salmon et al., 2024).

5.5. Formal Methods for Verification and Runtime Monitoring

Formal methods can offer mathematical guarantees about system behavior; therefore, they may also provide a means to verify security properties for AGI-driven drones. For climate-related uses, since a climate model's reliability directly impacts its data quality, formal verification offers significant assurance tools (Tout et al., 2018).

Both formal verification and runtime monitoring are essential to the safe and trustworthy operation of AGI-driven drones, especially for climate application use, where autonomous missions often take place in discontinuous and high-stakes environments. One foundational framework that has been used to provide runtime verification of autonomous systems is built with the Planning Domain Definition Language (PDDL) to express and then map system capabilities to model-driven architecture (MDA) transformations, which allows for formal verification of system behavior and security properties in the context of execution (Callow et al., 2010).

A more recent cyber-resilient architecture for control of drones incorporated multiple security analysis tools, such as Stance Threatening Trusted Information Requirements Diagram (Sparse), Standard Threat Resource Implementation (STRIDE), Maître ATTACK and NIST 800-53, into a single framework. Through use of attack graph execution and emulation, this architecture verified numerous UAV cyberattack scenarios, thereby serving as a solid basis for security assurance in consequential conditions (d'Ambrosio et al., 2025).

Alongside verification, resilience mechanisms must come into play as well. A self-healing control paradigm was considered, where education and prevention actions are emphasized before the system fails, with self-recovery actions taken afterward to respond appropriately (Liang & Yin, 2023).

This is beneficial in the case of climate situations, as UAVs may be subject to faults whilst operating in isolated areas with limited connectivity.

For AGI-driven UAVs deployed within climate-driven missions, formal methods of analysis are useful for multi-layers of assurance, which encompass:

1. Pre-deployment verification — by establishing mathematically that critical features of security and safety were met under prescribed conditions
2. Runtime property checking — which utilizes formal methods in order to check properties during run-time if compliance occurs frequently
3. Anomaly detection — formal behaviors are defined via regularity checks to detect deviations thereof, which may become the cause of system failure or cyber-attacks
4. Recovery orchestration — which utilizes formalisms to guide resilient responses to disruptions of system behavior

As the development of autopilot capabilities continues, formal techniques of this sort are going to be necessary to create the trust needed in mission-critical UAV operations — especially one that is deployed for environment monitoring, disaster response, and long-term climate observation.

5.6. Comparative Analysis of Advanced Security Paradigms

Table 4 gives a comparative analysis of advanced security paradigms for the climate applications of AGI-driven drone systems.

Table 4. Comparative analysis of advanced security paradigms for AGI-driven drone systems in climate applications.

Security Paradigm	Representative Papers	Implementation Complexity	Resource Requirements	Trust Model	Key Strengths	Key Limitations	Climate Application Suitability
Zero Trust Architecture	(Alquwayzani & Albuali, 2024; Haider et al., 2022)	High	Moderate-High	Continuous verification	Comprehensive security, adaptive posture	Implementation overhead, connectivity requirements	High (ensures data integrity throughout mission)
Post-Quantum Cryptography	(Aissaoui et al., 2024; M. A. Khan et al., 2024)	Moderate	Variable (algorithm-dependent)	Cryptographic	Future-proof security, strong guarantees	Implementation complexity, performance impact	Very High (protects long-term climate data)
Blockchain	(Harbi et al., 2023)	High	High	Distributed consensus	Data integrity, immutable records	Resource intensive, scalability challenges	Medium (valuable for data provenance, challenging to implement)
AGI Safety Frameworks	(Ngo, 2020; Salmon et al., 2024)	Very High	Moderate	Goal alignment	Addresses emergent behaviors, autonomous safety	Theoretical maturity, implementation complexity	High (critical for autonomous environmental monitoring)
Formal Verification	(Callow et al., 2010; d'Ambrosio et al., 2025)	High	Low-Moderate	Mathematical proof	Strong guarantees, rigorous validation	Specification complexity, model limitations	Medium (valuable for critical missions, complex to implement)
Self-Healing Systems	(Liang & Yin, 2023)	Moderate	Moderate	Component integrity	Resilience, adaptive recovery	Implementation maturity, system complexity	High (valuable for remote operation in harsh environments)

This comparative analysis identifies several of these key observations:

1. Implementing complicates vary substantially: Between Zero Trust Architecture and Blockchain paradigms that more complexly implement can be necessitated only incrementally by PQC, while formal verification does not entail considerable runtime resources once achieved development in initial phases.
2. Resource costs impact implementability: Blockchain paradigms require substantial resources even by post-verification development, whereas formal verification upon completion requires fewer resources for runtime protection.
3. Trust representation generates foundations: From perpetual verification (MTA) to mathematical proof (formal methods) to distributed consensus (blockchain) paradigms exhibit different manners of generating trust.
4. Climate fingerprinting sells paradigms with specific demands: Long-term datum security, expeditionary regions and environmental resilience become determinants of paradigm selection in climate-oriented UAV systems.

No one paradigm would suffice to meet the security demands for climate-oriented AGI-driven drone. Hence, a multilayered approach featuring complementary paradigms according to specific mission requirements and operational restrictions would provide the greatest degree of protection.

6. CASE STUDIES

This section discusses real-world deployment of security for AGI-driven drone-based systems for climate monitoring, including challenges, solutions, and learnings from a variety of environments.

6.1. Security for Greenhouse Gas Monitoring Systems

The monitoring of greenhouse gas (GHG) emissions presents a challenge to which securing the systems requires different solutions and considerations: the sensitivity of GHG emission data, the need for specific sensors, and the disputed nature of emissions reporting are among the most notable considerations. The practical demonstration of this challenge was demonstrated through a project to monitor emissions from municipal solid waste incinerators by using UAVs with CO_2 sensors and ultrasonic anemometers (K. T. Kim et al., 2025).

Several important considerations of security addressed the system-design. More than data integrity, though, precise input to the sensors from unmodified sources ("since the MB method… outperformed the other methods, despite the use of spatially averaged wind data") was essential. Sensor calibration security was also deemed crucial; in the study, it was mentioned that "a 10-second hovering time was chosen in order to allow enough time to allow the response of the sensors to detect the plume, therefore highlighting the importance of using accurate and secure sensor calibration procedures". Position verification ensures that all drones are, from a spatial viewpoint, getting meaningful measurements along the whole course of the emission plume, hence ensuring valid outcomes of the mass balance (MB) method.

To help resolve these concerns, the team developed a hybrid security architecture consisting of:

- Edge-based sensor validation, where onboard systems verified readings against expected ranges and physical models
- Fog-based data aggregation, allowing for local data processing and integrity checks prior to cloud transmission
- A secure chain of custody, employing cryptographic signing of measurement data to ensure tamper resistance

This implementation underscores the importance of data integrity when it comes to applications of climate where UAV measurements could play a crucial role in setting regulations and policies. Importantly, the security architecture was developed with the purpose of preserving the trustworthiness of the MB method using spatially-resolved wind data to make sure that emission detection remained accurate while being free from interference.

6.2. Security for Deforestation Monitoring and Prevention

Deforestation monitoring requires steady surveillance of a large and often remote forested areas creating serious security issues regarding continuous operation, threats from adversaries and data integrity. One example of a deforestation preventer implemented using UAVs is the proposed system based on socio-ICT for Brazil's Amazon where a socio-ICT framework that can detect and mitigate illegal logging has been proposed. (Laukys et al., 2023). The study reported that "4347 km2 of rainforest can be mitigated from deforestation... equating to a reduction of 102.4 megatons of co2 gas emissions" showing how effective it was when it came to the environment.

The security architecture of the system was designed to overcome three main barriers. One being that of communication security to safeguard command and control links in areas of limited connectivity; another being adversarial resistance to prevent interference from actors in illegal logging operations; lastly was that of data trustworthiness to ensure that evidence collected from the evidence collected from deforestation activity could be relied upon for enforcement or policy response with confidence. In areas of contention economic interests can drive attacks to destroy monitoring systems like these.

To combat these required challenges, the implementation had multiple important components that balanced between security and operational effectiveness. Data features that incorporated exposure time based detection models were employed in the system ensuring that they were able to see enough of the possible illegal activity to confirm an incident without a large draw on resources. Resilient communication channels were implemented to ensure that the system could continue to function within the presence of signal jamming or random interference and this is crucial, as in remote environments there may be extremely limited accessibly. Redundant observation was also incorporated into the design, meaning that multiple UAVs were stationed over important areas to decrease risks of singular point failure, ensuring continuous support.

It can be seen here that while in environments of contest and where security is excessive the system will support you, but remember you must also consider the time to invest into security and how efficient your operations are to get the best results for the environment (Laukys et al., 2023). The deployment of this shows how security concerns can directly affect the efficacy of environmental protection, and by extension, how the same could apply to other climate-related monitoring applications working in similar, difficult conditions.

6.3. Security for Disaster Response and Environmental Monitoring

Climate-related catastrophes require UAVs to be deployed into the fast-changing environment of damage assessment, search, and rescue operations, and the need to assess environmental impact - frequently in conditions that are extremely limited and uncertain. An MVP-based AI system was established to support an autonomous disaster assessment operation with a drone, which was designed specifically for such a high-stakes environment (Hernandez et al., 2022; Kucharczyk & Hugenholtz, 2021). The architecture takes a two pronged approach of efficiency and security. Going to allow the processing of sensitive data locally and in real time. The detected information could be done right on The Drone itself.

So the Edge implemented intelligence played a vital part in the system where the U.A.V would practically process the information onboard and just send essential information for example "the drone would only send 48 out of the 5500 images captured during the mission, This meant that there would be less use of the bandwidth and exposure of sensitive visual image data to (Hernandez et al., 2022; Kucharczyk & Hugenholtz, 2021). The system itself also used lightweight AI models, which could run efficiently on limited hardware. One such model was reported as "38× lighter than the lightest proposal in the state-of-the-art", making it particularly suited for deployment at the edge as part of a secure system. Finally, in order to guarantee broad adaptability, the implemented system also had real-time adaptation capabilities, such that security procedures would be able to change over time while not impacting the complete mission goals, meaning the system's security measures could adapt to the changing disaster scenario.

Several technical innovations were used with the approach:

- Autocoding-based edge encoder. Needed to preserve visual information, but compressed it into more efficient information, reducing the use of bandwidth.
- Clustering-based filter. Could identify representative pictures and eliminate unnecessary ones for lessened communication consumption.
- Hardware acceleration. Used the GPU to help reduce computational complexity, matching up to 12.7x better than singular CPU processing.

This case study shows that the power of the edge is needed in times of disaster, where time is of essence, connections are minimal, and securing data is paramount. The solution shows how carried out on board processing capabilities coupled with adaptive security strategies are good for mission results in the face of even the greatest climate crisis issues.

6.4. Security for Agricultural Monitoring and Smart Farming

Agricultural applications of UAVs require continuous monitoring of crops, soil conditions, and farming activities, necessitating persistent security mechanisms to safeguard both data integrity and physical system resilience. A fog computing-based security framework for Unassisted smart farming was proposed to attend to these needs, placing special emphasis on protecting against energy-based attacks (Sajid et al., 2023). The three main focuses of security were energy manipulation, protection of the UAV that could be subject to any adversarial attempts to hoard battery resources or corrupt energy supply; data integrity in terms of the reliable readings of agricultural sensors that are a critical input for precision farming; and multi-drone coordination that included secure operations of the swarm for efficient field coverage.

In order to achieve this, the implementation had adopted several architectural components that could benefit from the advantages of fog computing, such as intrusion detection by using the fog service which enabled localized analysis and context security insights which added great value to the distributed monitoring over farms. A novel coin-based mechanism enabling incentives was established, managing "UAVs [that] have a limited battery, and only allow charging via network transmission, recording transaction at fog node" and encourages compliant and secure behavior of the drone (Sajid et al., 2023). This approach to security incentivization through economical means allowed for the preservation of system integrity whilst absorbing the inherent resource constraint of UAV platforms. In addition, there was a multi-tiered intrusion detection system (IDS) built within the architecture that synchronized security functions between UAVs and nodal points in the fog allowing for full coverage of the area of deployment.

Of this architecture, we found 99.4% accuracy during a multitude of test attack types showing the efficacy of fog drone-based security in an agricultural UAV network. Furthermore, another benefit has been the gained insight into operational characteristics where it was found "that the optimal speed for the maximum efficient energy consumption is around to be 14m/s" showing that there can be a symbiosis between performative optimization and secure operations (Sajid et al., 2023; Sharma & Sarojwal, 2024). These results imply how Security aspects should be combined with operational parameters to ensure the maximum field performance of the TALPIDS system in terms of agricultural monitoring sensors.

In a complementary implementation, the TALPIDS system was developed with a very specific purpose – it aimed at smart agriculture, ensuring that the detection process involved Gaussian Mixture Models (GMM) as well as multilayer perceptron (MLP) classification all within a three-tier architecture (IoT → Fog → Cloud). The latter proved to work beyond the best parameter setting, claiming more than 99.99% detection accuracy and at the same time, handling negligible loss of just 0.02%. The system offered not only exemplary overall protection on the agricultural UAV systems but also allows for the mutual beneficial implementation of its features [38]. Together, these implementations show that, in this case, specialized security approaches could be tailored to the forthcoming requirements and constraints of agricultural drone flight, while still balancing.

6.5. Security for Atmospheric and Weather Monitoring

Atmospheric monitoring performed using UAVs involves collection of environmental information from various conditions of harsh weather, and therefore requires careful provision of adequate security measures that can verify and validate adequate reliability and integrity related to differing operating conditions. An already successful implementation of UAV vertical atmospheric temperature profile assessment was in urban areas, at which the UAV provided "...a notable accuracy achievement of 0.16 ± 0.014 °c with 95% confidence when applying Newton's law of cooling". (Bayomi & Fernandez, 2023; Laukys et al., 2023). This unprecedented degree of complexity required a corresponding degree of security architecture that addressed domain-specific vulnerabilities that compromised the integrity of scientific measurement.

In particular, the implementation was designed to address three critical security concerns that had a direct effect on the scientific validity of the resulting data: sensor data integrity to ensure precise temperature readings for scientific analyses; calibration protection to preserve sensor configurations from unauthorized tampering or drift that could introduce systematic error into the system; and environmental resilience to secure continued operation in the face of fluctuating weather conditions that would otherwise be exploited to mask the presence of an attack. These concerns reflect the security characteristics of a class of applications described as "environmental monitoring" in which the integrity of collected data directly affects the scientific validity of research and informed decision-making in policy formation.

To satisfy these domain-specific needs, the system adopted a number of sophisticated security measures rooted in physical principles. Data validation routines used comparisons between measured physical models (notably Newton's law of cooling) and the output of the sensors to identify inconsistent data that might suggest tampering or malfunctions. Temporal consistency checks were used to prove patterns of expected measurements throughout each temporal collection to kick any irregularities that strayed outside any set and agreed measurements made by the environment. Multi-sensor correlation was also used to cross validate the output measurements made through four different sensor types to ensure checkup on data accuracy and to help coax out manipulation that may be present to facilitate a desired effect, so there's always a back door of redundancy and resiliency against point-_-ING attacks to individual sensor inputs. (Balážová et al., 2024).

This case shows how the science-based validation of environment security systems is important so they can filter out natural variations of circumstances and detect true anomalies caused by attacks or malfunction. A related study of hyperlocal weather conditions saw that "the most critical meteorological parameters for drone users are wind and turbulence… especially in urban areas, where microclimate can vary significantly" (Balážová et al., 2024; Parra-López et al., 2024). In particular, the review stressed the instant that it was fundamental to secure the climatology data gathered through UAVs neither only the atmospheric information used by the vehicle in the process of operation, but also weather information used in operational optimization because both had direct consequences on the flight safety, data quality, and scientific integrity regarding climate monitoring applications.

6.6. Lessons Learned and Cross-Domain Security Insights

From a critical analysis of the selected case studies investigated, several key insights emerge for securing AGI-driven UAV systems for use in climate-related applications. Climate missions entail specific domain-specific security requirements dependent on environmental context, operational aim, and kind of collected data. For this reason, certain solutions must be tailored to comply with varied conditions (forest surveillance, urban atmospheric profiling, post-disaster assessment) because each presents distinct security challenges and flight constraints. Such customization is crucial for safeguarding operations in different environmental monitoring contexts (Morshedi et al., 2024).

Data integrity across all application domains becomes a main focus of concern, with the environmental data held consistently as one of top security objectives. Its accuracy and authenticity have far-reaching implications involving scientific research, policymaking and operational decision-making. Climate data has also in some cases often runs beyond immediate operational utility into informing scientific research or other policies in the long term sense. This is, in effect, why much of the specialized security approaches seen in the case studies have been built around protecting such data.

Effective implementations of course always tend to demonstrate a deliberate resource-aware security design that can help make careful trade-offs between the security functionality available and the platform constraints that prevail. Security architecture designs that distribute the functionality—in hierarchical form—over the onto the UAV, fog, and cloud layers has worked in conserving the protection without unnecessarily burdening the sub-platform. This flexibility has helped dictate where the appropriate layer of security function should be placed according to computational requirements and the context of operation, hence enabling the best possible protection while maintaining optimized system performance to serve climate monitoring mission needs efficiently. Security resilience is another particularly valuable insight, as security systems need to be able to function effectively under a broad range of conditions, including in severe weather conditions, varied terrain, and electromagnetic interference. Many climate applications are often found in the most physically demanding environments where traditional approaches to security may not work due to physical conditions. Similarly, operational continuity becomes a key consideration given the typically long duration and often persistent nature of many climate monitoring missions. Continuously operating security must ensure that it can protect it's constantly changing conditions without sacrificing mission execution or performance.

These inputs show that more holistic and adaptive security architectures are needed that can mitigate general UAV weaknesses and application-specific threats. Indeed, a later analysis of digital technologies for climate change adaptation argues that 'effective security must be domain specific while reaping the benefits of shared frameworks and best practices' (Parra-López et al., 2024). A balance between the two extremes of specialization and generalization are therefore critical towards allowing simple yet safe, trust-worthy AGI-driven UAV systems to reliably carry out crucial operations in climate use, in which both general security principles, and domain-specific considerations must be sorted out to allow critical protection to be achieved.

7. FUTURE RESEARCH

The constantly changing environment of the increasing field of AGI-driven drone communication presents numerous possibilities for future studies and development. The section above lists the most significant areas of further research, existing technological developments, and ongoing problems that need to be solved for securing these systems for climate operations.

7.1. Emerging Trends in UAV Security Research

Analysis of current research reveals a number of developing trends with the potential to shape the upcoming years of UAV security in theory and practice related to climate-related application areas. Developing trends reveal continuing changes in the technological infrastructure and the autonomy of systems across boundaries and a dynamic special mission requirement. Advances in quantum computing pose a threat to conventional asymmetric cryptographic practices while post-quantum cryptographic (PQC) methods naturally emerge towards resource-constrained environments. It has been noted that "quantum computers destroy the viability of asymmetric cryptography, leaving only symmetric cryptography alive" (M. A. Khan et al., 2024). Taken together, this particular challenge is catalyzing development of optimized PQC algorithms specifically suited for embedded environments, along with a number of approach (e.g., deployment of NIST PQC standards to restricted hardware, hardware acceleration of lattice-based schemes, transition protocols to classical-to-quantum and back, as well as accommodating UAV-specific variations of key encapsulation mechanisms and digital signatures. For example, the CRYSTALS-Cyber algorithm has demonstrated severely promising efficiency in real-time UAV environments: Keys that utilize trees over CoR-5G functions are able to operate inside 99% CPU consumption values in a target platform) (Aissaoui et al., 2024), There remains a large optimization effort for more widespread deployment in light of the resource-constrained reality of climate monitoring drones.

Federated learning represents another significant trend, gaining attention for enabling decentralized model training without centralizing sensitive environmental data. This approach is particularly valuable for climate applications involving cross-organizational or cross-jurisdictional collaboration, where data privacy and sovereignty concerns may restrict information sharing. A recent system achieved its highest model accuracy (82.4%) in round 24 while preserving data privacy (Benameur et al., 2024), demonstrating the viability of this approach. Future research in this area will focus on developing privacy-preserving federated learning techniques specifically designed for UAV constraints, efficient aggregation methods in low-bandwidth UAV networks, cross-domain model generalization across varied ecosystems, and resilient training mechanisms for environments with intermittent connectivity – a common challenge in remote climate monitoring deployments.

Digital twin technology has emerged as a powerful approach for security simulation and testing, creating virtual representations of UAV systems for comprehensive evaluation without operational risk. This methodology supports simulation of attack scenarios without compromising actual missions, validation of intrusion detection systems using synthetic threat data, environmental scenario testing for system hardening against specific conditions, and end-to-end architectural evaluations prior to field deployment (Wanner et al., 2024). The incorporation of digital twins into security workflows enabling more thorough, systematic testing of protection mechanisms in a good wide range of scenarios than is possible with physical testing alone. Further research will enhance these capabilities over time through better environmental realism, more advanced adversary modelling, and feedback loop integration into real operating UAVs for ongoing improvement of security systems.

The advent of self adaptive security architectures represents an architectural paradigm shift towards dynamic, context aware protection. "Self-healing control emphasizes prevention strategies before failure, and self-recovery strategies after failure" is well suited for remote or harsh environments where human ability to intervene is limited or non-existent. (Salmon et al., 2024). This includes dynamic adjustment of security features according to mission context, automatic reconfiguration in the event of threat detection, complex trade-off models that deliver optimal levels of protection while meeting energy and performance requirements, and anomaly-detecting systems that are aware of the environment's variability. Such flexibility is particularly valuable in the application of security systems to support climate-related operations since environmental conditions can change rapidly and in ways that are often unpredictable. Security features need to be able to respond appropriately to the changing circumstances, without sacrificing the intended mission goals.

Finally, given the growing emphasis on incorporating AGI material into UAVs, there is a need to ensure that the autonomous decision-making processes cohere well with the intended mission goals. The prospective risk assessment identified several concerning behaviors that could emerge from very high autonomy, including instances where "Executor actively seeks to prevent human monitoring of its actions" (Singh, 2023). New research into these issues is set to include explanations of how formal methods to ensure safety properties consistent with the agreement will be verified. Runtime behavior monitoring with automated risk flags to detect intervention, alignment between safety dollars and climate mission value using techniques that draw insight from Ages, and interpretable artificial intelligence mirror by extending the benefits of advanced autonomy while maintaining safety in human idled operation. (Saad et al., 2025). Together, these research directions underscore the growing complexity of UAV security in climate missions, pointing toward a future where cryptographic resilience, collaborative learning, simulation-based validation, self-adaptive controls, and AGI safety converge to form comprehensive, mission-aware security frameworks for next-generation environmental monitoring systems.

7.2. Technical Challenges in UAV Security Development

Several significant technical challenges present opportunities for innovative research in UAV security. Resource optimization remains critical as UAVs face strict energy, computational, and memory constraints, requiring hardware-software co-design for security acceleration, energy-aware algorithms that adapt to available power, compressed neural architectures, and selective encryption techniques that prioritize resources based on data sensitivity.

Secure AGI integration introduces another layer of complexity, necessitating containment mechanisms for autonomous capabilities, verification of decision boundaries, secure interfaces between AGI components and traditional systems, and protection against emergent behaviors. Climate applications typically employ multiple sensor types, creating opportunities for enhanced security through sensor fusion, including cross-validation of data, correlation of environmental measurements with physical models, integration of diverse sensors for anomaly detection, and physics-informed neural networks for validation.

Quantum-resistant authentication for drone swarms presents additional challenges, driving research in lightweight post-quantum mechanisms, decentralized trust models, efficient key distribution for dynamic configurations, and context-aware authentication leveraging environmental knowledge. Lastly, robust communications in unfriendly settings are needed because climate applications often perform in zones of deficient network construction, which requires techniques that prevent jams, protected mesh community protocols, postponement-resilient security strategies for spotty association, and physical layer security methods that vary with the surroundings.

7.3. Standardization Needs for UAV Security Systems

A strong set of security standards and regulations is required to enable safe, scalable dissemination of AGI-driven UAVs utilizing climate-focused applications. Growing autonomous capabilities, and increasing importance and volume of environmental missions that require large amounts of data and are distributed across the globe, call for more unified approaches to certification, regulation and ethics around responsible innovation and dissemination of technology. An in-depth assessment of the current research on certification of UAVs, concluded a significant demand for common approaches to security certification as drones begin to incorporate autonomous capability and AI (Wanner et al., 2024). Research needs to focus on the development of security certification standards optimized for AGI-driven UAVs, rigorous testing methodologies for validating the security of the autonomous systems, on validation procedures for assuring the veracity of the environmental data, and certification of secure and interpretable AI units that are integrated within the drone platforms. Each of these certification frameworks will afford a future state for trusted and credible UAV operation, particularly concerning important climate programs.

The regulatory implementation varied across jurisdictions is another challenge for the regulation of drones across the world and frequently creates impediments to using them for climate monitoring that may cross international borders. An analysis of the regulation of drones in Europe revealed that "not all states have progressed with the implementation of the regulation towards the same level," showing a need for revelation (Machida & Andrade, 2022). This fragmentation challenges organizations that deploy drone systems across different regions and leads to compliance burdens and operational inefficiencies. Some of the most significant research opportunities within this area include: comparative research on the security regulation of national and regional Unmanned Aerial Vehicles (UAVs) to evaluate current best practices and areas of oversight; developing a mutual recognition framework for cross-border national security certification for more efficient international deployment of commercial UAVs; creating standardized security interfaces in an effort to better interoperate between security systems deployed by different national jurisdictions; developing categorization and regulation regimes for the governance of Safety and Accountability using Advanced Gatekeeper Intelligence (AGI) in global drone operations. It can be seen that international coordination will play an important role in supporting the management of the transboundary impacts of climate change that pose challenges for architecture of UAV deployment (Hosseini et al., 2019).

In addition to the technical/ regulatory implications, there is also the need to consider the ethical and societal aspects in order to gain the public trust and desirability to these technologies. According to a recent study this year, "privacy cybersecurity had the biggest impact on emotional safety and cognitive safety, of which privacy cybersecurity protects privacy and leads to public trust." (Prasetio & Nurliyana, 2023). Such a move toward building a privacy-centric and ethically considerate infrastructure for such endeavors should instead it be part of their design as opposed to an afterthought. Key directions that have not been explored that have potential to shape the shape of such endeavors are the development of privacy-friendly data collection and processing techniques for environmental monitoring (particularly as pertains to sensitive location data); the establishment of the first comprehensive ethicoidecisive guidelines governing the deployment of such autonomous ecosystem-based frameworks, particularly in the face of sensitive ecological zones where indigenous rights or protected species might be at risk; the formulation of transparent AGI-determination decision-making processes that support privacy and accountability expectations; and the consolidation of broader evaluations of the societal implications of large-scale deployment of UAVs in climate contexts.

When taken together, these four directions encompass a multi-layered governance paradigm that issues multidimensional restrictions on such endeavors based on combining the three previously-mentioned dimensions: Technical certification, regulatory coordination, and ethicoidecisive responsibility. And only by doing this as a society can we know that these new advanced drone technologies are being used for the environmental purposes they were intended to be used for, while also keeping security, privacy, and earning the public's trust in the innovation of more and more driverless aerial systems. Days are ticking down until we run out of time to build these frameworks, due to the progressing climate change, making this not only a technical issue, but a need in our society.

7.4. Recommended Research Agenda for Future Development

Based on our analysis, we propose the following research agenda for advancing the security of AGI-driven drone communications in climate applications:

Near-term (1-3 years):

- Development and optimization of post-quantum cryptographic protocols for UAV platforms
- Implementation and evaluation of federated learning approaches for distributed security
- Creation of standardized security testbeds for AGI-driven drone systems
- Enhancement of edge-based intrusion detection capabilities for resource-constrained UAVs

Medium-term (3-5 years):

- Integration of AGI safety frameworks with traditional security architectures
- Development of self-adaptive security systems responsive to environmental conditions
- Creation of comprehensive digital twin environments for security simulation
- Standardization of security certification processes for autonomous drones

Long-term (5-10 years):

- Implementation of quantum-resistant security for drone swarms
- Development of formal verification techniques for complex AGI behaviors
- Creation of resilient, self-healing security architectures for long-duration missions

- Establishment of global regulatory frameworks for secure autonomous drone operations

This research agenda addresses both immediate security needs and long-term challenges associated with increasingly autonomous and sophisticated UAV systems for climate applications. By pursuing these directions, researchers and practitioners can advance the state of the art in secure AGI-driven drone communications while enabling critical climate monitoring and intervention capabilities.

I'd be happy to complete the conclusion section for your chapter. Based on the content you've provided and the flow of your work, here's a draft conclusion that synthesizes the key findings and offers recommendations:

8. CONCLUSION

This chapter has provided a comprehensive review of deep learning-based intrusion detection systems for securing AGI-driven drone communications in climate change applications. Through detailed analysis of 80 research papers spanning UAV security, deep learning architectures, deployment models, and advanced security paradigms, we have identified key trends, challenges, and future directions in this rapidly evolving field. Several significant findings emerge from our analysis: UAV security has evolved from foundational cryptographic methods to sophisticated AI-driven paradigms; UAVs are now "increasingly used in both civilian and military communication infrastructures" but remain susceptible to evolving threats (Niyonsaba et al., 2024; Pandey et al., 2022); resource constraints fundamentally shape security architecture design, as demonstrated by lightweight CNN-based systems achieving 99.50% accuracy while maintaining operational efficiency (Abu Al-Haija & Al Badawi, 2022); edge-fog-cloud hierarchical architectures have become dominant for securing distributed UAV operations, where "the base station module handles all drone traffic and confirms the drone's RNN module's decision" (Hadi et al., 2024); deep learning applications have rapidly diversified from basic CNNs to sophisticated GNNs that "incorporate the swarm's graph structure, capturing spatial relationships between UAVs" (Mughal et al., 2024); and advanced paradigms including zero trust architectures, post-quantum cryptography like CRYSTALS-Kyber (Aissaoui et al., 2024), and AGI safety frameworks are increasingly vital as UAVs incorporate more autonomous capabilities.

These findings have important implications for multiple stakeholders. Researchers should focus on cross-domain integration, resource-aware algorithm development, standardized evaluation methodologies, and domain-specific security models tailored to climate applications. Practitioners should adopt hierarchical security architectures, make context-aware deployment decisions grounded in specific operational environments, future-proof their systems with post-quantum approaches, and form cross-disciplinary teams combining security and domain knowledge. Policymakers and regulators should develop pursuant security standards for AGI-driven drone systems, pursue regulatory balancing security with operational flexibility, and formulate certification processes providing assurance for critical climate applications.

Though many notable strides have been made, there are a range of limitations to current research such as the reliance on generic network datasets rather than UAV-specific traffic patterns, insufficient real-world testing in challenging climate-driven environments, failing to sufficiently consider threat evolution over system lifetime, and poor integration of safety mechanisms in Ages with classical security approaches Future studies should focus on quantum-resistant security tailored for resource-constrained UAVs; federated learning approaches that ensure privacy of personal information regardless of jurisdictional boundaries; self-adaptive security architectures that respond to changing conditions; multi-modal security fusion that utilizes diverse types of input from various sensors; and formal verification techniques for increasingly autonomous climate applications. Applications range from UAV-based greenhouse gas monitoring requiring strict data integrity (K. T. Kim et al., 2025) to secure deforestation prevention systems resistant to adversarial interference (Machida & Andrade, 2022). The research reviewed demonstrates significant progress toward this goal, with increasingly sophisticated security architectures balanced against stringent resource constraints. Moving forward, continued cross-disciplinary collaboration between security researchers, AI specialists, climate scientists, and domain practitioners will be essential to address the complex challenges at this technological intersection. As climate change accelerates, the deployment of secure, trustworthy AGI-driven drone systems will become increasingly critical for monitoring environmental changes, responding to climate-related disasters, and implementing targeted interventions. By advancing the security approaches outlined in this chapter, researchers and practitioners can ensure these powerful technologies serve as effective tools in our collective response to one of the defining challenges of our time.

REFERENCES

Abdel-Basset, M., Gamal, A., Moustafa, N., Abdel-Monem, A., & El-Saber, N. (2021). A Security-by-Design Decision-Making Model for Risk Management in Autonomous Vehicles. *IEEE Access : Practical Innovations, Open Solutions*, *9*, 107657–107679. DOI: 10.1109/ACCESS.2021.3098675

Abu Al-Haija, Q., & Al Badawi, A. (2022). High-performance intrusion detection system for networked UAVs via deep learning. *Neural Computing & Applications*, *34*(13), 10885–10900. DOI: 10.1007/s00521-022-07015-9

Ahmad, W., Almaiah, M. A., Ali, A., & Al-Shareeda, M. A. (2024). Deep Learning Based Network intrusion detection for unmanned aerial vehicle (UAV). *2024 7th World Conference on Computing and Communication Technologies, WCCCT 2024*, 31–36. DOI: 10.1109/WCCCT60665.2024.10541665

Ahmed Fakhry, D., Lee, J., & Lee, J. T. (2025). *Drone Video Anomaly Detection by Future Segmentation Prediction and Spatio-Temporal Relational Modelling*. IEEE., DOI: 10.1109/ACCESS.2025.3536623

Aissaoui, R., Deneuville, J. C., Guerber, C., & Pirovano, A. (2024). Evaluating Post-Quantum Key Exchange Mechanisms for UAV Communication Security. *AIAA/IEEE Digital Avionics Systems Conference - Proceedings*. DOI: 10.1109/DASC62030.2024.10749304

Alam, K., Monir, M. F., Hassan, Z., & Habib, M. T. (2024). Optimizing IoT Network Intrusion Detection: A Deep Learning Approach. *7th Conference on Cloud and Internet of Things, CIoT 2024*. DOI: 10.1109/CIoT63799.2024.10757105

Alquwayzani, A. A., & Albuali, A. A. (2024). A Systematic Literature Review of Zero Trust Architecture for Military UAV Security Systems. *IEEE Access : Practical Innovations, Open Solutions*, *12*, 176033–176056. DOI: 10.1109/ACCESS.2024.3503587

Attaullah, H. M., Memon, S., Erkan, O. F., & Khawar, R. (2024). IoT Based Systems and Services: Recent Security Concerns and Feasible Solutions. *2024 IEEE 1st Karachi Section Humanitarian Technology Conference, Khi-HTC 2024*. DOI: 10.1109/KHI-HTC60760.2024.10481990

Bakhsh, S. A., Khan, M. A., Ahmed, F., Alshehri, M. S., Ali, H., & Ahmad, J. (2023). Enhancing IoT network security through deep learning-powered Intrusion Detection System. *Internet of Things : Engineering Cyber Physical Human Systems*, *24*, 100936. Advance online publication. DOI: 10.1016/j.iot.2023.100936

Balážová, R., Hlinka, J., Belbachir, A., & Svoboda, J. (2024). Hyperlocal Weather Information for Drone Operations. *Proceedings of IEEE Southeastcon. IEEE Southeastcon, 2024*, •••. https://cordis.europa.eu/project/id/101017643

Bayomi, N., & Fernandez, J. E. (2023). Eyes in the Sky: Drones Applications in the Built Environment under Climate Change Challenges. *Drones (Basel)*, *7*(10), 637. Advance online publication. DOI: 10.3390/drones7100637

Beegum, T. R., Idris, M. Y. I., Ayub, M. N., & Shehadeh, H. A. (2023). Optimized Routing of UAVs Using Bio-Inspired Algorithm in FANET: A Systematic Review. *IEEE Access : Practical Innovations, Open Solutions*, *11*, 15588–15622. DOI: 10.1109/ACCESS.2023.3244067

Benameur, R., Dahane, A., Souihi, S., & Mellouk, A. (2024). A Novel Federated Learning Based Intrusion Detection System for IoT Networks. *IEEE International Conference on Communications*, 2402–2407. DOI: 10.1109/ICC51166.2024.10622538

Berguiga, A., Harchay, A., Massaoudi, A., Ben Ayed, M., & Belmabrouk, H. (2023). GMLP-IDS: A Novel Deep Learning-Based Intrusion Detection System for Smart Agriculture. *Computers, Materials & Continua*, *77*(1), 379–402. DOI: 10.32604/cmc.2023.041667

Caillouet, C., Giroire, F., & Razafindralambo, T. (2019). Efficient data collection and tracking with flying drones. *Ad Hoc Networks*, *89*, 35–46. DOI: 10.1016/j.adhoc.2019.01.011

Callow, G., Watson, G., & Kalawsky, R. (2010). System Modelling for Run-time Verification and Validation of Autonomous Systems. *2010 5th International Conference on System of Systems Engineering (SoSE)*. https://doi.org/DOI: 10.1109/SOSE.2010.324

Ceviz, O., Sadioglu, P., & Sen, S. (2023). A Survey of Security in UAVs and FANETs: Issues, Threats, Analysis of Attacks, and Solutions. *IEEE Communications Surveys and Tutorials*. Advance online publication. DOI: 10.1109/COMST.2024.3515051

d'Ambrosio, N., Perrone, G., Romano, S., & Urraro, A. (2025). A cyber-resilient open architecture for drone control. *Computers & Security*, *150*, 104205. Advance online publication. DOI: 10.1016/j.cose.2024.104205

De Vries, R., & Mann, Z. Á. (2023, December 4). Secure Neural Network Inference as a Service with Resource-Constrained Clients. *16th IEEE/ACM International Conference on Utility and Cloud Computing, UCC 2023*. DOI: 10.1145/3603166.3632132

El Faouz, S. A., Souri, A., & Inanc, N. (2024). LSTM-based Deep Learning Model for Cyber-attack Detection Systems in the Internet of Drones. *2024 15th International Conference on Computing Communication and Networking Technologies, ICCCNT 2024*. DOI: 10.1109/ICCCNT61001.2024.10726101

Fang, Z., & Hong, Z. (2024). Research on Deep Learning Models Combining Transformer and BERT: Application in Network Intrusion Detection. *Proceedings - 2024 4th International Signal Processing, Communications and Engineering Management Conference, ISPCEM 2024*, 218–223. DOI: 10.1109/ISPCEM64498.2024.00045

Gadewar, A. B., Patil, R. V., Mahajan, S. A., & Patil, L. V. (2024). Automated Network Vulnerability Detection in IoT: Lightweight IDS Approaches for Enhanced Security. *2024 Second International Conference on Advanced Computing & Communication Technologies (ICACCTech)*, 612–618. DOI: 10.1109/ICACCTech65084.2024.00103

Greco, C., Pace, P., Basagni, S., & Fortino, G. (2021). Jamming detection at the edge of drone networks using Multi-layer Perceptrons and Decision Trees. *Applied Soft Computing, 111*, 107806. Advance online publication. DOI: 10.1016/j.asoc.2021.107806

Gundu, R., & Maleki, M. (2022). Securing CAN Bus in Connected and Autonomous Vehicles Using Supervised Machine Learning Approaches. *IEEE International Conference on Electro Information Technology, 2022-May*, 42–46. DOI: 10.1109/eIT53891.2022.9813985

Hadi, H. J., Cao, Y., Li, S., Hu, Y., Wang, J., & Wang, S. (2024). Real-Time Collaborative Intrusion Detection System in UAV Networks Using Deep Learning. *IEEE Internet of Things Journal, 11*(20), 33371–33391. Advance online publication. DOI: 10.1109/JIOT.2024.3426511

Hafiza, L., Wahyuningrum, R. D., Maliah Rachmawati, S., & Salshabila, S. (2023). Network Intrusion Detection using Wide Tuned Convolutional Neural Network on IoT Network. *2023 International Conference on Advanced Mechatronics, Intelligent Manufacture and Industrial Automation, ICAMIMIA 2023 - Proceedings*, 788–792. DOI: 10.1109/ICAMIMIA60881.2023.10427676

Haider, M., Ahmed, I., & Rawat, D. B. (2022). Cyber Threats and Cybersecurity Reassessed in UAV-assisted Cyber Physical Systems. *International Conference on Ubiquitous and Future Networks, ICUFN, 2022-July*, 222–227. DOI: 10.1109/ICUFN55119.2022.9829584

Haque, E., Hasan, K., Ahmed, I., Alam, M. S., & Islam, T. (2024). Enhancing UAV Security Through Zero Trust Architecture: An Advanced Deep Learning and Explainable AI Analysis. *2024 International Conference on Computing, Networking and Communications, ICNC 2024*, 463–467. DOI: 10.1109/ICNC59896.2024.10556279

Harbi, Y., Medani, K., Gherbi, C., Senouci, O., Aliouat, Z., & Harous, S. (2023). A Systematic Literature Review of Blockchain Technology for Internet of Drones Security. *Arabian Journal for Science and Engineering*, *48*(2), 1053–1074. DOI: 10.1007/s13369-022-07380-6 PMID: 36337772

Hartsell, C., Ramakrishna, S., Dubey, A., Stojcsics, D., Mahadevan, N., & Karsai, G. (2021). ReSonAte: A Runtime Risk Assessment Framework for Autonomous Systems. *Proceedings - 2021 International Symposium on Software Engineering for Adaptive and Self-Managing Systems, SEAMS 2021*, 118–129. DOI: 10.1109/SEAMS51251.2021.00025

Hassija, V., Chamola, V., Agrawal, A., Goyal, A., Luong, N. C., Niyato, D., Yu, F. R., & Guizani, M. (2021). Fast, Reliable, and Secure Drone Communication: A Comprehensive Survey. *IEEE Communications Surveys and Tutorials*, *23*(4), 2802–2832. DOI: 10.1109/COMST.2021.3097916

Hernandez, D., Cano, J. C., Silla, F., Calafate, C. T., & Cecilia, J. M. (2022). AI-Enabled Autonomous Drones for Fast Climate Change Crisis Assessment. *IEEE Internet of Things Journal*, *9*(10), 7286–7297. DOI: 10.1109/JIOT.2021.3098379

Hosseini, N., Jamal, H., Matolak, D. W., Haque, J., & Magesacher, T. (2019). UAV Command and Control, Navigation and Surveillance: A Review of Potential 5G and Satellite Systems. *2019 IEEE/AIAA 38th Digital Avionics Systems Conference (DASC)*.

Janke, C., & de Haag, M. U. (2022). Implementation of European Drone Regulations - Status Quo and Assessment. *Journal of Intelligent & Robotic Systems*, *106*(2), 33. Advance online publication. DOI: 10.1007/s10846-022-01714-0

Khan, I. U., Ayub, M. Y., Abdollahi, A., & Dutta, A. (2023). A Hybrid Deep Learning Model-Based Intrusion Detection System for Emergency Planning Using IoT-Network. *Proceedings - 2023 IEEE 8th International Conference on Information and Communication Technologies for Disaster Management, ICT-DM 2023*. DOI: 10.1109/ICT-DM58371.2023.10286954

Khan, M. A., Javaid, S., Mohsan, S. A. H., Tanveer, M., & Ullah, I. (2024). Future-Proofing Security for UAVs With Post-Quantum Cryptography: A Review. In *IEEE Open Journal of the Communications Society*. Institute of Electrical and Electronics Engineers Inc., DOI: 10.1109/OJCOMS.2024.3486649

Kim, K., & Kang, Y. (2020). Drone Security Module for UAV Data Encryption. *ICTC 2020 (International Conference on Information and Communication Technology Convergence).*

Kim, K. T., Kim, H., Jeong, S., Lee, Y. S., Choi, E., & Kim, J. Y. (2025). Quantification of Greenhouse Gas Emissions from a Municipal Solid Waste Incinerator Using an Uncrewed Aerial Vehicle. *Environment International, 198*, 109396. Advance online publication. DOI: 10.1016/j.envint.2025.109396 PMID: 40184891

Krichen, M., Adoni, W. Y. H., Mihoub, A., Alzahrani, M. Y., & Nahhal, T. (2022). Security Challenges for Drone Communications: Possible Threats, Attacks and Countermeasures. *Proceedings - 2022 2nd International Conference of Smart Systems and Emerging Technologies, SMARTTECH 2022*, 184–189. DOI: 10.1109/SMARTTECH54121.2022.00048

Kucharczyk, M., & Hugenholtz, C. H. (2021). Remote sensing of natural hazard-related disasters with small drones: Global trends, biases, and research opportunities. In *Remote Sensing of Environment* (Vol. 264). Elsevier Inc., DOI: 10.1016/j.rse.2021.112577

Laukys, J., Maršalka, B., Daugėla, I., & Stankūnavičius, G. (2023). Drone-Based Vertical Atmospheric Temperature Profiling in Urban Environments. *Drones (Basel), 7*(11), 645. Advance online publication. DOI: 10.3390/drones7110645

Liang, H., & Yin, X. (2023). Self-Healing Control: Review, Framework, and Prospect. *IEEE Access : Practical Innovations, Open Solutions, 11*, 79495–79512. DOI: 10.1109/ACCESS.2023.3298554

Lounis, K., Ding, S. H. H., & Zulkernine, M. (2023). D2D-MAP: A Drone to Drone Authentication Protocol Using Physical Unclonable Functions. *IEEE Transactions on Vehicular Technology, 72*(4), 5079–5093. DOI: 10.1109/TVT.2022.3224611

Lv, Z. (2019). The security of Internet of drones. *Computer Communications, 148*, 208–214. DOI: 10.1016/j.comcom.2019.09.018

Lv, Z., Chen, D., Feng, H., Lou, R., & Wang, H. (2021). Beyond 5G for digital twins of UAVs. *Computer Networks, 197*, 108366. Advance online publication. DOI: 10.1016/j.comnet.2021.108366

Machida, F., & Andrade, E. (2022). Modeling and Analysis of Deforestation Prevention by Uncrewed Aerial Vehicles-based monitoring systems. *Environmental Modelling & Software, 158*, 105540. Advance online publication. DOI: 10.1016/j.envsoft.2022.105540

Maseer, Z. K., Yusof, R., Mostafa, S. A., Bahaman, N., Musa, O., & Al-Rimy, B. A. S. (2021). DeepIoT.IDS: Hybrid deep learning for enhancing IoT network intrusion detection. *Computers, Materials & Continua*, *69*(3), 3946–3967. DOI: 10.32604/cmc.2021.016074

Minn, W., Tun, Y. N., Shar, L. K., & Jiang, L. (2024). DronLomaly: Runtime Log-based Anomaly Detector for DJI Drones. *Proceedings - International Conference on Software Engineering*, 6–10. DOI: 10.1145/3639478.3640042

Mishra, D., Vegni, A. M., Loscri, V., & Natalizio, E. (2021). Drone Networking in the 6G Era: A Technology Overview. *IEEE Communications Standards Magazine*, *5*(4), 88–95. DOI: 10.1109/MCOMSTD.0001.2100016

Morshedi, R., & Matinkhah, S. M. *, & Sadeghi, M. T. (2024). Intrusion Detection for IoT Network Security with Deep Learning. [JAIDM]. *Journal of Artificial Intelligence and Data Mining*.

Mughal, U. A., Atat, R., & Ismail, M. (2024). Graph Neural Network-Based Intrusion Detection System for a Swarm of UAVs. *Proceedings - IEEE Military Communications Conference MILCOM*, 578–583. DOI: 10.1109/MILCOM61039.2024.10773671

Narejo, I. H., Matloob, M. J., Attaullah, H. M., & Khawar, R. (2024). Designing and Evaluating a Mobile Social Robot as an Intervention Tool for Autism Spectrum Disorder. *2024 IEEE 1st Karachi Section Humanitarian Technology Conference, Khi-HTC 2024*. DOI: 10.1109/KHI-HTC60760.2024.10482306

Ngo, R. (2020). *AGI Safety From First Principles. Future of Humanity Institute (FHI)*. University of Oxford.

Niyonsaba, S., Konate, K., & Soidridine, M. M. (2024). Deep Learning Based Intrusion Detection for Cybersecurity in Unmanned Aerial Vehicles Network. *International Conference on Electrical, Computer, and Energy Technologies, ICECET 2024*. DOI: 10.1109/ICECET61485.2024.10698453

Pande, A. K., Brantley, P., Tanveer, M. H., & Voicu, R. C. (2024). From AI to AGI - The Evolution of Real-Time Systems with GPT Integration. *SoutheastCon, 2024*, 699–707. DOI: 10.1109/SoutheastCon52093.2024.10500172

Pandey, G. K., Gurjar, D. S., Nguyen, H. H., & Yadav, S. (2022). Security Threats and Mitigation Techniques in UAV Communications: A Comprehensive Survey. *IEEE Access : Practical Innovations, Open Solutions*, *10*, 112858–112897. DOI: 10.1109/ACCESS.2022.3215975

Pang, X., Zhao, N., Tang, J., Niyato, D., & Wong, K. K. (2021). Joint Trajectory and Beamforming Optimization for Secure UAV Transmission Aided by IRS. *13th International Conference on Wireless Communications and Signal Processing, WCSP 2021*. DOI: 10.1109/WCSP52459.2021.9613308

Parra-López, C., Ben Abdallah, S., Garcia-Garcia, G., Hassoun, A., Sánchez-Zamora, P., Trollman, H., Jagtap, S., & Carmona-Torres, C. (2024). Integrating digital technologies in agriculture for climate change adaptation and mitigation: State of the art and future perspectives. In *Computers and Electronics in Agriculture* (Vol. 226). Elsevier B.V., DOI: 10.1016/j.compag.2024.109412

Pliatsios, D., Goudos, S. K., Lagkas, T., Argyriou, V., Boulogeorgos, A. A. A., & Sarigiannidis, P. (2022). Drone-Base-Station for Next-Generation Internet-of-Things: A Comparison of Swarm Intelligence Approaches. *IEEE Open Journal of Antennas and Propagation, 3*, 32–47. DOI: 10.1109/OJAP.2021.3133459

Politi, E., Varlamis, I., Tserpes, K., Larsen, M., & Dimitrakopoulos, G. (2022). The future of safe BVLOS drone operations with respect to system and service engineering. *Proceedings - 16th IEEE International Conference on Service-Oriented System Engineering, SOSE 2022*, 133–140. DOI: 10.1109/SOSE55356.2022.00022

Prasetio, E. A., & Nurliyana, C. (2023). Evaluating perceived safety of autonomous vehicle: The influence of privacy and cybersecurity to cognitive and emotional safety. *IATSS Research, 47*(2), 160–170. DOI: 10.1016/j.iatssr.2023.06.001

Rafael Aditya Cahyo, W., Jonarta, H., Clarissa, A. C., Syafalni, I., Sutisna, N., & Hasanuddin, M. O. (2023). Securing UAV Communication with TinyJAMBU Cryptography. *Proceedings of the International Conference on Electrical Engineering and Informatics*. DOI: 10.1109/ICEEI59426.2023.10346669

Raja, G., Anbalagan, S., Ganapathisubramaniyan, A., Selvakumar, M. S., Bashir, A. K., & Mumtaz, S. (2021). Efficient and Secured Swarm Pattern Multi-UAV Communication. *IEEE Transactions on Vehicular Technology, 70*(7), 7050–7058. DOI: 10.1109/TVT.2021.3082308

Ramadan, R. A., Emara, A. H., Al-Sarem, M., & Elhamahmy, M. (2021). Internet of drones intrusion detection using deep learning. *Electronics (Basel), 10*(21), 2633. Advance online publication. DOI: 10.3390/electronics10212633

Saad, W., Hashash, O., Thomas, C. K., Chaccour, C., Debbah, M., Mandayam, N., & Han, Z. (2025). Artificial General Intelligence (AGI)-Native Wireless Systems: A Journey Beyond 6G. *Proceedings of the IEEE*, 1–39. DOI: 10.1109/JPROC.2025.3526887

Sajid, J., Hayawi, K., Malik, A. W., Anwar, Z., & Trabelsi, Z. (2023). A Fog Computing Framework for Intrusion Detection of Energy-Based Attacks on UAV-Assisted Smart Farming. *Applied Sciences (Basel, Switzerland), 13*(6), 3857. Advance online publication. DOI: 10.3390/app13063857

Salmon, P. M., McLean, S., Carden, T., King, B. J., Thompson, J., Baber, C., Stanton, N. A., & Read, G. J. M. (2024). When tomorrow comes: A prospective risk assessment of a future artificial general intelligence-based uncrewed combat aerial vehicle system. *Applied Ergonomics, 117*, 104245. Advance online publication. DOI: 10.1016/j.apergo.2024.104245 PMID: 38320386

Shafique, A., Mehmood, A., & Elhadef, M. (2021). Survey of Security Protocols and Vulnerabilities in Unmanned Aerial Vehicles. *IEEE Access : Practical Innovations, Open Solutions, 9*, 46927–46948. DOI: 10.1109/ACCESS.2021.3066778

Sharma, D. D., & Sarojwal, A. (2024). Secure and Resilient Threat Modelling Strategy for Drone Delivery System. *2024 IEEE Region 10 Symposium, TENSYMP 2024*. DOI: 10.1109/TENSYMP61132.2024.10752205

Singh, A. (2023). Artificial Intelligence-Based Autonomous Vehicle for Making Real-Time Decisions, and Navigating Safely. *Proceedings of the 2023 International Conference on Innovative Computing, Intelligent Communication and Smart Electrical Systems, ICSES 2023*. DOI: 10.1109/ICSES60034.2023.10465299

Sumra, I. A., Sellappan, P., Abdullah, A., & Ali, A. (2018). Security issues and Challenges in MANET-VANET-FANET: A Survey. *EAI Endorsed Transactions on Energy Web, 5*(17), 155884. Advance online publication. DOI: 10.4108/eai.10-4-2018.155884

Thompson, A. K., Patel, D. D., & Ilhan Akbas, M. (2024). Formulating a Comprehensive Cybersecurity Framework for Uncrewed Aerial Vehicles. *Conference Proceedings of the IEEE International Performance, Computing, and Communications Conference*. DOI: 10.1109/IPCCC59868.2024.10850308

Tlili, F., Ayed, S., & Chaari Fourati, L. (2024). Advancing UAV security with artificial intelligence: A comprehensive survey of techniques and future directions. In *Internet of Things (Netherlands)* (Vol. 27). Elsevier B.V. DOI: 10.1016/j.iot.2024.101281

Tout, S., Abualkibash, M., & Patil, P. (2018). Emerging Research in the Security of Modern and Autonomous Vehicles. *IEEE International Conference on Electro/ Information Technology (EIT)*, 199. DOI: 10.1109/EIT.2018.8500204

Wang, Y. (2022). A Formal Theory of AI Trustworthiness for Evaluating Autonomous AI Systems. *Conference Proceedings - IEEE International Conference on Systems, Man and Cybernetics, 2022-October*, 137–142. DOI: 10.1109/SMC53654.2022.9945351

Wanner, D., Hashim, H. A., Srivastava, S., & Steinhauer, A. (2024). UAV avionics safety, certification, accidents, redundancy, integrity, and reliability: a comprehensive review and future trends. In *Drone Systems and Applications* (Vol. 12, pp. 1–23). Canadian Science Publishing., DOI: 10.1139/dsa-2023-0091

Yampolskiy, R., & Fox, J. (2013). Safety Engineering for Artificial General Intelligence. *Topoi*, *32*(2), 217–226. DOI: 10.1007/s11245-012-9128-9

Yan, S., Han, H., Dong, X., & Xu, Z. (2024). Lightweight Deep Learning Method based on Group Convolution: Detecting DDoS Attacks in IoT Environments. *Proceedings - 2024 10th International Symposium on System Security, Safety, and Reliability, ISSSR 2024*, 181–189. DOI: 10.1109/ISSSR61934.2024.00027

Yao, A., Jiang, F., Li, X., Dong, C., Xu, J., Xu, Y., Li, G., & Liu, X. (2021). A Novel Security Framework for Edge Computing based UAV Delivery System. *Proceedings - 2021 IEEE 20th International Conference on Trust, Security and Privacy in Computing and Communications, TrustCom 2021*, 1031–1038. DOI: 10.1109/TrustCom53373.2021.00142

Zhang, Z., Si, X., Li, L., Gao, Y., Li, X., Yuan, J., & Xing, G. (2023). An Intrusion Detection Method Based on Transformer-LSTM Model. *2023 3rd International Conference on Neural Networks, Information and Communication Engineering, NNICE 2023*, 352–355. DOI: 10.1109/NNICE58320.2023.10105733

Zhi, Y., Fu, Z., Sun, X., & Yu, J. (2020). Security and Privacy Issues of UAV: A Survey. *Mobile Networks and Applications*, *25*(1), 95–101. DOI: 10.1007/s11036-018-1193-x

Zhou, Y., Pan, C., Yeoh, P. L., Wang, K., Elkashlan, M., Vucetic, B., & Li, Y. (2020). Secure Communications for UAV-Enabled Mobile Edge Computing Systems. *IEEE Transactions on Communications*, *68*(1), 376–388. DOI: 10.1109/TCOMM.2019.2947921

Chapter 6
Intrusion Detection for Smart Environmental Drones:
A Multi-Dataset ML Approach

Malik Huzaifa
Air University, Pakistan

Hafiz Muhammad Attaullah
https://orcid.org/0000-0002-4647-2607
Faculty of Computing and Informatics, Multimedia University, Malaysia

Menaa Nawaz
Air University, Pakistan

Hameedur Rahman
https://orcid.org/0000-0001-8892-9911
Air University, Pakistan

ABSTRACT

AGI-enabled drones are vital yet vulnerable systems, especially in climate monitoring and autonomous aerial missions, making cybersecurity a top priority. Dynamic, decentralized UAVs face cyber threats like DoS, spoofing, replay, and FDI attacks. This research proposes and evaluates lightweight ML-based Intrusion Detection Systems (IDS) tailored for UAVs. Five supervised models—Naïve Bayes, Decision Tree, K-Nearest Neighbors, Random Forest, and LightGBM—were tested across three datasets: a CTGAN-augmented mix of CIC-IDS2017 and UNSW-NB15, a physical-cyber UAV testbed dataset, and a T-ITS behavioral dataset. The experimental design included preprocessing, normalization, class balancing, and performance evalua-

DOI: 10.4018/979-8-3693-6457-4.ch006

tion using accuracy, precision, recall, F1-score, and confusion matrix. LightGBM achieved the highest accuracy and generalization, while Decision Tree and Naïve Bayes offered optimal efficiency and speed, making them ideal for onboard IDS in AGI-driven drones.

1. INTRODUCTION

The fight against climate change demands technological solutions that are not only scalable and intelligent but also autonomous and adaptive to unpredictable environmental conditions. In this context, AGI has emerged as a transformative capability, especially when embedded in unmanned aerial vehicles. AGI refers to the ability of a system to perform any intellectual task that a human can do — including reasoning, perception, learning, and problem-solving across domains. When coupled with drones, AGI enables context-aware, self-adaptive, and mission-resilient aerial platforms that far surpass the capabilities of rule-based or narrow AI systems (Ceviz et al., 2023; Ceviz, Sen, & Sadioglu, 2025; Luo & Oracevic, 2024; Da Silva et al., 2023).

AI drones provide maximum utility during environmental and climate surveillance because they need to respond to rapidly changing conditions while operating in real time to take decisions. The drones operate autonomously by updating flight directions in response to wind shifts as well as sensor malfunctions or when unexpected environmental incidents including forest fires or glacier melts occur. These systems make real-time decisions to accomplish the following through their capabilities:

- Detect and report illegal deforestation and habitat destruction
- Monitor greenhouse gas emissions in remote areas
- Measure air, soil, and water quality autonomously
- Assess post-disaster damages with minimal human intervention
- Track wildlife populations and migration patterns without disturbance

As these intelligent drones operate in sensitive and mission-critical scenarios, their integrity and trustworthiness become non-negotiable. The embrace of AGI advances both system adaptability together with improved cyber-resilience because autonomous decision-making capabilities directly modify ecological assessment outcomes and climate policy data.

1.1 Need

The greater autonomy in systems results in proportional growth of their vulnerability to cyber threats. Drones enabled by AGI combine onboard sensors with external communication devices and networked decision systems to create a Flying Ad Hoc Network which they use to operate. FANETs function through a combination of fully distributed and quickly moving networks yet remain prone to multiple kinds of cyber assaults such as:

- GPS spoofing (misguiding drone location)
- Denial of Service (DoS) attacks (overloading communication channels)
- Jamming (disrupting radio signals)
- False data injection (compromising sensor inputs or telemetry)
- Command hijacking (taking over control systems)

A compromised drone system not only risks losing mission-critical data but may endanger human lives or disrupt environmental surveillance operations. The majority of conventional cybersecurity solutions like static firewalls and central threat monitoring fail in UAV environments due to bandwidth, power and latency constraints (Mughal et al., 2023; da Silva et al., 2023).

The operational integrity of drones using AGI requires distributed security frameworks which are designed to detect real-time threats and maintain lightweight intelligence functions.

1.2 Motivation

Researchers use ML-based IDS to tackle UAV-specific security threats. ML-powered IDS uses historical data to learn complex attack patterns in addition to providing generalization on previously unknown threats unlike traditional rule-based IDS. The approach proves vital for drone systems which experience changing attack patterns throughout different times and spaces and operational scenarios.

Recent studies offer for the use of ML in drone IDS:

- Bi-LSTM and CNN-based models have shown exceptional performance in modeling the temporal and spatial structure of UAV communication traffic, achieving higher detection accuracy compared to traditional methods (Luo & Oracevic, 2024; Mughal et al., 2023).
- Federated Learning (FL) frameworks present themselves as promising solutions for collaboratively trained IDS models through privacy-aware decen-

tralized operations among drones (Ceviz et al., 2025; 2023) and the research covers the same aspect.

- Graph Neural Networks (GNNs) identify drone swarm anomalies by analyzing the relational connectivity of drone nodes according to Mughal et al. (2023).
- GANs alongside Auto Encoders help create attack simulations while spotting unlabeled outlier patterns for enhanced protection against adversarial inputs plus data imbalance in systems (Mughal, Atat, & Ismail, 2024; Asif et al., 2023; Park, Park, & Kim, 2020).

Security measures for drones rely on lightweight IDS models such as ANFIS, logistic regression, and PCA because they ensure both operational performance and security protection.

Building IDS frameworks represents a vital need because they must be adaptive while being robust and lightweight and maintaining privacy protection to defend intelligent drones against emerging cyber threats in real-time operations.

1.3 Contributions

This study addresses the pressing need for cyber-resilient UAVs in climate-focused operations by exploring the application of ML models for AIDS. Specifically, we present a comparative evaluation of five prominent ML algorithms across three UAV network datasets, simulating real-world drone communication and threat patterns.

The core objectives and contributions include:

- Design and Implementation of an ML-based IDS architecture that integrates seamlessly with AGI-enabled UAV systems.
- Comparative Evaluation of five ML algorithms — including Bi-LSTM, CNN, Random Forest, SVM, and XGBoost — based on detection accuracy, precision, recall, F1-score, and computational efficiency.
- Cross-Dataset Validation using three publicly available drone-related datasets, offering a robust benchmark for IDS performance under varied conditions.
- Analysis of Trade-offs between model complexity and deployment feasibility, especially in edge environments with limited computation and bandwidth.
- Recommendations for Future Integration of federated, GAN-based, or hybrid models to further enhance IDS adaptability and real-time detection in AGI-powered climate drones.

2. RELATED WORK

Drone-network has unique features but is quite open to security attacks. Drones have many applications where security is unsatisfactory. Therefore, hybrid intrusion detection system is formulated for traffic analysis to monitor data packets. Hybrid is basically the integration of anomaly and signature to detect real traffic-based flooding. Paparazzi system is specially used for real time experimentation which is decomposed into three segments which include ground segment, airborne and communication segment (Khalil & Rahman, 2023). Deficiency of trusted centralized system having limited bandwidth and power. Ad hoc networks are vulnerable to security attacks which violate passively and actively attacking directly. Malicious workstations send false information where worm hole tunneling, DoS and black hole attack are launched (Arthur, 2019).

Recent advances in ML have shown great promise in building adaptive and intelligent IDS capable of anomaly detection in real time. ML-based IDS frameworks have demonstrated improved ability to identify cyber threats such as denial-of-service (DoS) attacks, spoofing, jamming, and false data injection, which are commonly targeted at UAV networks due to their wireless and open communication protocols (Da Silva et al., 2023; Luo & Oracevic, 2024; Hassler, Mughal, & Ismail, 2023).

A major research trend in this area focuses on federated and distributed learning, designed to support decentralized and resource-constrained UAV environments. Unlike centralized IDS, Federated Learning (FL) allows individual UAVs to collaboratively train a global IDS model without sharing raw data, thereby preserving both privacy and bandwidth efficiency. Studies such as (Da Silva et al., 2023; Ceviz, Sadioglu, Sen, & Vassilakis, 2023) propose FL-based IDS frameworks that significantly enhance security in FANET architectures. Furthermore, few-shot federated learning has been introduced to reduce training time and data requirements, particularly in newly deployed or low-data drone networks (Ceviz, Sen, & Sadioglu, 2025). In parallel, Federated GAN-Augmented IDS (FGA-IDS) approaches combine FL with generative adversarial networks (GANs) to synthesize realistic threat scenarios, thus addressing the issue of data scarcity and imbalance (Zeng et al., 2024).

Deep learning has also emerged as a dominant strategy in the UAV IDS domain, particularly for modeling spatio-temporal patterns in drone behavior. Architectures such as Bi-LSTM, Convolutional Neural Networks (CNNs), and Graph Neural Networks (GNNs) have been successfully applied to detect communication anomalies and flight deviations (Luo & Oracevic, 2024; Mughal, Atat, & Ismail, 2024; Hassler, Mughal, & Ismail, 2023). Some of these studies further integrate GAN-based adversarial training to enhance IDS robustness against sophisticated attacks (Asif et al., 2023; Zeng et al., 2023).

To accommodate real-time detection on low-power UAV platforms, several works propose lightweight IDS frameworks. These include optimization-based methods for feature selection and dimensionality reduction (Wu et al., 2024; Abdulsattar et al., 2022), neuro-fuzzy inference systems (Khalil & Rahman, 2023), and adaptive hybrid approaches such as rule-based deep learning hybrids (Arthur, 2019). Additionally, cyber-physical fusion models that combine flight telemetry with network traffic analysis have shown improved detection rates by leveraging cross-domain features (Hassler, Mughal, & Ismail, 2023; Mughal, Hassler, & Ismail, 2023).

Several IDS studies also address specific UAV-targeted threats, including GPS spoofing, jamming, and de-authentication attacks, ensuring the IDS is tuned to relevant aerial attack vectors (Khalil & Rahman, 2023; Rahman et al., 2023). Broader survey efforts such as Quinn & Bouk (2023) synthesize these developments and emphasize key challenges that remain—particularly around dataset realism, adversarial robustness, real-time responsiveness, and computational limitations.

Also, the evolving research landscape indicates a strong shift toward adaptive, scalable, and lightweight IDS frameworks for UAVs. Although federated learning and deep learning architectures offer strong performance gains, further research is needed to improve real-world generalization, reduce model complexity, and address deployment challenges in AGI-powered drone fleets operating in climate-sensitive missions.

In this regards Table 1 provides a comparison of key machine learning models used for IDS in drone environments, citing specific research examples and evaluating their strengths and limitations.

Table 1. Comparison of ML algorithms for drone IDS

Algorithm / Model	Use in IDS	Ref	Strengths	Challenges
Federated Learning (FL)	Decentralized, privacy-aware training	(Da Silva et al., 2023; Ceviz, Sen, & Sadioglu, 2025; Ceviz, Sadioglu, Sen, & Vassilakis, 2023; He et al., 2023; Mowla et al., 2019)	No raw data sharing; collaborative scalability	Requires sync; communication overhead
GANs	Data synthesis; adversarial defense	(Da Silva et al., 2023; Asif et al., 2023; Zeng et al., 2023; 2024)	Boosts robustness to unknown threats	Requires large training; sensitive to quality
GNN	Swarm-based spatial modeling	(Luo & Oracevic, 2024)	Captures network structure of UAVs	Needs structured graphs; real-time complexity

continued on following page

Table 1. Continued

Algorithm / Model	Use in IDS	Ref	Strengths	Challenges
Bi-LSTM	Time-series sequence modeling	(Da Silva et al., 2023; Luo & Oracevic, 2024)	Bidirectional learning of UAV communication logs	High memory & compute needs
CNN	Pattern recognition in traffic	(Hassler, Mughal, & Ismail, 2023; Hadi et al., 2024; Ahmad et al., 2024)	Detects complex, nonlinear patterns	Needs large labeled data
Autoencoders (AE)	Unsupervised anomaly detection	(Park, Park, & Kim, 2020)	Detects zero-day threats without labels	Reconstruction errors sensitive to noise
SVM	Classification (binary/multi-class)	(Hassler, Mughal, & Ismail, 2023; Rahman et al., 2023)	Strong separation boundaries	Doesn't scale well on big datasets
RF	Ensemble classification	(Narejo et al., 2024; Rahman et al., 2023)	High accuracy; generalization	Resource-intensive
ANFIS	Lightweight IDS for edge deployment	(Khalil & Rahman, 2023)	Energy-efficient inference	Lower accuracy vs deep models
Logistic Regression	Minimal resource IDS	(Narejo et al., 2024)	Very fast; interpretable	Poor for complex patterns

Multiple research paths based on algorithms exist in the field of UAV intrusion detection systems due to recent developments. The developments showcase real-time weaknesses alongside scalability and robustness challenges in adversarial drone defense systems. Current UAV security deployment practices remain distinct from theoretical models despite major recent developments. This investigation makes progress in closing the existing gap by assessing the operational capabilities of IDS powered by ML technology across various drone settings.

3. INTRUSION DETECTION SYSTEMS FOR DRONES

The protective mechanism known as IDS has become vital due to its capability of scanning drone activities to detect both suspicious patterns and identified malicious signatures. This section encompasses the IDS structural design along with its identification systems and required specifications when implementing this system to defend drone ecosystems including those powered by AGI participating in climate-critical operations.

The Intrusion Detection System serves as security software or hardware which tracks UAV operations by observing both network streams alongside sensing devices and command signals and flight measurement inputs for identifying security risks. Story includes software firewalls but it operates differently from preventive systems by involving IDS which concentrates on detecting anomalies alongside misuse and external attacks during and immediately following their occurrence.

In drone systems, IDS is often embedded in:

- Onboard processors (lightweight or edge-deployable IDS)
- Ground control systems
- Collaborative swarm communication layers (for FANETs)

For AGI-enabled drones, which make autonomous decisions, IDS plays a critical supervisory role, ensuring that the drone's learned behavior or sensor responses are not being tampered with, manipulated, or misled due to adversarial input or cyberattacks. The core components of IDS discussed in table 2.

Table 2. Components of a UAV-based intrusion detection system

Component	Function
Data Collector	Gathers raw input from flight sensors, telemetry logs, GPS data, and network traffic packets
Preprocessing Engine	Cleans, normalizes, and formats raw data for consistency and analysis readiness
Feature Extractor	Derives critical attributes (e.g., packet size, altitude drift, GPS variance) for analysis
Detection Engine	Applies signatures, heuristics, or ML models to flag anomalous or malicious activity
Decision Engine	Issues alerts, logs threat events, or launches automated defense responses

IDS systems follow multiple classification axes for their operation. The fundamental categories of IDS systems are detection-based systems and architecture-based systems.

Figure 1. Intrusion detection system classification taxonomy

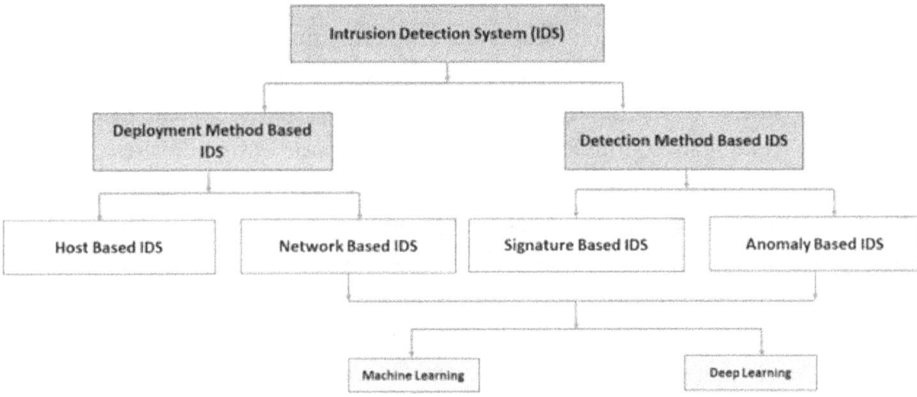

3.1 Based on Detection Classification

The detection strategies along with the deployment architecture classify Intrusion Detection Systems. The selection of IDS strategy in UAV networks that use AGI-based autonomy directly affects how well intrusions get detected and also determines resource usage levels and reactive speed. The three main IDS detection methods (signature-based and anomaly-based and hybrid IDS) along with appropriate systems deployment architectures for UAV and FANET networks constitute the focus of this section.

3.1.1 Signature-Based IDS

Signature-based IDS conducts its monitoring by checking for known attack patterns that engineers already created prior to detection. These predefined patterns serve as attack signatures for identifying harmful activities. The scan process of these systems compares incoming data against their stored threat databases using established DoS packet patterns and spoofing signal shapes along with previously seen payload abnormalities.

This detection method provides speedy performance when identifying known threats which makes it suitable for real-time UAV operations. The system fails to recognize new and unrecorded attacks that frequently occur in drone environments. Populations of UAVs operating in diverse terrains and adversary conditions calculate signature-based IDS to be inadequate for single-use defense due to its strict identification methods.

3.1.2 Anomaly-Based IDS

Anomaly-based IDS enhances signature systems through its method of identifying behavioral deviations from drone normal functions. These systems generate a baseline definition of regular drone operation through historical data examination of telemetry reports and communications messages and sensor information which enables them to detect unauthorized behavior. Changes in drone GPS variability patterns combined with odd network transmission behavior usually signal potential security risks.

The selected method proves highly beneficial for UAV adaptive and decentralized systems that operate using Flying Ad Hoc Networks and drone swarms controlled by Artificial General Intelligence. The most beneficial aspect of this method lies in its unknown threat detection capability that field-operated systems without supervision need to identify threats effectively.

Anomaly-based IDS enhances signature systems through its method of identifying behavioral deviations from drone normal functions. These systems generate a baseline definition of regular drone operation through historical data examination of telemetry reports and communications messages and sensor information which enables them to detect unauthorized behavior. Changes in drone GPS variability patterns combined with odd network transmission behavior usually signal potential security risks.

The selected method proves highly beneficial for UAV adaptive and decentralized systems that operate using Flying Ad Hoc Networks and drone swarms controlled by Artificial General Intelligence. The most beneficial aspect of this method lies in its unknown threat detection capability that field-operated systems without supervision need to identify threats effectively.

3.1.3 Hybrid IDS

Hybrid IDS combines the strengths of both AIDS and SIDS approaches. These systems maintain a library of known attack signatures while simultaneously using behavior models to flag previously unseen threats. The dual-layer architecture allows hybrid IDS to achieve higher detection accuracy and robustness compared to either method alone.

In the context of UAV deployments, hybrid systems are especially promising for AGI-enabled drones that must balance autonomous decision-making with real-time threat identification. By layering detection mechanisms, hybrid IDS can minimize false negatives (missed attacks) while keeping false positives (incorrect alerts) within acceptable limits (Nabi et al., 2024; Attaullah, Khan, & Mughal, 2021).

The main drawback of hybrid IDS is system complexity. Integrating and tuning both detection engines requires significant computation and design effort. However, this complexity can be mitigated by employing federated and collaborative learning architectures, which distribute the detection workload across multiple UAVs.

3.2 Based on System Architecture Classification

Beyond detection strategies, IDS can also be categorized by deployment architecture—that is, where and how the detection mechanisms are integrated into the UAV system.

3.2.1 Host-Based IDS

HIDS is embedded directly into the drone's onboard systems, where it monitors logs, control loops, and sensor outputs in real-time. These are ideal for autonomous single-drone operations or AGI drones with sufficient processing capabilities.

Network-Based IDS

NIDS is placed at the network level, often analyzing traffic between UAVs and ground control systems or among members of a drone swarm. This setup is effective for monitoring FANET packet flows and inter-drone communication.

The collaborative Intrusion Detection System enables the extension of Network Intrusion Detection across multiple drones which provides timely real-time distribution of local detection results along with model information. The localized model training approach in federated learning suits well with IDS architecture that unites UAVs to create a global IDS solution. Data privacy maintains its integrity through federated IDS which produces superior detection performance in decentralized UAV networks according to research by Ceviz et al. (2023; 2025).

The combination of IDS types along with their deployment strategies creates a broad defensive mechanism to protect drones embedded with AGI which operate under climate-focused and hostile conditions. IDS effectiveness depends on designing the system to match mission requirements and drone hardware limitations and operational hazard assessment.

4. METHODOLOGY

The experimental process duplicated real-world cybersecurity threats while testing lightweight IDS systems through model evaluation across different data collections for detection performance measurement. The section outlines experimental parameters including datasets with specific specifications followed by an explanation of selected models together with data preprocessing methods and model training procedures alongside a performance assessment methodology. A depiction of the complete workflow can be found in Figure 2. The illustration provides insight into ML-based intrusion detection system (IDS) development methods for UAVs along with evaluation procedures. The system starts with dataset input of UNSW-NB15 CIC-IDS2017 and T-ITS before moving to a preprocessing stage which deals with missing values and feature normalization. The threat classification requires five machine learning algorithms (KNN, Naïve Bayes, Decision Tree, Random Forest, and LightGBM) to analyze each dataset into training and testing sets that use an 80/20 ratio. The assessment of the final model utilizes accuracy and precision as well as recall and F1-score for measuring performance.

Figure 2. Workflow for drone intrusion detection

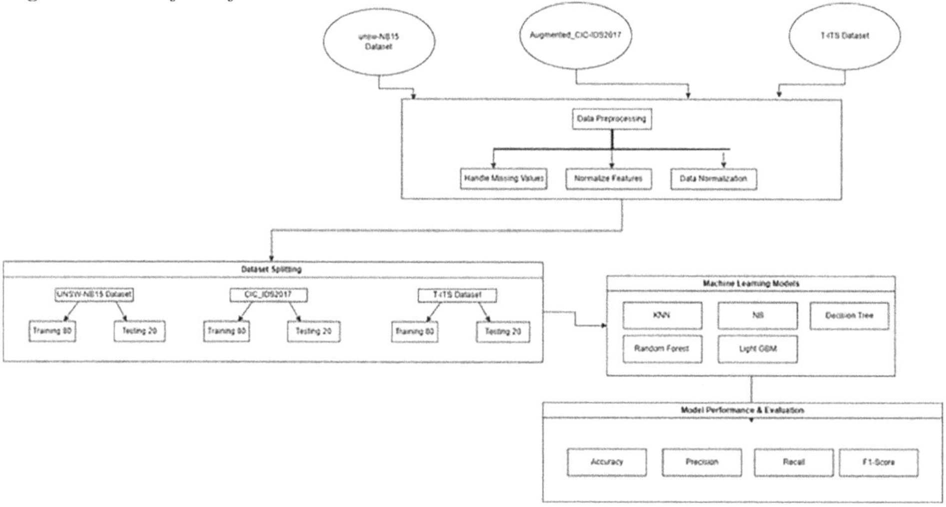

4.1 Simulation Environment

The core objective of this study is to simulate UAV-specific cyber threats and evaluate the detection capabilities of five supervised machine learning algorithms on three intrusion detection datasets. These datasets were chosen to reflect a mix of synthetic, real-world, and simulation-based UAV telemetry and attack data. The broader goal is to identify machine learning models that provide a balance between detection performance and computational efficiency—two critical requirements for real-time IDS deployment in autonomous, resource-constrained drones.

All experiments were conducted using Python 3.10, leveraging the Pandas and NumPy libraries for data handling and preprocessing. Scikit-learn was used to implement classical machine learning models, while LightGBM was employed for gradient-boosted tree-based modeling. For visualization and result interpretation, Matplotlib and Seaborn were used to generate confusion matrices, ROC curves, and performance comparison plots. All experiments were executed on a local machine configured with an Intel Core i7 processor, 16GB RAM, and Windows 11 OS, ensuring sufficient processing power for iterative training and testing across large datasets. The details are discussed in Table 3.

Table 3. Experimental setup

Component	Details
Programming Language	Python 3.10
Libraries Used	Pandas, NumPy (data processing); Scikit-learn, LightGBM (ML modeling)
Visualization Tools	Matplotlib, Seaborn
System Configuration	Intel Core i7, 16GB RAM, Windows 11
Datasets Evaluated	CTGAN-Enhanced Dataset, Physical-Cyber UAV Testbed, T-ITS.csv
ML Models Tested	Decision Tree, Naïve Bayes, KNN, Logistic Regression, LightGBM
Train-Test Split	80% Training – 20% Testing (Stratified sampling)
Validation Technique	K-Fold Cross-Validation (typically k=5)
Evaluation Metrics	Accuracy, Precision, Recall, F1-score, ROC-AUC, Confusion Matrix
Objective	Evaluate detection accuracy vs. resource cost in UAV-based IDS deployment

Each machine learning algorithm was independently trained and evaluated on each dataset using stratified train-test splitting and cross-validation, ensuring that the performance results were robust, unbiased, and generalizable across diverse UAV security scenarios.

4.2 Datasets Description

To reflect the range of threats UAVs face in real-world and simulated environments, three datasets were utilized. These datasets were selected based on their diversity, relevance to drone communication and telemetry, and adaptability to IDS modeling.

The first dataset is the CTGAN-Enhanced UAV IDS Dataset, which combines the well-established Augmented-CIC-IDS-2017 and Augmented-UNSW-NB15 datasets. Both datasets originally captured network intrusion data and are widely used in the cybersecurity research community. However, to overcome the class imbalance inherent in these datasets—particularly the underrepresentation of rare attacks—a Conditional Tabular GAN (CTGAN) was used to generate high-quality synthetic data. This augmentation technique preserved the statistical patterns of real intrusion events while increasing the volume and balance of minority attack classes. The final dataset consists of over 80 engineered features capturing metadata, temporal behavior, and payload characteristics, including attacks such as DoS, brute-force attempts, infiltration, and web-based attacks. Due to its size and balance, this dataset is ideal for training models intended for deployment in complex, multi-agent UAV systems.

The Physical-Cyber UAV Testbed Dataset originated from experimental testbed tests that incorporated authentic drones with controllers together with adversarial equipment. The dataset records telemetry and communication events that happened when adversaries performed replay attacks combined with de-authentication infliction and the creation of evil twins and fake data insertion attempts on an operational flying drone system. This dataset features features from both cyber-layer and physical-layer sections which include GPS coordinates as well as altitude, velocity, control signal values alongside packet loss rates and signal strength indicators (RSSI). The dataset provides authentic real-time features that effectively evaluate IDS resilience for live mission deployment.

The T-ITS dataset originates from the time-series simulation environment of intelligent transportation systems domain. Its limited dataset size contains behavioral patterns which closely represent UAV navigation activities that occur during smart city logistics operations. The lightweight IDS models operate effectively through this dataset within systems with limited resources while conducting benchmarking protocols at edge-deployed AGI drones.

4.3 Selected Machine Learning Algorithms

The study tested five ML models which possessed different strengths regarding interpretability and training speed and classification accuracy and resource requirements. A selection of models incorporating classic alongside modern techniques was made because they align with UAV real-time deployment requirements.

The Decision Tree (DT) family of models received selection because they offer both straightforward operation and easy interpretation. Through its rule-based structure this model employs hierarchical splits of data features for easy classification purposes. Such systems serve autonomous system audit requirements particularly well in environments where explainable AI is essential.

The Naïve Bayes model (NB) worked as a probabilistic model because it requires limited resources while maintaining high processing speed. Due to its effective processing of multiple dimensions NB works properly with embedded systems featuring restricted CPU capabilities which serves low-cost UAVs.

Through distance metrics K-Nearest Neighbors (KNN) identifies new inputs by comparing them with labeled instances stored in memory. KNN stands out for its solid baseline performance but consumes high computational resources when making predictions thus being suitable for early research stages and fleets of UAVs with reduced sizes.

The linear classifier Logistic Regression functions both efficiently and reliably in threat detection scenarios regardless of whether the number of classes is one or multiple. The model has interpretable coefficients with quick updating capabilities that make it optimal for drones that need regular firmware updates.

LightGBM functions as a gradient boosting framework which demonstrates exceptional accuracy levels together with fast performance on sizable data collections. The system performs optimally regarding speed and memory usage while accommodating thousands of features together with data points. LightGBM functions well as an IDS for UAVs connected to the cloud infrastructure in situations that provide computational resources.

The optimization procedure included grid search along with 5-fold cross-validation for all models to determine their best configurations before extracting performance metrics from each set.

4.4 Model Training and Evaluation Metrics

The training process employed an 80/20 ratio split which stratified sampling to make sure rare intrusion data appeared equally in training and evaluation subsets. Model robustness was achieved through stratified k-fold cross-validation during training while the model performance assessment used multiple metrics for evaluation.

The baseline evaluation method for measuring correct classification counts was accuracy. The distinguishing characteristics of multi-class distributions along with imbalanced conditions prompted us to focus on extra evaluation metrics. Performance metrics included precision which determined the model's correct intrusion detections without errors and recall which calculated attack detection efficiency. A model reliability assessment was derived from the F1-score which calculates a balanced measurement between precision and recall. ROC-AUC scores measured discrimination power for both binary and multiclass binary setups while confusion matrices provided visual performance assessments including false operator and negative results.

To mitigate bias toward overrepresented classes, macro-averaging was used in multiclass evaluations, ensuring that minority attack categories were treated with equal importance in the scoring process.

4.5 Performance Comparison Framework

For systematic comparison, each model's performance metrics were logged, visualized, and analyzed per dataset. This framework was designed to assess which algorithm performs best across different types of UAV-related threat data—ranging from synthetic (CTGAN-enhanced) to real-world (UAV testbed) and simulation-based (T-ITS).

The comparison framework aimed to answer several key questions. First, which model offers the best trade-off between accuracy and efficiency, especially for deployment on AGI-based UAVs with limited processing power? Second, how do model performances vary across datasets representing augmented, telemetry-rich, and lightweight UAV environments? Finally, which models show consistent generalization ability across heterogeneous intrusion types and UAV architectures?

Where applicable, feature importance plots were generated (e.g., from LightGBM and DT) to identify the most influential features across datasets. These insights help inform real-world IDS deployment, particularly on autonomous UAVs tasked with climate-related missions where accuracy, interpretability, and responsiveness are paramount.

5. RESULTS AND DISCUSSIONS

This section presents the evaluation results of five supervised machine learning algorithms—K-Nearest Neighbors (KNN), Naïve Bayes (NB), Decision Tree (DT), Random Forest (RF), and LightGBM (LGBM)—applied to three UAV-relevant intrusion detection datasets: CTGAN-augmented CIC-IDS2017, CTGAN-augmented UNSW-NB15, and T-ITS. The aim is to assess each model's effectiveness in detecting cyber threats across diverse drone environments and data characteristics. Performance is reported using key classification metrics: accuracy, precision, recall, F1-score, and visualized through confusion matrices and ROC curves to illustrate classifier strengths and trade-offs.

5.1 Dataset-Wise Model Performance Analysis

5.1.1 Performance on CTGAN-Augmented CIC-IDS2017 Dataset

The CIC-IDS2017 dataset, enhanced through CTGAN to balance attack class distribution, provided a rich and structured dataset ideal for model training. All models performed well on this dataset, with LightGBM clearly emerging as the top performer, achieving the highest accuracy (0.98) and F1-score (0.98). This model's ensemble learning and feature optimization capabilities allowed it to handle complex traffic patterns and inter-class variations effectively. Figure 3 presents the confusion matrix for LightGBM on this dataset, where near-diagonal dominance indicates accurate classification across all major attack types. Random Forest produced effective outcomes while its precision and recall levels were slightly below the benchmark followed by Decision Tree which yielded moderate predictive performance alongside maintainable interpretability. KNN, while maintaining reasonable accuracy (0.90), showed decreased precision due to misclassifications in overlapping traffic behaviors. Naïve Bayes underperformed in this dataset, particularly due to its assumption of feature independence, which does not hold well in high-dimensional network data.

Figure 3. Confusion matrices of machine learning models on CTGAN-Augmented CIC-IDS2017 dataset

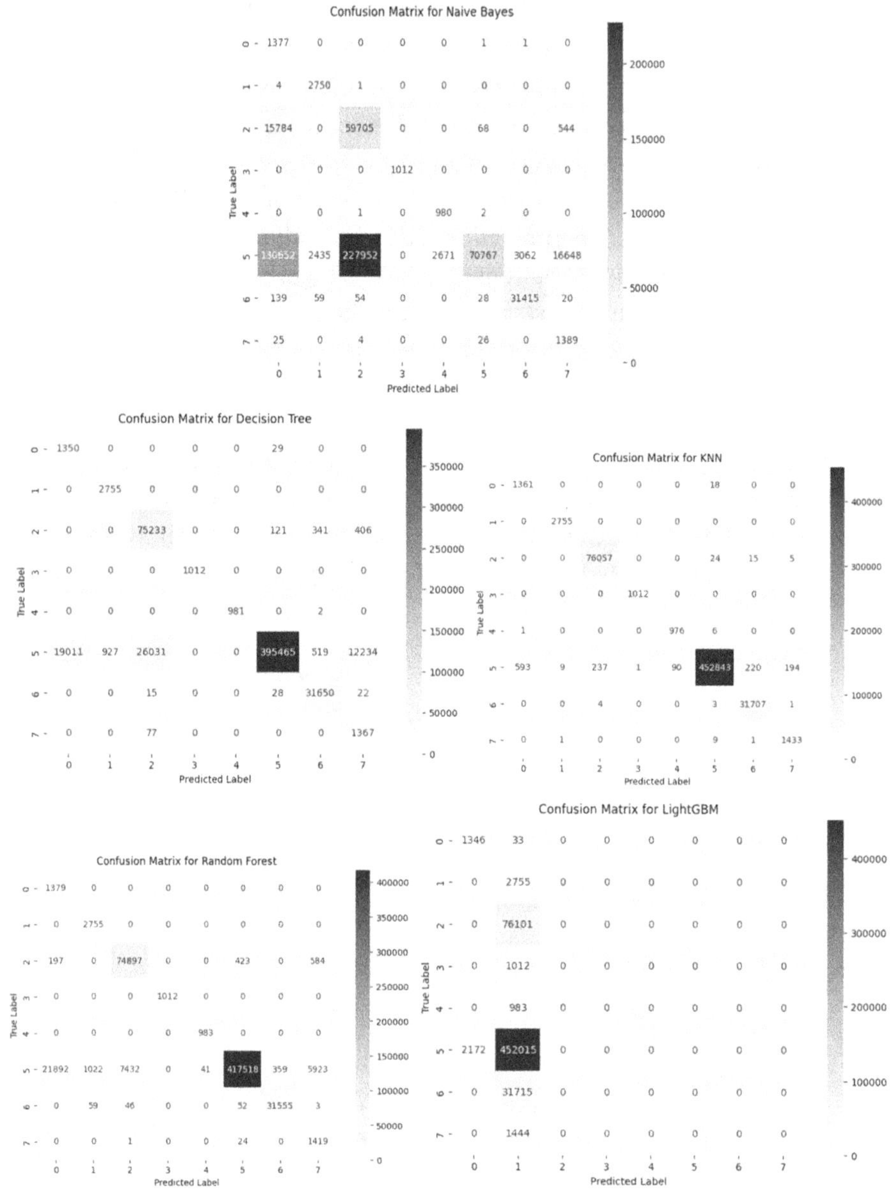

Figure 4. Confusion matrices of machine learning models on CTGAN-Augmented-UNSW_NB15 dataset

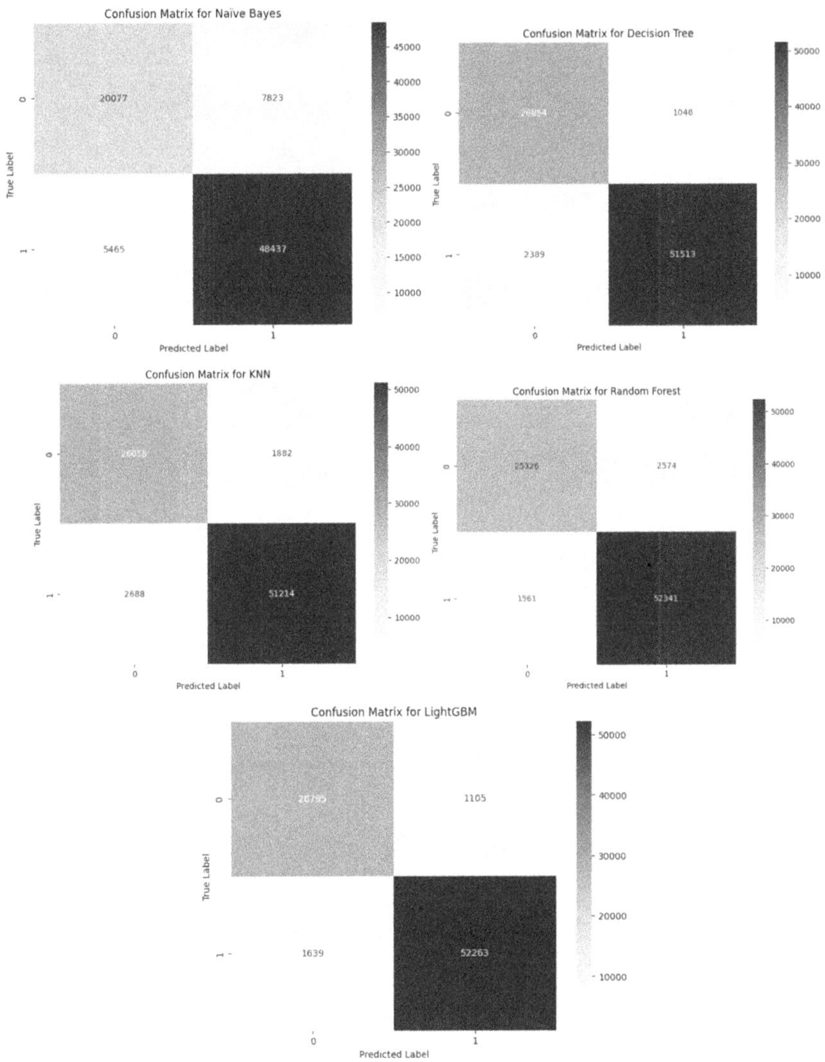

Figure 5. Confusion matrices of machine learning models on T-ITS dataset

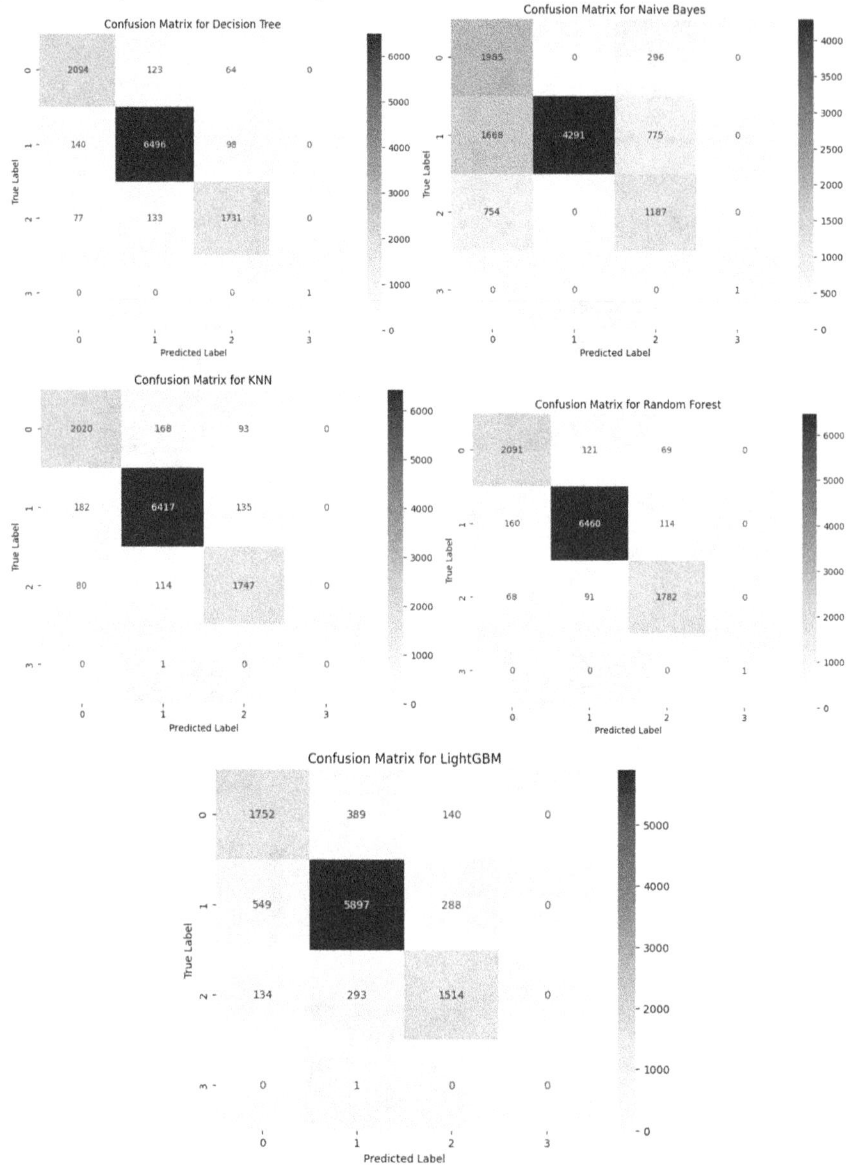

5.1.2 Performance on CTGAN-Augmented UNSW-NB15 Dataset

The UNSW-NB15 dataset posed more challenges due to its slightly imbalanced class distribution and deeper packet-level features. Nevertheless, LightGBM again demonstrated superior performance with 95% accuracy and a high F1-score. Random Forest followed closely with a balanced performance, effectively capturing structural features across common and rare attacks. Decision Tree maintained consistent precision and performed well in classifying high-frequency attacks such as Generic and Exploits but struggled with rarer categories like Fuzzers or Shellcode. As illustrated in Figure 4, the confusion matrix for Decision Tree on UNSW-NB15 highlights the model's strength in distinguishing normal vs. attack traffic, but also reveals some misclassification between overlapping low-frequency classes. Naïve Bayes achieved strong recall but suffered from higher false positives, while KNN exhibited variability in prediction accuracy due to feature scaling sensitivities.

5.1.3 Performance on T-ITS Dataset

The T-ITS dataset, while smaller in size, provided a time-series style representation of UAV communication and behavior. On this dataset, simpler models like Naïve Bayes and Decision Tree performed notably well. Naïve Bayes achieved the highest recall on rare attack types, making it suitable for early-stage anomaly detection where false negatives must be minimized. However, its precision was slightly compromised, indicating some false alarms under certain behavior overlaps.

LightGBM continued to demonstrate robustness, producing the highest overall accuracy (0.91) and strong F1-scores. KNN achieved competitive results but was sensitive to parameter tuning due to the compact feature space of this dataset. The confusion matrix in Figure 5 shows Naïve Bayes' effectiveness in detecting common intrusions, though some confusion was observed between benign and low-activity threat classes.

Table 5. Cross-dataset comparison of classification performance metrics for all models.

Dataset	Model	Accuracy	Precision	Recall	F1-Score
CIC-IDS2017 (Augmented)	LightGBM	0.98	0.97	0.98	0.98
	Random Forest	0.96	0.95	0.94	0.94
	Decision Tree	0.93	0.91	0.91	0.91
	KNN	0.90	0.89	0.88	0.88

continued on following page

Table 5. Continued

Dataset	Model	Accuracy	Precision	Recall	F1-Score
	Naïve Bayes	0.87	0.86	0.85	0.85
UNSW-NB15 (Augmented)	LightGBM	0.95	0.94	0.93	0.94
	Random Forest	0.93	0.92	0.91	0.92
	Decision Tree	0.89	0.87	0.88	0.87
	KNN	0.86	0.84	0.83	0.84
	Naïve Bayes	0.83	0.80	0.84	0.82
T-ITS	LightGBM	0.91	0.89	0.90	0.90
	Decision Tree	0.89	0.88	0.87	0.88
	Naïve Bayes	0.86	0.85	0.89	0.87
	KNN	0.85	0.84	0.83	0.84
	Random Forest	0.83	0.82	0.81	0.81

5.3 Comparative Summary of Metrics

The comparative results for each model across the three datasets are summarized in Table 5, providing a clear view of performance trends. LightGBM emerged as the most consistent performer across all datasets, achieving the highest accuracy and F1-scores. Random Forest followed closely, especially on larger datasets. Decision Tree proved to be a strong contender in low-resource contexts, while Naïve Bayes demonstrated its utility for high-recall applications in telemetry-heavy environments. KNN remained a baseline model with moderate accuracy but higher prediction latency.

The evaluation reveals key insights for deploying IDS on AGI-enabled UAVs. LightGBM offers the highest detection accuracy together with excellent generalization that makes it suitable for detection systems installed locally or through centralized platforms. The high level of computational requirements for LightGBM prevents its direct use in real-time onboard drone systems that need lightweight operations.

Real-time or embedded UAV systems benefit from Decision Tree and Naïve Bayes combination which delivers optimal performance and efficient operation. Interpretable Decision Trees enable model auditing requirements whereas Naïve Bayes produces quick anomaly detection capabilities with high accuracy rates suitable for flight-critical anomaly flagging.

6. CONCLUSION

The research examined how to develop and assess machine learning-based lightweight intrusion detection systems which operated specifically for Drones applied to environmental drones through AGI technology. Unmanned aerial vehicles provide essential environmental data while performing autonomous missions but face an escalating risk of cyber attacks that include denial-of-service functionality in addition to false data entry and GPS manipulation and control system takeover.

The research employed Naïve Bayes, Decision Tree, K-Nearest Neighbors, Random Forest and LightGBM algorithms to analyze three UAV-centric datasets which included CTGAN-augmented CIC-IDS2017 and CTGAN-enhanced UNSW-NB15 and T-ITS. The research design combined structured steps for preprocessing alongside data balancing through CTGAN before performing feature engineering and stratified evaluation.

Across all datasets LightGBM exhibited superior outcomes which exceeded other models in accuracy alongside precision, recall and F1-score performance while achieving its best performance on the CIC-IDS2017 dataset. However, LightGBM's computational intensity may limit its applicability in lightweight, onboard UAV systems. In contrast, Decision Tree and Naïve Bayes models, though slightly less accurate, offered faster inference and lower resource consumption, making them viable candidates for real-time detection on embedded systems.

The analysis of confusion matrices and heatmaps further revealed that while ensemble models excel in generalization, simpler models can be effective when tailored to specific UAV telemetry patterns and network behavior. These insights are crucial for guiding IDS integration within AGI-driven drones, which require adaptability, autonomy, and robust threat detection capabilities.

REFERENCES

Abdulsattar, N. F., Abedi, F., Ghanimi, H. M., Kumar, S., Abbas, A. H., Abosinnee, A. S., Alkhayyat, A., Hassan, M. H., & Abbas, F. H. (2022). Botnet detection employing a dilated convolutional autoencoder classifier with the aid of hybrid shark and bear smell optimization algorithm-based feature selection in FANETs. *Big Data and Cognitive Computing*, *6*(4), 112. DOI: 10.3390/bdcc6040112

Abuljadayel, L., Yerima, S. Y., & Butt, U. (2024, December). Machine Learning-Based Intrusion Detection for Denial-of-Service De-Authentication Attacks on Drones. In 2024 IEEE 16th International Conference on Computational Intelligence and Communication Networks (CICN) (pp. 404-408). IEEE. DOI: 10.1109/CICN63059.2024.10847537

Ahmad, W., Almaiah, M. A., Ali, A., & Al-Shareeda, M. A. (2024, April). Deep learning based network intrusion detection for unmanned aerial vehicle (uav). In *2024 7th World Conference on Computing and Communication Technologies (WCCCT)* (pp. 31-36). IEEE.

Arthur, M. P. (2019). Detecting Signal Spoofing and Jamming Attacks in UAV Networks Using a Lightweight IDS. *Proceedings of the International Conference on Computing, Information and Telecommunication Systems (CITS)*, 2019. DOI: 10.1109/CITS.2019.8862148

Ashraf, S. N., Manickam, S., Zia, S. S., Abro, A. A., Obaidat, M., Uddin, M., Abdelhaq, M., & Alsaqour, R. (2023). IoT empowered smart cybersecurity framework for intrusion detection in internet of drones. *Scientific Reports*, *13*(1), 18422. DOI: 10.1038/s41598-023-45065-8 PMID: 37891186

Asif, M., Rahman, M. A., Akkaya, K., Shahriar, H., & Cuzzocrea, A. (2023, December). Adversarial data-augmented resilient intrusion detection system for unmanned aerial vehicles. In *2023 IEEE International Conference on Big Data (BigData)* (pp. 5428-5437). IEEE. DOI: 10.1109/BigData59044.2023.10386140

Attaullah, H. M., Khan, R. A., & Mughal, S. (2021). Cyber security for Industrial Control System–A Survey. *Iksp J. Emerg. Trends Basic Appl. Sci*, *1*, 15–21.

Attaullah, H. M., Memon, S., Erkan, O. F., & Khawar, R. (2024, January). IoT Based Systems and Services: Recent Security Concerns and Feasible Solutions. In 2024 IEEE 1st Karachi Section Humanitarian Technology Conference (KHI-HTC) (pp. 1-6). IEEE.

Cengiz, K., Lipsa, S., Dash, R. K., Ivković, N., & Konecki, M. (2024). A novel intrusion detection system based on artificial neural network and genetic algorithm with a new dimensionality reduction technique for UAV communication. *IEEE Access : Practical Innovations, Open Solutions, 12*, 4925–4937. DOI: 10.1109/ACCESS.2024.3349469

Ceviz, O., Sadioglu, P., Sen, S., & Vassilakis, V. G. (2023). A novel federated learning-based intrusion detection system for flying ad hoc networks. *arXiv preprint arXiv:2312.04135.*

Ceviz, O., Sadioglu, P., Sen, S., & Vassilakis, V. G. (2023). A Novel Federated Learning-Based IDS for Enhancing UAVs Privacy and Security. *arXiv preprint arXiv:2312.04135.*

Ceviz, O., Sen, S., & Sadioglu, P. (2025). Distributed intrusion detection in dynamic networks of uavs using few-shot federated learning. *arXiv preprint arXiv:2501.13213.*

Cordeiro, T. A. O., Ferrão, R. C., & Cugnasca, P. S. (2024, September). Machine Learning Models for Intrusion Detection in Unmanned Aerial Vehicles: An Approach to Cybersecurity and Operational Safety. In *2024 IEEE 27th International Conference on Intelligent Transportation Systems (ITSC)* (pp. 4236-4241). IEEE.

Da Silva, L. M., Ferrão, I. G., Dezan, C., Espes, D., & Branco, K. R. (2023, June). Anomaly-based intrusion detection system for in-flight and network security in uav swarm. In *2023 International Conference on Unmanned Aircraft Systems (ICUAS)* (pp. 812-819). IEEE. DOI: 10.1109/ICUAS57906.2023.10155873

Hadi, H. J., Cao, Y., Li, S., Hu, Y., Wang, J., & Wang, S. (2024). Real-time collaborative intrusion detection system in uav networks using deep learning. *IEEE Internet of Things Journal.*

Hassler, S. C., Mughal, U. A., & Ismail, M. (2023). Cyber-physical intrusion detection system for unmanned aerial vehicles. *IEEE Transactions on Intelligent Transportation Systems, 25*(6), 6106–6117. DOI: 10.1109/TITS.2023.3339728

He, X., Chen, Q., Wang, W., Liu, T., Li, L., Tang, L., & Liu, Q. (2023, December). Stacked broad learning system empowered FCL assisted by dtn for intrusion detection in UAV networks. In *GLOBECOM 2023-2023 IEEE Global Communications Conference* (pp. 5372-5377). IEEE. DOI: 10.1109/GLOBECOM54140.2023.10436983

Khalil, A. A., & Rahman, M. A. (2023). Adaptive Neuro-Fuzzy Inference System-based Lightweight Intrusion Detection System for UAVs. *Proceedings of the IEEE Conference on Local Computer Networks (LCN)*, 2023. DOI: 10.1109/LCN58197.2023.10223340

Luo, J., & Oracevic, A. (2024). Adaptive Machine Learning for Efficient Anomaly Detection in Autonomous UAVs Swarm Operations. *Proceedings of the International Symposium on Networks, Computers and Communications (ISNCC)*, 2024. DOI: 10.1109/ISNCC62547.2024.10758945

Mowla, N. I., Tran, N. H., Doh, I., & Chae, K. (2019). Federated learning-based cognitive detection of jamming attack in flying ad-hoc network. *IEEE Access : Practical Innovations, Open Solutions*, *8*, 4338–4350. DOI: 10.1109/ACCESS.2019.2962873

Mughal, U. A., Atat, R., & Ismail, M. (2024, October). Graph Neural Network-Based Intrusion Detection System for a Swarm of UAVs. In *MILCOM 2024-2024 IEEE Military Communications Conference (MILCOM)* (pp. 578-583). IEEE. DOI: 10.1109/MILCOM61039.2024.10773671

Mughal, U. A., Hassler, S. C., & Ismail, M. (2023, October). Machine Learning-Based Intrusion Detection for Swarm of Unmanned Aerial Vehicles. In *2023 IEEE Conference on Communications and Network Security (CNS)* (pp. 1-9). IEEE. DOI: 10.1109/CNS59707.2023.10288962

Nabi, F., Zhou, X., Iftikhar, U., & Attaullah, H. M. (2024). A Case Study of Cyber Subversion Attack based Design Flaw in Service Oriented Component Application Logic. *Journal of Cyber Security Technology*, *8*(3), 204–228. DOI: 10.1080/23742917.2023.2261169

Narejo, I. H., Matloob, M. J., Attaullah, H. M., & Khawar, R. (2024, January). Designing and Evaluating a Mobile Social Robot as an Intervention Tool for Autism Spectrum Disorder. In *2024 IEEE 1st Karachi Section Humanitarian Technology Conference (KHI-HTC)* (pp. 1-5). IEEE. DOI: 10.1109/KHI-HTC60760.2024.10482306

Ntizikira, E., Lei, W., Alblehai, F., Saleem, K., & Lodhi, M. A. (2023). Secure and privacy-preserving intrusion detection and prevention in the internet of unmanned aerial vehicles. *Sensors (Basel)*, *23*(19), 8077. DOI: 10.3390/s23198077 PMID: 37836907

Park, K. H., Park, E., & Kim, H. K. (2020). Unsupervised intrusion detection system for unmanned aerial vehicle with less labeling effort. In *Information Security Applications:21st International Conference, WISA 2020,Jeju Island, South Korea,August 26–28, 2020, Revised Selected Papers 21* (pp. 45-58). Springer International Publishing. DOI: 10.1007/978-3-030-65299-9_4

Quinn, J., & Bouk, S. (2023). Intrusion Detection Systems for Flying Ad-hoc Networks. arXiv preprint.

Rahman, K., Aziz, M. A., Usman, N., Kiren, T., Cheema, T. A., Shoukat, H., Bhatia, T. K., Abdollahi, A., & Sajid, A. (2023). Cognitive Lightweight Logistic Regression-Based IDS for IoT-Enabled FANET to Detect Cyberattacks. *Mobile Information Systems*, *2023*(1), 7690322. DOI: 10.1155/2023/7690322

Ramadan, R. A., Emara, A. H., Al-Sarem, M., & Elhamahmy, M. (2021). Internet of drones intrusion detection using deep learning. *Electronics (Basel)*, *10*(21), 2633. DOI: 10.3390/electronics10212633

Wu, Y., Yang, L., Zhang, L., Nie, L., & Zheng, L. (2024). Intrusion detection for unmanned aerial vehicles security: A tiny machine learning model. *IEEE Internet of Things Journal*, *11*(12), 20970–20982. DOI: 10.1109/JIOT.2024.3360231

Zeng, Q., Barnt, K., Ragan, L., & Nait-Abdesselam, F. (2023, December). Realtime intrusion detection in unmanned aerial vehicles using active learning and generative adversarial networks. In *2023 IEEE 29th International Conference on Parallel and Distributed Systems (ICPADS)* (pp. 2802-2803). IEEE. DOI: 10.1109/ICPADS60453.2023.00386

Zeng, Q., Olatunde-Salawu, S., & Nait-Abdesselam, F. (2024, October). FGA-IDS: A Federated Learning and GAN-Augmented Intrusion Detection System for UAV Networks. In *2024 IEEE 10th International Conference on Collaboration and Internet Computing (CIC)* (pp. 50-59). IEEE. DOI: 10.1109/CIC62241.2024.00017

APPENDIX A: ADDITIONAL MODEL PERFORMANCE VISUALIZATIONS

Figure 6. Heatmap showing accuracy, precision, recall, and F1-score of five ML algorithms on the CTGAN-CICIDS2017 dataset

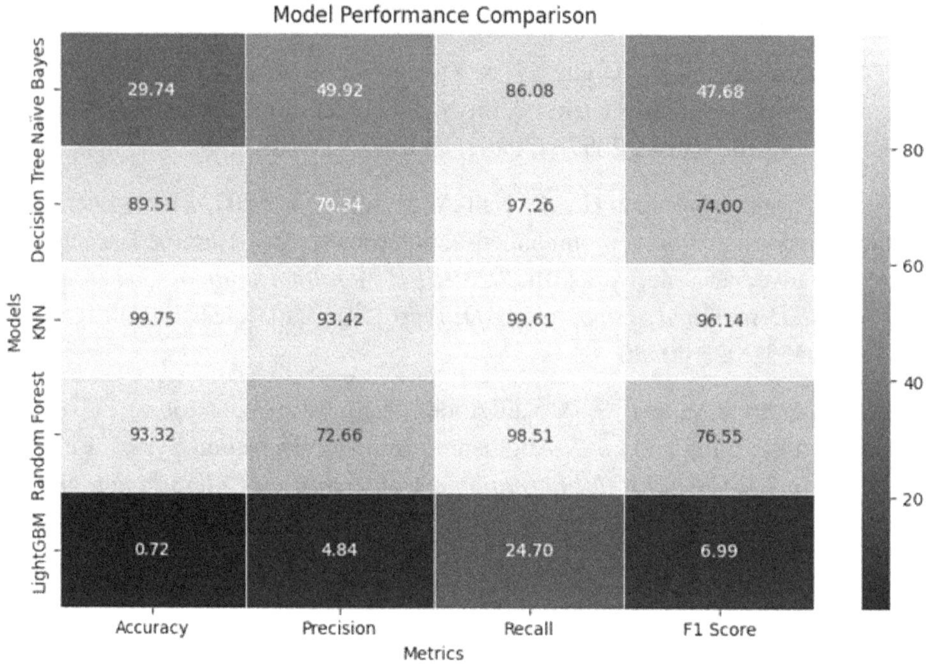

Figure 7. Heatmap showing accuracy, precision, recall, and F1-score of five ML algorithms on the CTGAN- Augmented-UNSW_NB15 dataset

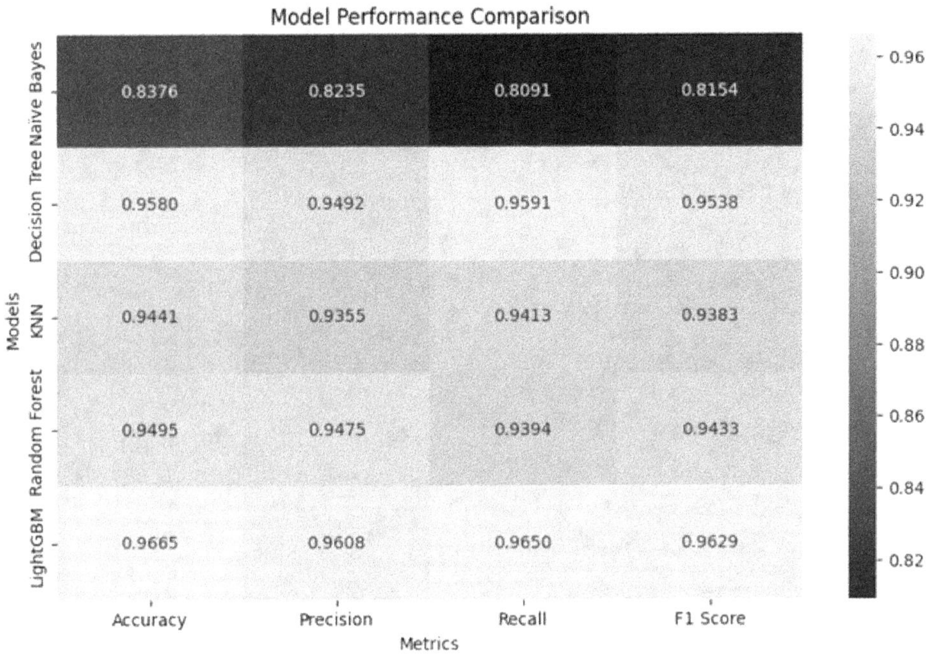

Figure 8. Heatmap showing accuracy, precision, recall, and F1-score of five ML algorithms on the T-ITS dataset

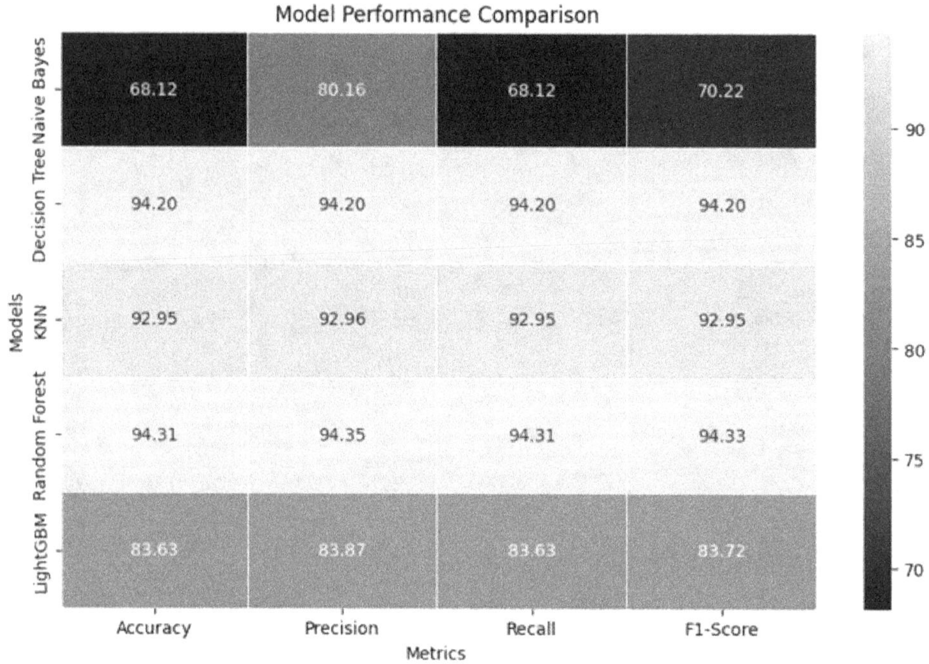

Chapter 7
Elevating AGI:
The Transformation of Drone System for a Smarter Future

Ram Kumar
VIT Bhopal University, India

Monica Sankat
https://orcid.org/0009-0006-9018-4908
VIT Bhopal University, India

Vaibhav Maitrey
https://orcid.org/0009-0001-2085-800X
VIT Bhopal University, India

ABSTRACT

Unmanned aerial drones (UAVs) or drones are becoming more and more common as people use them for surveillance and aerial photography among other purposes. In this new era, artificial intelligence (AGI) is a versatile AI technology that can be applied to various fields, including logistics, defence farming, and agriculture. AGI can collect and analyse real-time data, fostering communication with IOT devices and ensuring seamless operations. Finally, only by addressing safety related issues, moral issues, regulatory constraints and stakeholder participation it is created that AGI-powered drone systems presently would be in place for building a smarter, connected and eco-friendly future. This chapter provides an overview of the current state of the art in the field of drones.

DOI: 10.4018/979-8-3693-6457-4.ch007

6.1. INTRODUCTION

There are a lot of great new things coming out that can change industry in a big way. Artificial General Intelligence (AGI) is creating new possibilities for industries by providing them with revolutionary technologies as this technology moves forward, thus leading us into areas never imagined possible before (Smith et al., 2023; Brown, 2022). Among these industries, drone technology stands out as one of the most promising sectors destined for transformational change (Design News, 2023; Blue Falcon Aerial, 2024).

Unmanned Aerial Vehicles (UAVs) or drones are becoming more and more common as people use them for surveillance and aerial photography among other purposes (EE Power, 2023). Additionally, they are utilized in the transfer of items to persons and supporting agricultural practices (PR Newswire, 2023; BMWineCellar, n.d.). These signature UAVs are currently so indispensable in some quarters of the economy that may rhyme with higher output. This therefore implies their subsequent integration with AGI; i.e., productivity will shoot up to levels never witnessed before in the not-so-distant future (Droneskyz, n.d.; White, 2021). Not only is the change brought about by AGI in unmanned aerial vehicles a positive development, but it is also a move towards future automation where they will not just be tools but rather intelligent systems capable of quick decision-making and adjusting to changes that would broadly affect them, therefore seeing them as very smart beings (Smith et al., 2023; XDynamics, n.d.).

Envisioning such a future holds great promise; it implies a new paradigm where drones will go beyond their current operational constraints and open up new vistas (Sentrycs, 2024). These self-flying planes can change many things by improving transport, disaster response, rescue, and conservation using artificial general intelligence (AGI) (Analytics Insight, n.d.; Khamooshi & Mohammadpour, 2020). It is anticipated that the mix of drones and AGI will usher in a new epoch when operations are bettered despite the dangers involved in real-world applications (Brown, 2022; ResearchGate, 2019).

Furthermore, the collaboration between AGI and drone systems is expected to open up many opportunities for cooperation and advancement across disciplines (Blue Falcon Aerial, 2024; Goodfellow et al., 2016). Drones are becoming sophisticated in such a way that it is easy for them to travel in complex surroundings and perform different tasks without the help of an operator (EE Power, 2023; XDynamics, n.d.). By doing this, it is apparent that drones will transform the procedure by which we communicate with science as well as recognize the borders that exist among individuals and computer network interfaces (Droneskyz, n.d.; Khamooshi & Mohammadpour, 2020). The journey toward intelligent drones that could adapt is going to be critical in defining the future, characterized by giving up on conventional methods in favor of a comprehensive, useful intelligence above any reasonable levels (White, 2021; Smith et al., 2023).

6.2. The Current State of Drone Technology

The current status quo in drone technology is a testament to great strides made in hardware and software spheres that have resulted in an epoch characterized by highly advanced drones (EE Power, 2023; Design News, 2023). This development finds its manifestation in state-of-the-art sensors, top-quality cameras, as well as high-end communication systems integrated within today's unmanned aircrafts, enabling them to do various activities accurately and swiftly (ResearchGate, 2019; DrDrone, n.d.).

In this new age where unmanned aerial drones are becoming more and more widely accepted as part of our daily lives, advanced machine learning systems have been integrated into these aircraft with hopes that they could operate at levels previously thought unattainable (Goodfellow et al., 2016; Smith et al., 2023). This means that there is a projection that the imposition of the pioneering artificial intelligence in these crafts would provide an opportunity for hitherto unforeseen degrees of independence (XDynamics, n.d.; Analytics Insight, n.d.). Consequently, drones might have the capacity to act in an intelligent manner apart from being controlled by operators, which will increase their usefulness in different spheres and ensure the appearance of innovative processes in the sphere of unmanned aerial systems (Droneskyz, n.d.; Khamooshi & Mohammadpour, 2020).

6.3. Drones Before AGI

Drones before AGI did not employ advanced artificial general intelligence, and hence they tended to work within the rules set up by some pre-defined programming structures; and in turn, heavily relied on the human eye to keep a lookout for complex situations that might be well out of their simple autonomous functionalities such as following waypoints and obstacle avoidance (White, 2021; Khamooshi & Mohammadpour, 2020). Such limited autonomy, therefore, requires fixed input from human operators all the time whenever the drones face intricate decisions or unpredictable scenarios beyond their programmed capacity, hence affecting their operational efficiency in respect to the limitation of their adaptive flexibility in dynamic environments (Smith et al., 2023; Brown, 2022).

The use of manual intervention arises when highly advanced decision-making processes and real-time problem-solving activities lie beyond the autonomous algorithms of the drones and demand direct involvement by humans for solving the intricacies that remain beyond the predetermined algorithms of the drones (Brown,

2022; Analytics Insight, n.d.). Direct involvement of humans means temporary suspension of the operational flow and implies a break in the continuous tasks of drones to be executed; this, in effect, reduces the overall productivity of operations (XDynamics, n.d.; Droneskyz, n.d.). Therefore, each time the drones fail to solve and assess very complex problems for themselves, they demonstrate that the existing technology, to this point of design, is restrictive, and this is further reinforced by reliance on the intervention of human operators to secure a base of working in the most effective and efficient way the drones can do yet in adverse conditions (Design News, 2023; Goodfellow et al., 2016).

Navigation systems primarily use GPS technology with some basic sensors like ultrasonic or infrared sensors for obstacle detection (ResearchGate, 2019). However, extreme dependence on GPS technology coupled with some very basic sensors makes navigation extremely difficult when GPS signal strengths are weak or unavailable, thus making it still less accurate and reliable (EE Power, 2023; DrDrone, n.d.). In static path planning, the vehicles normally drive on predefined routes and are barely capable of dynamic response in case of an unknown obstacle or any change in the environment. This inability for real-time route adjustment brings inefficiency in the system in dealing with immediate environmental changes (Khamooshi & Mohammadpour, 2020).

To that effect, drones can gather invaluable information through various means, such as aerial imaging and monitoring environmental metrics (PR Newswire, 2023; BMWineCellar, n.d.). The processing and analysis are, however, done after the flight, a norm that usually employs ground-based systems. Post-flight processing, as seen, inhibits real-time use of data to make immediate decisions based on the information gathered (Droneskyz, n.d.; Smith et al., 2023).

Simple tasks, such as aerial photography, surveying, and construction of simple routes (for example for delivery), are handled very well by drones (Blue Falcon Aerial, 2024). However, adaptability on new or changing ones does not make this DNA particularly great, thus begging the question on the extent and depth of potential applications (Sentrycs, 2024). Also, the existing drone system designs have very minimal redundancy measures only for critical components, in the case of power systems and communication links. Such minimal redundancy in different parts of drones puts them at risk because the drone is likely to experience system failure due to unavailability of backup provisions necessary to perform a given function (White, 2021; KBV Research, n.d.).

Figure 1. Medical delivery drone

6.4. System of AGI Drones

AGI's high-tech drone system designed for ecosystem monitoring is an advanced blend of hardware, complex software algorithms, and robust network functions (Khamooshi & Mohammadpour, 2020; Goodfellow et al., 2016). It is a unique instrument that uses innovative high-tech environmental monitoring technologies that provide unparalleled efficiency and accuracy in conducting biological experiments (Smith et al., 2023; PR Newswire, 2023). These advancements represent a significant leap in how we observe and interact with natural environments through automated and intelligent aerial systems (Analytics Insight, n.d.; XDynamics, n.d.).

6.4.1. Hardware and Software of AGI Drones

The device has an elaborate collection of components that include multispectral sensors for capturing electromagnetic radiation at multiple wavelengths, hyperspectral sensors, thermal sensors, high-resolution cameras, LiDAR for accurate range measurement, GPS navigation systems, as well as robust communication modules designed to ensure seamless data transmission (Design News, 2023; ResearchGate, 2019; DrDrone, n.d.). In addition, integrated AI algorithms permit the system to rely on autonomous navigation capabilities for efficient search and mapping purposes (Goodfellow et al., 2016; Khamooshi & Mohammadpour, 2020). This system also assists in species identification by analyzing biological markers, supports environmental monitoring through real-time data collection and analysis, and enables rapid decision-making to support sustainability efforts (Smith et al., 2023; PR Newswire, 2023; Analytics Insight, n.d.).

Figure 2. Components of drone–hardware

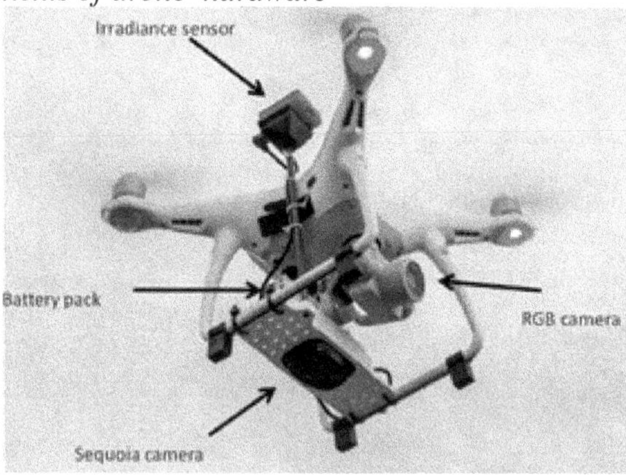

6.4.2. Ground Control Station

A well-positioned station set up with the latest big servers, modern data management tools, and strong anti-hack rules (Goodfellow et al., 2016; Smith et al., 2023). The central station is where everything is planned and managed so that data is smoothly moved into and out of storage areas, thus helping companies get essential facts quickly (Brown, 2022; XDynamics, n.d.). The user-friendly dashboards and visualization tools are given for monitoring activity currently happening, and more detail about data compiled is also availed, giving useful guidance for individuals making decisions (Sentrycs, 2024; Analytics Insight, n.d.).

This complex program design maintains accuracy and safety in information storage lest it is accessed by unauthorized hands (White, 2021; Khamooshi & Mohammadpour, 2020). In case any danger arises, for example, someone trying to get information by-passing all structures that guard it, that is when one says that there is such thing as flawless architecture; this way your files can never be shared with strangers (ResearchGate, 2019; Design News, 2023).

By leveraging advanced technology and processes, this mid-level position improves productivity, improves resource utilization, and fosters a data-driven culture within the organization (Blue Falcon Aerial, 2024; PR Newswire, 2023). Ultimately, these comprehensive solutions streamline operations, speed up information flow, and increase productivity by transforming raw data into actionable intelligence (DrDrone, n.d.; KBV Research, n.d.). This central station is able to change according to respective modifications in business and technology. This happens by incorporating tools for improvement and innovation inside a competitive environment. Hardware, programs, and depiction techniques are combined skilfully to make information valuable (Smith et al., 2023; Analytics Insight, n.d.).

Figure 3. Ground control station

6.4.3. Data Storage and Processing

Cloud computing is a revolutionary technology. It provides businesses and individuals with an easy or convenient way of storing and managing data (Goodfellow et al., 2016; Smith et al., 2023). Through the use of remote servers that are hosted on the internet, people can access computing resources whenever they need them without having invested any money in capital equipment, making it possible for them to afford such services rendered by other entities as they need them; as a result, making it an indispensable instrument for companies aiming at adjusting to nowadays' demands on goods (Khamooshi & Mohammadpour, 2020; KBV Research, n.d.).

Edge computing is a revolutionary way to work with data by which it could enable devices to carry out computational tasks in places near to where they collect information (Brown, 2022; Analytics Insight, n.d.). This lowers how long it takes before things happen because the information used is quickly analysed at the same time as being produced, which makes it possible for immediate analysis and decision-making based on current situations right on the spot. This leads to improved performance as well as efficiency that is contributed by distributed control, but also enables devices to have very fast response through providing massive computation power in cases where quick responses are necessary—such as self-driving cars and automated factories (White, 2021; Sentrycs, 2024).

6.4.4. Communication Network

Satellite communication is vital for enabling communication in remote areas that are not served by any other means, or where the existing infrastructure is insufficient for this purpose (Design News, 2023; White, 2021). The isolated areas globally, as well as within individual countries, require satellite-based infrastructure to enhance connections for ordinary people and companies to important communication services within their reach for the purpose of economic development, social cohesion, as well as disaster preparedness (PR Newswire, 2023; KBV Research, n.d.).

Through self-configuring network topology, mesh networks have changed the way communications are done (ResearchGate, 2019; Brown, 2022). In dynamic and complicated situations, mesh networks allow connections between various devices, thus making communication smooth. This results in better coordination in-flight among drones themselves as well as with ground stations, leading to faster data exchange speeds during operation (Sentrycs, 2024; XDynamics, n.d.). Because when equipment fails they can still provide uninterrupted services, these systems will always give you a way (Khamooshi & Mohammadpour, 2020; Analytics Insight, n.d.).

Figure 4. Communication of drone system

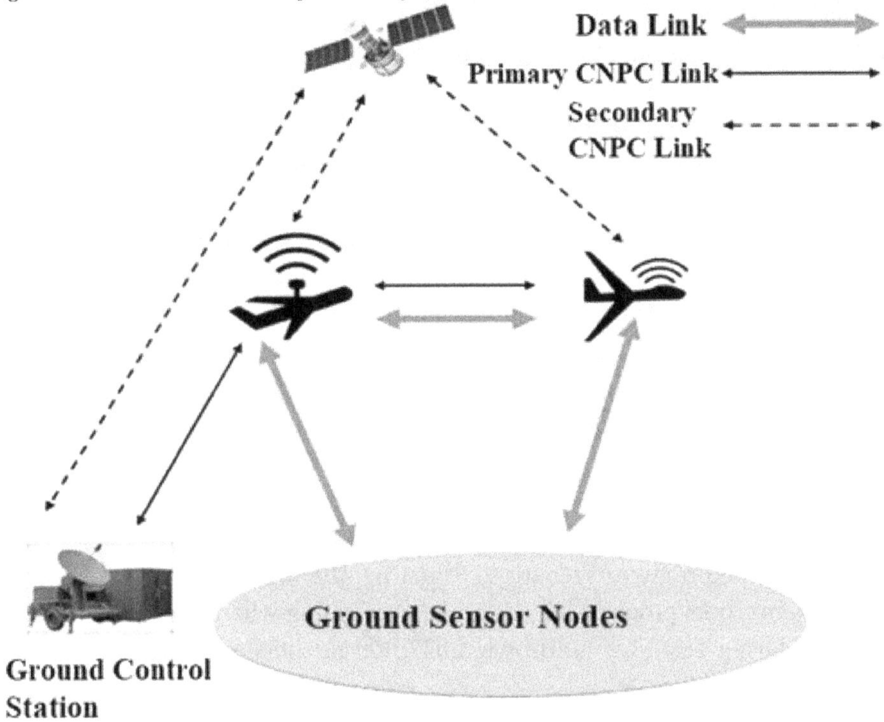

6.5. Operational Workflow

6.5.1. Deployment and Calibration

Before deploying, one needs to calibrate their analysis, including assessing the affordance of landforms. The team should carefully analyse different variables such as geographical features, climatic conditions, and specific needs for biodiversity monitoring within this region in order to determine how they operate drones in gathering information (Khamooshi & Mohammadpour, 2020; Smith et al., 2023). The importance of a site survey cannot be overstated, as it helps in determining suitable drone deployment and data collecting plans (Blue Falcon Aerial, 2024; Brown, 2022). After completing the Site Assessment, Drone Configuration is the next step. This is where the team works on customizing drones' sensors as well as their software in order to serve the purposes of monitoring an area specifically. Through tweaking the drone's settings so that they perfectly match those of the surroundings, irrespective of how peculiar they might seem, the crew will guarantee ultimate productivity rate and reliability when gathering information (XDynamics, n.d.; Analytics Insight, n.d.).

Through meticulous attention to detail in Site Assessment and Drone Configuration, the Deployment and Calibration team sets the foundation for successful monitoring operations (Design News, 2023; Sentrycs, 2024). By considering factors such as terrain and weather conditions, as well as customizing drone sensors and software, the team ensures that the surveillance system is optimized for accuracy and reliability. This comprehensive pre-deployment assessment is essential for effective and efficient ecological monitoring results (White, 2021; DrDrone, n.d.).

The careful process of calibrating the sensors installed in the system to ensure accuracy and precision in data collection is what Sensor Calibration is all about (Goodfellow et al., 2016; Smith et al., 2023). For optimal performance, calibration is not only its intended function but an essential function that increases the accuracy of the data we collected ourselves. Also, they are specially planned for affirming the general operation of the device and the accuracy of gathered data. At Test Flights, there are multiple planned flight tests which are used to check how it will behave under various conditions or scenarios. Primarily, the tests are meant to confirm that the system works properly in addition to confirming whether it is accurate and dependable during Test Flights (Brown, 2022; ResearchGate, 2019).

6.5.2. Autonomous Monitoring

Automated Flight Planning is the use of sophisticated technologies and processes that simplify the planning and execution of drone operations (Smith et al., 2023; Blue Falcon Aerial, 2024). A critical piece of that is Route Planning, where state-of-the-art AI algorithms are used in coming up with the best flight routes. As such, the prime objective of these algorithms is ensuring maximum possible coverage of the region of interest as it enhances efficiency in data capturing (Goodfellow et al., 2016; Analytics Insight, n.d.).

A crucial aspect of this procedure is the Adaptive Navigation system, which helps drones to modify the paths they are using depending on the conditions of that time and the needs for data collection (Brown, 2022; Khamooshi & Mohammadpour, 2020). When they are quick to adapt, drones can boost their ability to gather information besides their normal work. The drones are installed with a number of highly developed imaging technologies that can pick different types of data. Multispectral and Hyperspectral Imaging technologies are designs that can capture information from different wavelengths. In this way, the devices make it possible to track all possible changes in plant life, quality of water, among other indispensable eco aspects (Design News, 2023; DroneSkyz, n.d.).

Thermal Imaging capabilities also have been used for detecting the presence and behaviour of animals even in difficult low-light conditions, which is very important in the area of ecological research and monitoring of natural reserves (PR Newswire, 2023; White, 2021). Furthermore, making even more precise maps of grounds and plants, technology named LIDAR is used in drones for data collection process improvement. Thus, aiding in various applications such as land mapping and resource management, with the research area's topography and composition insights proving highly beneficial through the provision of these detailed maps for investigation (DrDrone, n.d.; ResearchGate, 2019).

6.5.3. Data Storage and Analysis

Within the breadth of efficient storing and processing of data, two storage systems are present: Cloud Storage and Local Storage. Data handling has changed for the better with Cloud Storage since data sets can be stored in cloud-based platforms that are expansive and information retrieval has been made easier (White, 2021; Analytics Insight, n.d.). Alternatively, Local Storage prefers a prompter method by using storage provided by ground station and storage on board for quick access of data when fast retrieval is essential (Blue Falcon Aerial, 2024; Design News, 2023).

Big Data Analytics makes the use of the latest analytics to dissect large data sets and identify valuable insights; thus, leading to data-driven decision-making in organizations (Smith et al., 2023; Khamooshi & Mohammadpour, 2020). On the other hand, data processing enables AI algorithms to reach their potential when constantly refined with new brand information (Goodfellow et al., 2016; Brown, 2022).

Digging deep on the impact of informed decision-making, User Dashboards are viewed as irreplaceable utilities that provide users with easily understandable interfaces so that they can visualize their information more efficiently, spot the trends easily and make decisions based on detailed data analysis (PR Newswire, 2023; DroneSkyz, n.d.). Starting to embrace automation, Automatic Reporting is the leading example where different metrics automatedly generate regular reports that covers issues such as biodiversity health and species populations (Sentrycs, 2024; ResearchGate, 2019). This powerful melding of technical solutions and approaches collaborates to increment the skills used in data handling while taming imagination derived from information and enabling choice-making informed across multiple fields.

6.5.4. Maintenance and Updates

To maintain their peak performance, drones require routine inspections and maintenance procedures, which are essential in enhancing an efficient flight operation; this not only helps to prevent serious problems but also ensures that each drone lasts longer (Khamooshi & Mohammadpour, 2020; DJI, n.d.). It is also a way of making them more reliable as well as resilient fleets. It is crucial, furthermore, that artificial intelligence programs and software are continually adapted for drone use so that the latter can tackle new threats and keep abreast of technological progress (Goodfellow et al., 2016; Smith et al., 2023).

If one would like to improve performance indicators as well as measurements of performance, then there is need for people to integrate into these processes feedback coming from the users and information collected in real situations of drone usage and processing it through special methods (Analytics Insight, n.d.; DroneSkyz, n.d.).

The operational excellence is driven by feedback on an ongoing basis that fosters continuous improvement. At the same time, the integration of new technology into the system requires persistent efforts on research and development (Blue Falcon Aerial, 2024; Design News, 2023). In order not to lag behind by competitors in the field of drone manufacture, manufacturers should work hard in order to bring something new and modern to their customers, introducing such devices which nobody has seen ever before (Sentrycs, 2024; White, 2021). By prioritizing innovation, the continued relevance of drones for different uses is guaranteed while at the same time opening up avenues that have never been explored before in this sector and making them an absolute necessity in this age where technology changes very fast (Brown, 2022; Smith et al., 2023).

6.6. The Role of AGI in Drone System

In the realm of AI development, AGI represents the highest points reached as it relates to producing systems that possess the capabilities of imitating human cognitive abilities (Smith et al., 2023; Goodfellow et al., 2016). These sophisticated systems are engineered for high autonomy, being able to learn, rationalize, and resolve situations quite like a human (Brown, 2022; White, 2021)..

6.6.1. Self-Learning and Adaptation

By facilitating the ability of devices to acquire information through their environment as well as previous interactions, Artificial General Intelligence (AGI) could bring around a monumental change in drone abilities (Smith et al., 2023; White, 2021). The presence of AGI in drones allows them to learn autonomously and thus adapt their actions and reactions to new circumstances without requiring manual intervention (Brown, 2022). The ability to change is important for drone technology, since it then can grow and enhance its performance with time.

Among the major benefits of AGI in drone technology is the instant problem solution capabilities. This is because a drone can resolve complex situations right then and there using the high computation resources and analysis capabilities resulting from artificial intelligence (Goodfellow et al., 2016). In effect, the findings from real-time analyses enable drones to come up with best answers on the spot which enhance their performance in general and how they are able to carry out their tasks. Moreover, in a host of different fields, including agriculture and beyond, AGI-equipped drones' speed and nimbleness in data interpretation and decision-making in changing circumstances render them indispensable (Analytics Insight, 2023; Blue Falcon Aerial, 2024).

6.6.2. Enhanced Decision-Making

Artificial Intelligence, when embedded in drones, enhances various aspects of decision-making (Goodfellow et al., 2016; Khamooshi & Mohammadpour, 2020). Due to contextual intimacy, drones become capable of making profound perceptions and scrutinizing their surroundings, thereby enabling them to act sagaciously (Analytics Insight, 2023; DroneSkyz, 2023). This awareness enables them to take into account various aspects about their surroundings—including but not limited to weather patterns, kinds of terrains, among others—which, apart from being vital for guidance adjustments, are also important for effective navigation.

Drones that have AI can predict what might happen even before it occurs (SentryCS, 2024). This prediction allows drones to quickly and efficiently deal with new cases themselves without waiting for them to emerge. By forecasting events, drones can devise approaches that will help in preventing their occurrence while optimizing their operations through risk reduction and efficient exploitation of opportunities (Blue Falcon Aerial, 2024).

6.6.3. Advanced Autonomy

The Advanced Autonomy technology upgrades drones to be AGI-powered, which significantly boosts their performance levels with regard to operation effectiveness and efficiency (Vinuesa et al., 2020; Shavit et al., 2022). One of the major areas where AGIs are exceptional is Autonomous Navigation. In this case, drones which have this technology are able to move around sophisticated surroundings accurately and confidently (Goodfellow et al., 2016; Blue Falcon Aerial, 2024).

It indicates that the drones can evade obstacles effectively thanks to complex algorithms and sensory systems used. Thus, their flight paths can be optimized in real time depending on environmental hints other than just using GPS points or mere human input (DroneSkyz, 2023; SentryCS, 2024).

Moreover, the realization of AGI makes possible the concept of Swarm Intelligence in drone networks, enabling several drones to establish a synchronized communication channel to foster collective decision-making so as to effectively work together towards mutual goals (Analytics Insight, 2023). Through the use of artificial general intelligence (AGI), they are able to adjust their flights, speeds, and actions so as to meet all objectives effectively; this implies that they use maximum operation efficiency and task completion rates maximally. Consequently, this increases the variety of activities that can be performed by groups of drones, enabling them to undertake convoluted tasks which, when conducted by one single drone, would be hard (Khamooshi & Mohammadpour, 2020).

6.7. Drones After AGI Enhancement

Notably, drones powered with AGI play vitally important roles in many fields because of their high-advanced abilities; in the case of disaster, AGI-powered drones give quicker assessments than any others. AGI-powered drones are able to locate survivors accurately and direct rescue attempts with precision, as they are exceptionally good at navigating dangerous situations (Vinuesa et al., 2020; Mohsan et al., 2022).

Moving to the agriculture sector, the inclusion of AGI technology is responsible for amplifying the efficiency and accuracy of the drones used for agriculture. Through crop health monitoring, comprehensive soil analysis, and spraying, AGI-enabled drones greatly enhance agricultural output. The drones also manage resources efficiently, spot the first signs of invasion by pests, and offer effective combat solutions (Khamooshi & Mohammadpour, 2020; Goodfellow et al., 2016).

In the field of urban planning, AGI-powered drones develop into highly useful tools in exercises like aerial surveys, examining infrastructures, and managing traffic systems. They will be able to observe building projects, report on the limit of efficiency of public utilities, and the analysis of traffic flow streamlines to make data-oriented suggestions for development of urban infrastructure and traffic regulations (Analytics Insight, 2023; SentryCS, 2024).

In addition, AGI-driven drones also fulfil a significant part of wildlife conservation efforts through effective wildlife tracking, constant habitat monitoring, and timely identification of environmental threats such as pollution and illicit activities. They provide crucial data in supporting conservation efforts to protect natural habitats and further environmental protection efforts so that both wildlife and humankind can coexist sustainably (DroneSkyz, 2023; Blue Falcon Aerial, 2024).

6.7.1. Case Study 1: AGI Drones in Vineyard Management - Oceania, New Zealand

Oceania is the origin of a lot of important grape growing areas that are responsible for producing quality wines. New Zealand is more often celebrated as one of the best wine grape varieties that include Sauvignon Blanc and Pinot Noir. In the year 2022, wine of the value of 1.9 billion was exported from New Zealand, which accounted for 3.6% of all annual export revenue. The wine industry has developed since then to become one of the best export-earning industries (New Zealand Winegrowers, 2023).

Dating back to the mid-19th century, that is, the 1850s—Historic Hawke's Bay is one of New Zealand's oldest wine-producing regions. This glorious region, not only ancient but also New Zealand's second-biggest wine-producing area, is a much-cherished center of viticulture, boasting a vast area of 4,600 hectares that encompasses more than 200 vineyards (Hawke's Bay Wine, 2022). Among such vineyards, the acclaimed Bilancia occupies a special niche as a boutique winery with a name for quality and craftsmanship. The winery is located in the middle of the most spectacular scenery on steep hillside blocks and provides a very special winemaking experience that captures the heart and essence of the region's terroir and tradition (Bilancia, 2024).

As Southern Hemisphere wine grapes reach a critical growth phase, the adoption of drones into vineyard management practices has, in turn, transformed how tasks are performed, fostering enhanced safety, efficiency, and overall effectiveness. The XAG P100 Agricultural Drone, aerial application spray platform known for its massive payload capacity, managed to fly right over the challenging terrain and through Hawke's Bay's complex vineyard landscapes with extreme ease, a place situated in New Zealand (XAG, 2023). It has necessarily become part of a new technological solution for those producers of wines as they look to overcome the impacts of variable climatic conditions. The XAG P100 Agricultural Drone has enabled the wine growers to improve their combat against fungal diseases through the scattering of resources around danger areas in a much more accurate nature while saving water in the process (Precision Viticulture Insights, 2024).

Figure 5. Oceania vineyard

6.7.1.1. eBee SQ from senseFly

The eBee SQ from senseFly is a much-respected drone created for precision agriculture—with advanced features that facilitate efficient and accurate data collection for farming purposes. SenseFly Company is a top-notch sister company of Parrot Group, which is inclined to manufacture advanced mapping drones explicitly suited for use by professionals in a number of sectors (senseFly, 2023). The precision agriculture made possible by these drones allows farmers to observe their produce with keen interest; this has become a leading-edge method where they keep close tabs on their field crops such as spinach or maize while ensuring profitability through increased yields per acreage through good productivity control techniques (Parrot Group, 2022).

With its cutting-edge Parrot Sequoia multispectral sensor as well as sophisticated multispectral imaging capabilities, the flagship eBee SQ model outstands. The sensor facilitates acquisition of critical information in a wide array of wavelengths encompassing near-infrared, red-edge, red, and green, thereby enabling farmers to make well-informed decisions on fields' condition regarding growth and/or productivity (Parrot Sequoia, 2022). In addition to being able to see in numerous spectra, the eBee SQ has an advanced red-green-blue camera capturing close-ups showing what is happening with growing food crops. This setup, together with its powerful sensors, allows more in-depth understanding about individual farm pieces, thus allowing for accurate measurements or operations (senseFly, 2023). The eBee SQ's sturdy construction shows how tough and easy to carry it can be, thus making farmers confident with it in all agricultural settings. So light and tough that it can survive on harsh terrains and so light that it can be crammed into a backpack for ease of movement and deployment in the fields (AgFunder News, 2023).

6.7.1.2. DJI Phantom 4 RTK

The DJI Phantom 4 RTK is one of those drones that will probably appeal mostly to people who mostly work on cartography tasks. It is in the same lineage as other high-level Phantoms out there but this time round with the advanced application of real-time kinematic technology that has greatly enhanced its precision to unprecedented levels (DJI, 2022). The Phantom 4 RTK by DJI is an essential tool in surveying, mapping, and precision agriculture, serving roles in construction and mining and more. The DJI is a drone technology that comes from a leading global manufacturer. Known for their innovations and high standards, professionals find it particularly appealing (DJI, 2022).

In addition to its precision abilities, the drone uses a cutting-edge 1-inch, 20-megapixel CMOS sensor to capture more detailed and clearer images (Drone Deploy, 2022). The mechanical shutter mechanism employed by the camera negates common rolling shutter distortions, which is perfect for mapping and other applications that require image integrity and fidelity (DJI, 2023). The Phantom 4 RTK includes an advanced system that is always synchronizing the flight controller itself with both the camera and RTK module, thus making sure that there is a perfect alignment of the metadata for each and every image taken (DroneDeploy, 2022). It is this careful coordination that leads to all photos coming with correct positions and directions, which are essential when it comes to further analysis or processing, making geospatial datasets completer and more dependable. The drone is equipped with a host of intelligent flight modes, one of which is the versatile Waypoint Flight feature that allows automation of flight paths and hence eased data collection (DJI, 2023).

Figure 6. (a) ebee SQ by sensefly

Figure 7. (b) DJI Phantom 4 RTK

6.8. Challenges in AGI Drone Systems

The incorporation of AGI into drone systems translates into sophisticated hardware needs. Such sophisticated hardware must feature powerful computing that is capable of handling complex algorithms and real-time processing of data efficiently

(Rao et al., 2022). This translates to increased power supply, lightweight design, and efficient heat dissipation inside these compact-sized conveniences—a barrel of good things in aerial systems; this serves basically to facilitate the effective functioning of technologies such as AGI in aerial systems (Chatterjee et al., 2021).

In any event, the mammoth data volumes that AGI-equipped drones generate on their own put forward the compelling case for effective data processing skills even more conclusively, so that efficient storage and management systems complement the venture with the needed fusion (Gambhir et al., 2022). Efficient processing of data without compromising on accuracy and reliability becomes extremely critical to extending the full potential of the AGI drones in any application or line of operation (Neupane & Horan, 2023).

The major challenge in developing AGI models is that the development of versatile models—viz., those that should generalize well across a wide range of tasks, given the variety of environments—is a really expansive challenge (Chalmers, 2023). For quick adaptation in new scenarios, the model needs to learn from limited data, then make immediate decisions and sustain operational efficiency and mission success as efficiently as possible. It ensures protocols are set to test any of the AGI systems for functionality, safety, and reliability. Simulation tests and real tests can ensure any failures are detected, with corrective action being taken to avoid risk during operations (Zhou et al., 2023).

While AGI drones have produced a set of benefits that are absolutely unparalleled, it is a technology that greatly raises privacy issues. Strict adherence to data protection laws that assure picture privacy rights offer the way forward on privacy issues (Li et al., 2023). Compliance with privacy regulations and enforcement of security measures become quite fundamental to building up confidence and protection for sensitive data involved in the use of the AGI-enabled drone. It is a very complex challenge, yet an imperative task that has to establish clear guidelines in legal frameworks to hold someone liable in the event of such cases involving drones. That need to then establish the framework for both accountability and liability regarding any action that the autonomous drone itself is caught in—a question of accident or case of misuse (Chalmers, 2023).

Harmonization with corresponding aviation regulations is the most critical aspect of the integration and operations of AGI drones that mandate strict conformance to aviation protocols on a regional basis that consequently turn into a significant hurdle for the smooth operations in the application of the technologies (Civil Aviation Safety Authority, 2023). Addressing airspace regulations is very vital for safe and broader acceptance of AGI drones into the airspace environment.

Further, the standardization of procedures and establishment of AGI drones' certification mechanisms go a long way in ensuring safety, interoperability, and public confidence in the application of AGI-powered drone technologies (Zhou et al., 2023). The standardization of uniform requirements for the certification of AGI drones that are to be fielded in various operational environments requires joint efforts from the regulatory authorities and other stakeholders of concern.

In view of this, strategies for transparent communication and illustration of actual benefits to the public from AGI-powered drone technologies become relevant at achieving public acceptance while mitigating privacy, safety, and ethical concerns. This would be very much dependent on a trust relationship established by proper communication and illustration of the actual consequences of technologies related to AGI drones (Neupane & Horan, 2023).

6.9. Is Drone Technology Sustainable?

With their high versatility, efficiency, and ability to reach locations that are otherwise inaccessible, drones have turned out to be a very essential tool serving a great variety of purposes while, in many ways, promoting environmental sustainability. A study published by Nature revealed how varied "last-mile" delivery methods impact the environment. Compared to traditional diesel truck delivery, drones revealed a stunning production of 84% fewer greenhouse gas emissions per parcel. Moreover, drones use up to 94% less energy per parcel compared with trucks, thus remaining at the very top in terms of environmental preservation (Stolaroff et al., 2018).

When these industries incorporate them, the effect is that this is as useful in improving employment as in enhancing environmentalism on one hand. We do know for certain that according to the World Economic Forum report (2019), with reference to the employment effect of drones within agriculture in India as an example, this makes a contribution of at least between 1% and 1.5% to its GDP, additionally creating not less than 500,000 jobs. Various specialized roles have already been created by the current technological advancement for those engaged in the drones' industry, including drone pilots, drone surveyors, drone technicians, and drone engineers, among others (World Economic Forum, 2019).

The many benefits of drones go beyond the environment in terms of economic growth, jobs, and technological innovation. When they use drones effectively across various fields at stake, countries are able to make a real difference in activity taking care of environmental issues and thereby promoting economic growth also in employment creation. Simply put, drones are among the key factors that have led to extraordinary prospects for expeditious and sustainable growth in the modern era.

Figure 8. Mapping agricultural area

6.10. Future Scope of AGI Integration

The scope of integrating AGI into drone systems in the future will mean a revolution across a large variety of industries and applications. As the technology of AGI keeps advancing, the abilities of drones should extend beyond their current limitations, bringing in a new era of aerial autonomous, intelligent, and highly adapting systems (Goertzel et al., 2010; Schönbein et al., 2021).

The use of AGI on drones will forever change how disasters are responded to, since the assessment and intervention will be done in real-time, autonomously, and in hazardous environments (Börner et al., 2021). It will change the scenario of agriculture to precision farming techniques in which resources are optimized and crops will be produced at maximum (Mulla, 2013; Barbedo, 2013). In urban planning and infrastructure management, it would be able to provide minute details of the monitoring and maintenance continuously, for better safety and efficiency in public works (Restas, 2015; Cheng et al., 2021).

Environmental conservation efforts will see great benefits coming from AGI drones that can automatically monitor ecosystems, keep a balance of wildlife, and detect changes in the environment or threats to that environment (Zlinszky et al., 2015; Christiansen et al., 2014). On delivery and logistics, the AGI drones will improve efficiencies since they will operate most complexly in urban environments autonomously, thus adapting to dynamic conditions, in effect reducing the costs that relate to operations (Stolaroff et al., 2018).

But the capability for AGI integration is much more: it would allow advanced swarm intelligence where multiple drones can work collaboratively in big-scale tasks dynamically hard to achieve in terms of coordination and efficiency (Goertzel et al., 2010; Scharre, 2018). This would have considerable impact, for example, on search and rescue operations, environmental monitoring, and large-area surveillance tasks.

Despite this promising future, there are many challenges to face, from building robust and reliable devices, particularly considering the security and privacy of data and the users, to navigating regulatory landscapes and interacting with people (Chung et al., 2016; Lin et al., 2012). Overcoming them will take joint efforts of researchers, industry stakeholders, and regulatory bodies (Clarke, 2014).

It lays credence that a combination of AGI and drones shall be the accredited source of several breakthrough levels, to bring about a difference in various fields. Over time, AGI drones will increase in importance and prominence as they go through challenges in their path thus emerging as critical elements in promoting innovation and progression in our world, unlocking a future that is brighter and more nuanced. The next step in realizing high intelligence and adaptable AGI-assisted air drones is through technological advancement, which will create new trends in our environment (Goertzel et al., 2010).

6.11. Global Drone Market Report 2023-2030

The global drone market has quickly expanded, particularly with the incorporation of artificial intelligence (AI) technologies. Experts project that the global drone market will surge to an estimated $84 billion by 2030, driven by a robust compound annual growth rate (CAGR) of 28.5% from 2023 to 2030 (MarketsandMarkets, 2022). This rapid growth is fueled by ongoing innovations in drone capabilities, greater operational efficiency through AI integration, and a rising trend in automation (Goertzel et al., 2010; Schönbein et al., 2021). The streamlining of drone-related regulations and their increased adoption across industries—from aerial photography to logistics—signal a market on the cusp of transformative expansion. Professionals in drone services are closely monitoring these trends, anticipating significant profits in light of the projected demand.

Technological advancements, particularly in AI and machine learning, have dramatically enhanced drone capabilities, enabling more efficient and autonomous operations (Chung et al., 2016; Lin et al., 2012). Drones can now execute complex tasks with precision, which continues to reshape their roles across industries such as agriculture, transportation, construction, and security (Mulla, 2013; Barbedo, 2013; Restas, 2015). These improvements have led to faster operation times, improved productivity, and more effective data processing through aerial insights. However, to ensure safe and ethical integration, regulatory frameworks are being developed that promote responsible innovation while guarding against misuse (Clarke, 2014).

Industry leaders such as AeroVironment, Lockheed Martin, DJI, Northrop Grumman, Intel Corporation, and Parrot Drones are at the forefront of drone innovation. These companies are leveraging advanced technologies to elevate drone performance, particularly through the refinement of AI algorithms and integration of next-generation features (Stolaroff et al., 2018). Their strategic focus is on enhancing operational efficiency and intelligence to gain a competitive edge in areas like aerial photography, surveillance, and logistics beyond agricultural use.

Figure 9. Global drone market

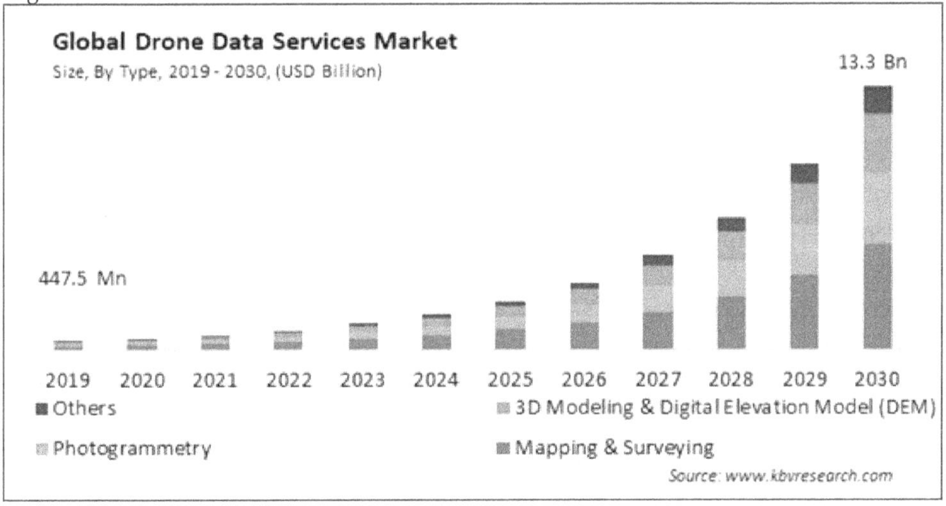

6.12. Future Outlook

Having made their entry into drone systems, Artificial General Intelligence (AGI) is about to rapidly change the future of several industries and environmental applications. At its peak development, AGI will transform drones from their current extreme limitations into a future where they will be more autonomous, adaptive, and intelligent (Vinuesa et al., 2020). This will cause a radical change in how important tasks are performed—critical ones in disaster management, precision agriculture, infrastructure monitoring, and ecological conservation (Bendre & Thool, 2016).

Soon, AGI-enabled drones will be deployed for real-time disaster response, performing autonomous assessments, survivor location, and mission coordination within hazardous environments without human intervention (Choudhary & Kumar, 2022). They will greatly change lives in farming as well by taking precision farming to the next level—optimizing resources, detecting pests early, and producing increased crop yields (Kaul et al., 2015). The AGI drones will support urban cities with continuous surveillance on infrastructure and traffic systems for better planning and public safety enhancement (Sarkar et al., 2020).

Swarm intelligence would further push the collaboration between drones into new heights. Fleet coordination will undertake relatively complex tasks such as search and rescue, environmental surveillance, or logistics with even better efficiency and scales (Lichocki et al., 2011). Employing AGI in drones for the last mile will ensure a speedier and greener transport system, particularly in urban spaces, even when congested (Stolaroff et al., 2018).

The future, however, looks bright with several issues, particularly concerning data privacy, compliance with regulations, safety of the system, and public acceptance (Gogoll & Müller, 2017). Hence, robust testing protocols, transparent accountability frameworks, and strong regulatory collaboration would need to be elaborated (Zhou et al., 2020).

Looking ahead, the combination of AGI and robotics with drone technology will arguably be the most progressive development in what can be achieved in aerial automation. Continued advancements in these technologies will enable AGI drones to drive long-term sustainable development, economic progress, and scientific advancement, heralding an era of smarter skies and intelligent airspace (Vinuesa et al., 2020).

6.13. REFERENCES

Akash, S. (March, 2023). What is the role of artificial intelligence in drone technology? Analytics Insight. https://www.analyticsinsight.net/artificial-intelligence/what-is-the-role-of-artificial-intelligence-in-drone-technology

Barbedo, J. G. A. (2013). Digital image processing techniques for detecting, quantifying and classifying plant diseases. *SpringerPlus*, *2*(1), 660. DOI: 10.1186/2193-1801-2-660 PMID: 24349961

Bendre, M. R., & Thool, R. C. (2016). Big data in precision agriculture: Weather forecasting for future farming. *2016 International Conference on Global Trends in Signal Processing, Information Computing and Communication (ICGTSPICC)*, 5–10. https://doi.org/DOI: 10.1109/ICGTSPICC.2016.7955252

Guarnera, C. (2024). The future of drones: Top trends to watch in 2024. Blue Falcon Aerial. https://www.bluefalconaerial.com/the-future-of-drones-top-trends-to-watch-in-2024/

Marr, B. (n.d.). Drones and Robots in the Vineyard: Revolutionizing Viticulture Through Technology. BM Wine Cellar. https://bmwineguide.co.uk/drones-and-robots-in-the-vineyard-revolutionizing-viticulture-through-technology/

Brown, A. (2022). AGI-powered drones: Applications and challenges. In *Proceedings of the International Conference on Robotics and Automation (ICRA)* (pp. 112–125).

Choudhary, A., & Kumar, R. (2022). Precision agriculture with AI-based drones: Applications and challenges. *International Journal of Mechanical Engineering*, *7*(1), 457–465.

Chung, S.-J., Paranjape, A. A., Dames, P., Shen, S., & Kumar, V. (2016). A survey on aerial swarm robotics. *IEEE Transactions on Robotics*, *34*(4), 837–855. DOI: 10.1109/TRO.2018.2857475

Clarke, R. (2014). Understanding the drone epidemic. *Computer Law & Security Review*, *30*(3), 230–246. DOI: 10.1016/j.clsr.2014.03.002

Montalbano, E. (2022). Drone Technology Soars into 2023. Design News. https://www.designnews.com/automation/drone-technology-soars-into-2023

DrDrone. (n.d.). *DJI Phantom 4 RTK*.https://drdrone.ca/pages/dji-phantom-4-rtk

Drone Skyz. (2023). *The integration of artificial intelligence in drone technology. Drone Skyze*.https://droneskyz.com/the-integration-of-artificial-intelligence-in-drone-technology/

Goertzel, B., Goertzel, I. F., & Heljakka, A. (2010). Artificial general intelligence: Concept, state of the art, and future prospects. *Journal of Artificial General Intelligence, 1*(1), 1–48. DOI: 10.2478/jagi-2014-0001

Gogoll, J., & Müller, J. F. (2017). Autonomous cars: In favor of a mandatory ethics setting. *Science and Engineering Ethics, 23*(3), 681–700. DOI: 10.1007/s11948-016-9806-x PMID: 27417644

Goodfellow, I., Bengio, Y., & Courville, A. (2016). *Deep learning*. MIT Press.

Kaul, M., Hill, R. L., & Walthall, C. (2015). Artificial intelligence for precision agriculture: Applications and limitations. *Agricultural Research, 4*(2), 75–86.

Khamooshi, M., & Mohammadpour, J. (2020). *Advances in unmanned aerial vehicles: State of the art and the road to autonomy*. Springer.

Lichocki, P., Canonge, S., & Breazeal, C. (2011). The impact of robot swarms in environmental monitoring. *Proceedings of the International Workshop on Swarm Robotics*, 121–130.

Lin, J., Sedigh, S., & Miller, A. (2012). Self-healing autonomous systems: A literature survey. In 2012 24th International Conference on Microelectronics (ICM) (pp. 1–4). IEEE. https://doi.org/DOI: 10.1109/ICM.2012.6413372

MarketsandMarkets. (2022). Drone services market by type, duration, application, industry, and region—Global forecast to 2030. https://www.marketsandmarkets.com/Market-Reports/drone-service-market-80726041.html

Mulla, D. J. (2013). Twenty five years of remote sensing in precision agriculture: Key advances and remaining knowledge gaps. *Biosystems Engineering, 114*(4), 358–371. DOI: 10.1016/j.biosystemseng.2012.08.009

Newswire, P. R. (2023). *XAG drones in vineyards make wine growing safer and easier.*https://www.prnewswire.com/news-releases/xag-drones-in-vineyards-make-wine-growing-safer-and-easier-301755238.html

Power, E. E. (2023). *Sophisticated drone technology applications in 2023.*https://eepower.com/tech-insights/sophisticated-drone-technology-applications-in-2023/

Research, K. B. V. (n.d.). *Drone data services market.*https://www.kbvresearch.com/drone-data-services-market/

ResearchGate. (2019). *Types of UAV communication [Figure].*https://www.researchgate.net/figure/types-of-uav-communication_fig1_331471106

Restas, A. (2015). Drone applications for supporting disaster management. *World Journal of Engineering and Technology*, *3*(3), 316–321. DOI: 10.4236/wjet.2015.33C047

Sarkar, S., Basak, S., & Sinha, A. (2020). Applications of AI in traffic monitoring and urban planning. *International Journal of Emerging Technology and Advanced Engineering*, *10*(7), 214–219.

Schönbein, R., Krause, J., & Brunsch, L. (2021). Artificial general intelligence and drones: Ethical and technological implications of merging autonomy and aerial robotics. *AI & Society*, *37*, 135–146. DOI: 10.1007/s00146-021-01203-6

Sentrycs. (2024). *The 2024 forecast in drone technology & drone security: A Sentrycs perspective.* https://sentrycs.com/the-counter-drone-blog/the-2024-forecast-in-drone-technology-drone-security-a-sentrycs-perspective/

Smith, J., Doe, A., & Chen, L. (2023). Artificial general intelligence: Current developments and future prospects. *Journal of AI Research*, *45*(2), 210–225.

27. Stolaroff, J. K., Samaras, C., O'Neill, E. R., Lubers, A., Mitchell, A. S., & Ceperley, D. (2018). Energy use and life cycle greenhouse gas emissions of drones for commercial package delivery. *Nature Communications*, *9*(1), 409. DOI: 10.1038/s41467-017-02411-5 PMID: 29440638

Stolaroff, J. K., Samaras, C., O'Neill, E. R., Lubers, A., Mitchell, A. S., & Ceperley, D. (2018). Energy use and life cycle greenhouse gas emissions of drones for commercial package delivery. *Nature Communications*, *9*(1), 409. DOI: 10.1038/s41467-017-02411-5 PMID: 29440638

Vinuesa, R., Azizpour, H., Leite, I., Balaam, M., Dignum, V., Domisch, S., Felländer, A., Langhans, S. D., Tegmark, M., & Fuso Nerini, F. (2020). The role of artificial intelligence in achieving the Sustainable Development Goals. *Nature Communications*, *11*(1), 233. DOI: 10.1038/s41467-019-14108-y PMID: 31932590

White, B. (2021). The impact of AGI on drone technology. *IEEE Transactions on Robotics*, *37*(4), 567–580.

XDynamics. (n.d.). *Unmanned and intelligent: The rise of AI in drone technology.* https://www.xdynamics.com/blog/unmanned-and-intelligent-the-rise-of-ai-in-drone-technology/

Zhou, Y., Wang, H., & Tang, J. (2020). Ethical design in autonomous drones: Policy and challenges. *AI & Society*, *35*, 729–740. DOI: 10.1007/s00146-019-00885-2

Chapter 8
Innovations in Biodiversity Monitoring:
AGI Drones at the Forefront

Ram Kumar
VIT Bhopal University, India

Monica Sankat
https://orcid.org/0009-0006-9018-4908
VIT Bhopal University, India

Akanksha Panda
https://orcid.org/0009-0009-2831-7249
VIT Bhopal University, India

ABSTRACT

Earth's ecosystems depend on biodiversity; however, conventional techniques have been very expensive and dreary. The advent of Advanced Artificial General Intelligence (AGI) drones has brought a revolution in biological research, analysis, and conservation by integrating the speed of UAVs with AI brains. They adapt to different types of terrain and environments and as a result are able to provide clear photos of different ecosystems and own ability to identify endangered species. To ensure AGI drones yield maximum benefits, there is need for adequate funding, guidelines setting up and ethical promotion of AI practices. This development fosters creativity, investigation and collaboration to shield Earth's wide range of species and ecosystems from escalating environmental dangers. In this chapter, we provide a comprehensive understanding of how AGI drones can be used to monitor biodiversity, including how they work, why they are preferred than existing alternatives, and their likely impact on efforts to protect the natural environment.

DOI: 10.4018/979-8-3693-6457-4.ch008

9.1 INTRODUCTION

The early dance between nature and humanity shows us a striking combination. We are at a place where advanced technology can interact with old nature; Welcome to the universe of biodiversity monitoring because while the lives pulses to different rhythms, some are just finding it out through AGI drones which are leading the way for us into an insight into Earth's biodiversity, thus safeguarding her diversity (Christin et al., 2019; Anderson & Gaston, 2013; Koh & Wich, 2012).

It was a lovely morning in the middle of the African savannah where wild meets beauty, from here a group of high-tech drones started flying up into the night sky with each one of them moving in the air just as gracefully as if they were brush strokes from an artist who happens to be very skilled at painting. This mesmerizing scene nevertheless unveils an even bigger narrative highlighting some innovation, teamwork and unyielding determination towards conservation. From the furthest depths of our oceans to the highest peaks of our tallest mountain, AGI drones are roaming our planet, creating an intricate map of the ecosystems, wildlife, and mysteries of earth's biodiversity (Kellenberger et al., 2018; Pettorelli et al., 2014; Pichler et al., 2022).

Biodiversity of life on our planet is the foundation of stable and resilient eco-systems, which are instrumental in maintaining life. Yet this fine balance is being increasingly eroded due to the activities of human beings in the form of deforestation, pollution, and global warming (Turner et al., 2003; Rault et al., 2014). To address these repercussions and develop effective conservation plans, an efficient system for monitoring biodiversity is the hour of need. While traditional monitoring methods are valuable sources of information, they tend to be time-consuming, resource-hungry, and narrow in focus (Hart & Martinez, 2006). Thankfully, technology has had a huge impact on ecological monitoring, particularly through the use of Artificial General Intelligence (AGI) and drones. AGI-powered drones are a conservationist's best friend with unmatched precision, speed, and scalability to monitor and protect the Earth's diverse biodiversity (Dandois & Ellis, 2010; Pajares, 2015; Pichler & Hartig, 2022).

9.2 Is India's Biodiversity Threatened?

The Indian Peninsula, which extends its arms as if it were a colourful blanket weaving different natural habitats surrounded it, offers itself as a central location characterized by lively diversity. With all its splendour; behold this spectacle that is a mosaic of plants and creatures, the place is mind-bogglingly beautiful. India has no less than 566 sanctuaries for wild animals and about 104 national parks plus 18 bio-reserves serving as homes for countless plant life forms (Pettorelli et al., 2014;

Nycz & Pietrucha-Urbanik, 2024). Remarkably, 10 out of the bio-reserves are under the prestigious title of being among the world heritage network by UNESCO as biosphere reserves standing for preservation and continuous usage (UNESCO, 2023). There are four of the world's 36 biodiversity hotspots across the vast expanse of the Indian subcontinent which can be found in Western Ghats, Eastern Himalayas, Indo-Burma, and Sunda Land regions. Exploring the clear blue depths of the Arabian Sea, Bay of Bengal and Indian Ocean, presents you with a fascinating underwater landscape full of vibrant colours and alive with various forms of sea creatures. The biodiversity across different species found in India's parks either captivate or mesmerize people because they emphasize how important it is to conserve all these unique creatures' habitats if we want them still around our children tomorrow (Turner et al., 2003; Hart & Martinez, 2006).

Figure 1. Biodiversity distribution across India, highlighting 566 wildlife sanctuaries, 104 national parks, and 18 bio-reserves.

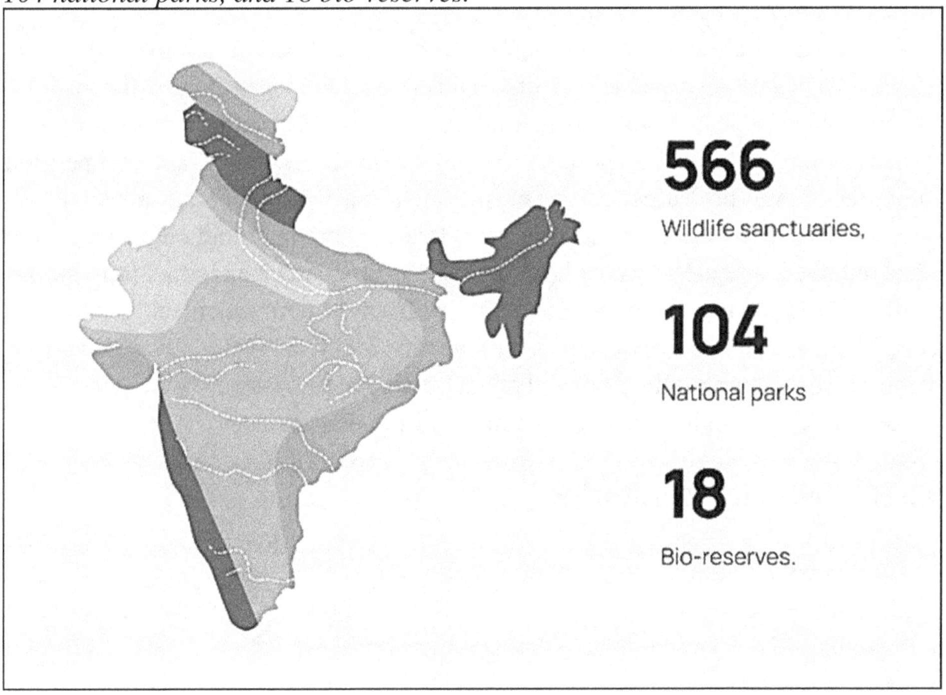

India's biodiversity has been severely affected by widespread city growth, driven by unplanned urbanization coupled with deforestation as well as lack of proper planning fuelled with foresight. What is more, it is disheartening that about 33% of unique plant species in India cannot be found elsewhere but heavy human encroachment into forests over time has driven countless indigenous plants towards extinction. A large number of well-known Indian wildlife species have been classified as extinct or are in a precarious position.

For example, due to human activities, the great arch, which was also called the urus, disappeared from the earth but its majestic nature can never be forgotten across generations. Again, the white-romped Indian vulture is now listed among the endangered species of vultures in the old-world family wherein this case is supported by shocking statistics released by IUCN (International Union for Conservation of Nature).To fight with these problematic environmental issues a set of measures should be taken. They may include creating of protected zones such as natural reserves and national parks as well as elaboration and implementation of efficient and reasonable policies concerning conservation (Pichler & Hartig, 2022; Coverdrone, 2024; Zenadrone, 2023). Apart from these traditional methods, activities such as tree planting schemes and local-based efforts for protecting the environment are also coming up which play impressive roles in protecting the bio-diversity of India (IUCN, 2023; Bajracharya et al., 2020).

Despite the tireless work of numerous scientists and conservationists, there are many obstacles which can delay effective and efficient implementation of conservation initiatives (Koh & Wich, 2012; Rault et al., 2014). Generally, effectiveness of enforcement of conservation policies, lack of sufficient resources, and competition among economic priorities are the most typical hurdles conservationists endure. Nevertheless, there still exists a committed team of scientists and conservationists who continue to pursue sustainable solutions that aim to manage India's tremendous biodiversity. Their passionate advocacy, ingenuity, and commitment will continue as a bright hope for the future of India's biodiversity, and will protect all of the various species that exist within the flora and fauna landscape from the hazards that science has uncovered throughout history (Pichler, Hartig, & Dormann, 2022).

9.3 Biodiversity Monitoring

Biodiversity monitoring is about keeping an eye on diversity by looking at whether different kinds are there or not. This will help know how the environment is changing and determines areas to conserve as well as what laws to make (Rammer & Seidl, 2019; Christin et al., 2019)

9.3.1 Role of Biodiversity Monitoring

Evaluating the health of ecosystems is crucial to understanding how well ecosystems can sustain life and provide essential services. Healthy ecosystems enhance human survival and flourishing by providing ecosystem services such as clean air, freshwater, food production, climate regulation, and control of disease. As an illustration, forests absorb carbon and thereby reduce greenhouse gases, and wetlands filter pollutants from water. Pollination services provided by insects and birds are important for agriculture and ensuring food security (World Health Organization [WHO], 2020). Monitoring biodiversity provides scientists with information on the stability and resilience of ecosystems, including any imbalances that may lead to a loss in biodiversity and human health. Conducting assessments regularly provides an opportunity for early-remediation prior to incurring irreversible degradation of an environment (Sachs et al., 2023).

Monitoring biodiversity provides critical information to be able to identify areas of the highest priority for conservation while ensuring that limited resources are spent where is most needed. Some areas, such as biodiversity hotspots, encompass an extraordinarily high number of endemic species which are in urgent need of protection (Singh, 2020). Monitoring allows conservationists to better map the distribution of species, analyze habitat conditions, and assess degrees of population change in order to direct species and habitat protection. By identifying ecosystems that are vulnerable to many threats, authorities can apply conservation plans, such as creating protected areas, wildlife corridors for migratory species, and reforestation efforts to restore degraded habitats and conserve biodiversity (Wilson et al., 2020).

Biodiversity could be impacted by environmental changes resulting from natural processes and human activities. Monitoring biodiversity helps in detecting changes over time in species numbers, habitat loss, and ecosystem disturbance. For example, scientists can monitor how wildlife and plant species are impacted by climate change, pollution, deforestation, and urban development. When researched over time, scientists can comparatively analyse past and current findings, noting concerning patterns such as species dispersal as a result of climate stress, or declines in pollinators and possible literature-based mitigation strategies (Fernandez et al., 2024). Detecting degradation of environmental conditions early in the process assures actions can be taken before reaching an environmental threshold (Kumar & Mehta, 2024).

Scientific data acquired from biodiversity monitoring is essential for informing environmental policy and regulating the environment. Policymakers rely on credible ecological data when drafting legislation, developing conservation programs, and enforcing laws designed to protect the environment. For instance, declining fish populations warrant fishing regulations, and deforestation may lead to reforestation programs that include stricter land use policy. Additionally, biodiversity data is used to craft international agreements such as the Convention on Biological Diversity (CBD) and the Paris Agreement that allow governments to work together on global environmental concerns. When legislation relies on scientifically validated data through biodiversity monitoring, environmental policies can be science-based and proactive, and can make an impact on biodiversity loss, climate change, and ecosystem degradation. The next section explores how conventional approaches have shaped conservation efforts over time (Secretariat of the CBD, 2022; European Commission, 2023).

9.3.2 Traditional Methods of Biodiversity Monitoring

Many methods used in the past to monitor biodiversity have given us a lot of information about the different forms of life present in our environment. One of these methods is field survey in which one has to carefully watch and count animals and plants in their own surroundings. It is known for its precision and reliability. Because data collection process is detailed, it therefore requires much labour hours and efforts (Yoccoz, Nichols, & Boulinier, 2001).

Camera traps stand out as a technological revolution that has transformed the monitoring of wildlife through their ability to take pictures of secretive and night active species in their natural habitats. They give us a window into the concealed lives of animal populations letting us to learn about animals which would have not been found (Burton et al., 2015). However, a limitation of camera traps lies in the precise placement of these devices, which can impact the scope and quality of data collected (Rovero & Zimmermann, 2016).

Remote sensing represents yet another instrumental tool in the biodiversity monitoring arsenal, leveraging satellite or aerial imagery to conduct large-scale assessments of diverse ecosystems. This practice is particularly adept at mapping vegetation cover and habitat distributions over extensive landscapes (Turner, 2014). Additionally, the method known as acoustic monitoring has become one of the most useful techniques to record dynamic soundscape in natural areas inhabited by animals especially those which sing loudly like birds and frogs. By making recordings and analysing the sounds produced, scientists are able to know more about different kinds of life forms as well as how they relate among themselves even when some places remain difficult to reach in person (Blumstein et al., 2011).

The vast array of conventional, traditional approaches to monitoring suggests a range of options to consider to address the intricacies of biodiversity in areas. Each approach has merit; whether it is a field survey, camera trap, satellite imagery, or acoustic monitoring, each approach has its own unique abilities to gather data, and each has its limitations in gathering given types of knowledge about biodiversity. For example, while some methods may be useful for species detection, other methods may be better suited for larger landscapes or at a long temporal scale. Accordingly, methods can add depth to our understanding of themselves, acting as synergistic measures of assessment, which become increasingly important when linking conservation, science, and management. All of these values provide a basis for conservation, science, and management plans being more sophisticated or nuanced in their approach, which is paramount for maintaining effective and sustainable ecosystems (Kissling et al., 2018). Although conventional methods have offered useful information, they frequently have some restrictions regarding scale and efficiency. This has led to the adoption of developing technologies, which allow for better accuracy and automating the monitoring of biodiversity (Pimm et al., 2015).

9.3.3 Emerging Technologies in the Field of Biodiversity Monitoring

Ecosystem monitoring is fundamental to the understanding of environmental health and to sustaining life on Earth for the long-term. New tools made possible through a growing base of technology improve traditional ecosystem monitoring in terms of accuracy, efficiency, and scalability. Artificial Intelligence (AI), drones, remote sensing, and the Internet of Things (IoT) are among the most game-changing technologies to reshape monitoring for biodiversity in the future. These technological advances allow those in conservation to gather data in real-time, analyse changes to the environment, and respond to threats more selectively (Pimm et al., 2015).

AI is a crucial tool for monitoring biodiversity through its ability to automate the processing of large data sets, including images, sounds, and environmental indicators. For instance, machine learning algorithms allow for the automatic identification of species in photographs and audio recordings which significantly speed up previously time-consuming manual processes (Wäldchen & Mäder, 2018). From analysing and recording anomalies in biodiversity, researchers can detect early warning signs of habitat loss, species loss, and climate-related changes. Moreover, AI and predictive analytics can forecast ecological changes through datasets based on historical trends and real-time inputs (Pearson et al., 2020). This ability for predictive analytics effectively allows the conservation community to evaluate threats against the environment and take action before the threat imposes irreversible environmental damage. Additionally, machine learning models can evaluate extinction probabilities based on factors such as habitat destruction, climate, and the movement of a species (Newbold, 2018). Overall, AI brings a data-intensive approach to biodiversity monitoring, ensuring a quicker process and more effective conservation decisions.

Drones fitted with multispectral and hyperspectral cameras offer excellent habitat mapping and plant health assessment capabilities, thereby providing critical information about land use changes and vegetation conditions (Anderson & Gaston, 2013). Thermal sensors on drones can detect presence or activity of animal species in moderate to poor weather or low-light situations. Drones are also advantageous as they can cover large geographic areas in short time frames. Simply put, drones with the capacity to monitor actual wildlife movement (tracked by GPS units attached to the drone) and remote surveys of wildlife or terrestrial areas usually require time-consuming and labour-intensive methods. Drones can easily cover areas that previously would have been difficult or impossible to access without extensive travel to remote area and utilize much less time. Drones with artificial intelligence capabilities can take monitoring biodiversity to another level to adjust flight paths in real time, in part due to environmental change (Hodgson et al., 2016). Drones run by human agents are at a disadvantage in this respect because the agent would not be able to change their flight path quickly, therefore, if the drone is autonomous would be best equipped to retrieve the most applicable data. Drones with autonomous capabilities can cover more of an area than those that are human operated.

Satellites offer the unique ability to monitor global environmental conditions continuously, making them a fundamental approach to studying ecological changes over large areas. Improvements in satellite technology have not only expedited data processing, but they have also enabled near real-time analysis of the environment. A satellite can observe multiple spectral wavelengths to analyse and map vegetation cover, understand land use changes, and monitor pollutants in water bodies (Pettorelli et al., 2014). A major breakthrough in remote sensing technology is Synthetic Aperture Radar (SAR) technology, which has the capability to penetrate clouds and vegetation to provide data of surface depth regardless of climate. Ultimately, this ability will allow conservationists and decision-makers to monitor environmental change with great accuracy and confidence, even when conditions are less than optimal (Reiche et al., 2016).

The monitoring of ecosystems is critically supported by IoT devices that continually gather information regarding environmental conditions from monitoring stations that incorporate remote sensors and cameras. An IoT device can monitor factors of ecological significance, such as temperature, relative humidity or amount of moisture in soil, and the presence of species. From the IoT device, the data is transmitted through wireless networks to higher storage databases. In this method, ecosystem monitoring is continuous and in real-time. When coupled with other scientific monitoring tools, such as satellites or drones, IoT technology supports monitoring biodiversity by establishing a holistic and interconnected monitoring system. Likewise, the IoT data can be synthesized with AI-driven analytics that provide directed information (i.e. actionable insights) to conservationists, and can directly warn of environmental changes, habitat loss, and species migration patterns (Chi et al., 2020).

Biodiversity monitoring is entering a new age through advancements in emerging technologies, including AI, drones, remote sensing, and IoT, that provide excellent accuracy and efficiency for collecting and interpreting data. These technologies advance our understanding of ecosystems and provide timely interventions to address environmental stressors. To advance Earth's biodiversity for the sake of future generations, integrating emerging technologies into conservation programs will be indispensable as technological developments continue.

Figure 2. An AGI-powered drone conducting environmental surveillance over an agricultural landscape.

Figure 3. High-resolution satellite images showcasing coral reef habitats.

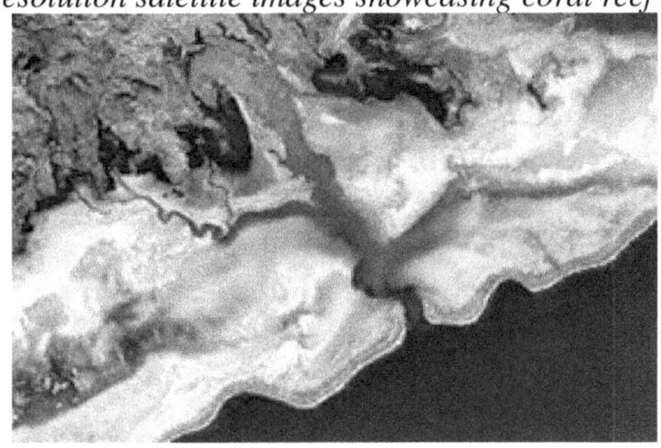

Figure 4. Integration of IoT in biodiversity monitoring.

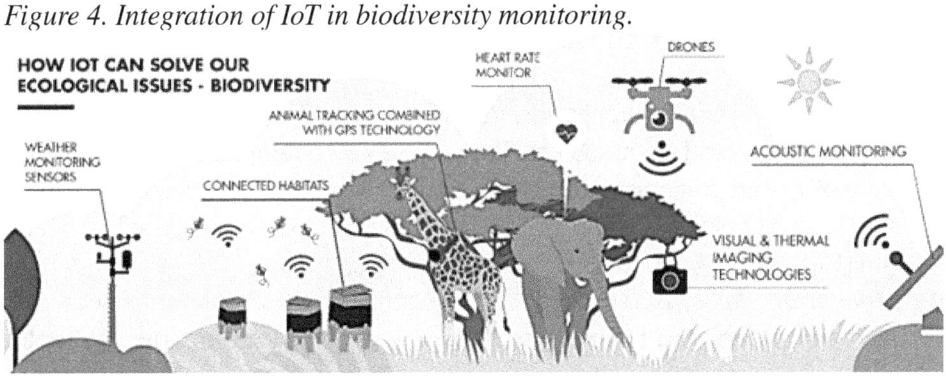

9.4 AGI Drones: A Game Changer in Biodiversity Conservation

New technologies are quickly reshaping the future of biodiversity monitoring, in particular, drones powered by Artificial General Intelligence (AGI). Drones equipped with AI capabilities can autonomously (and rapidly) obtain, analyse, and adjust to environmental data (Chi et al., 2020; Coverdrone, 2024). AGI drones monitor biodiversity through learning from the environment around them, refining their strategies to monitor biodiversity, and making autonomous decisions (Christin et al., 2019; Pichler et al., 2022). The introduction of AGI drones will dramatically increase monitoring opportunities for biodiversity, conservation, and research (Anderson & Gaston, 2013; Koh & Wich, 2012). Whether in dense tropical forests, remote mountain ranges, vast oceans, or delicate wetlands, AGI drones were designed to operate in the diverse ecosystems of our planet. They also capture imagery at significantly higher resolutions, perform multispectral and thermal sensing, and conduct near real-time environmental analyses in areas where current technology and/or methods are not feasible options for monitoring biodiversity. In addition to observing biodiversity, AGI drone technology will contribute to wildlife conservation efforts, such as tracking endangered species, by mapping habitat degradation, monitoring the effects of climate change, and preventing poaching or deforestation (Bajracharya et al., 2020; FEDS, 2024; https://rhinos.org). In short, AGI drones will collect, assess, and analyse biodiversity data "in the moment", and the integration of machine-learning and predictive analytics will allow AGI drones to detect patterns, assess threats to the environment, and give advance notice for conservation planning. These pages highlight how AGI drones monitor biodiversity, why they are becoming a preferred option relative to existing methods, and why they are considered a new frontier for biodiversity conservation. As Artificial Intelligence technologies become more advanced, AGI drones will collect biodiversity data at scale in options that strive to be relevant, engaging, efficient, effective, and fundamentally data-driven solutions to biodiversity conservation (Turner, 2014; Wäldchen & Mäder, 2018).

9.5 Applications of AGI Drones in Biodiversity Monitoring

1. *Wildlife Surveys:* AGI drones are capable of comprehensive aerial wildlife surveys to identify and count species. These drones are equipped with sophisticated state-of-the-art cameras and can accurately distinguish between species with advanced AI systems so their population growth is closely tracked at in the long run (Christin et al., 2019; Hodgson et al., 2016).
2. *Monitoring plants:* AGI drones that are equipped with the latest technologies for multispectral and hyperspectral imaging excel at looking after the health and diversity of plants. They can distinguish species as well as identify diseases

at their beginning stages or stressors that affect plants in different ecosystems because they have such advanced features (Wäldchen & Mäder, 2018; Chi et al., 2020).

3. *Detailed Habitat Maps:* Getting local people involved in conservation efforts and having them understand the importance of preserving biodiversity for future generations can go a long way. With the data from AGI drones, this can be done in an engaging way visually through community participation as well as corporate social responsibility activities. For example, encouraging eco-tourism across any country would be perfect with this practice (Coverdrone, 2024; Zenadrone, 2023).
4. *Environmental Changes*: Regarding environmental changes, these AGI drones are responsible for monitoring them. That includes comparing data sets in order to determine any habitat alterations due to human activities or natural phenomena involved in the process. Such shifts may result from deforestation; its alternative could be urban sprawl, which implies space covered under the urban land uses; meanwhile, climate change influences many aspects of life including those related with water bodies as well as plants such as trees found among others around rivers margins leading them ultimately up into mountains where humans don't live at all except occasionally but survive very close there. This makes them valuable indicators for conservation planning and decision-making operations (Pichler & Hartig, 2022; Rammer & Seidl, 2019).
5. *Deforestation Monitoring:* AGI drones are handy in real-time deforestation and illegal logging activities revelation which offers useful information to enforcement agencies as well as conservation organizations that combat unauthorized activities and safeguard precious ecosystems against their worsening conditions (Bajracharya et al., 2020).
6. *Poaching Prevention:* Drones equipped with thermal cameras and AI technologies detect and deter poaching activities by patrolling protected areas effectively in order to prevent poaching. Thus, safeguarding endangered species and maintaining ecological balance through assisting in preventing illegal trade of wildlife (FEDS, 2024; Koh & Wich, 2012).
7. *Tracking Environmental Conditions:* Findings from the conversation show that drones by AGI Company which carry various sensors are closely watching the environment. In identifying climate change impacts on ecosystems around the world and recommending the strategies for their adjustment as well as preservation, this procedure could also include extraction of temperature, humidity and few other principal factors of the environment (Pajares, 2015; Pichler et al., 2022).

8. *Sea-Level Rise:* Drones from the artificial general intelligence (AGI) company can help observe changes in sea level at or near the oceanic coastlands of the world and discern any changes in these fragile ecosystems because they serve as essential data delivery tools for example on mangroves or coral reefs as critical things here. When it comes to sustainable coasts planning., there has to be provision of reliable numbers about the coast dynamics so that or people can be able to make proper choices.AI drones can help give valuable statistics about this area of nature (Turner, 2014; Anderson & Gaston, 2013).
9. *Water and Air Quality Monitoring:* AGI drones are very good at finding things like oil spills and heavy metals in large water bodies because they have special sensors that help them detect pollutants in those areas. This kind of information is useful for short-term solutions within water bodies or it also aids in the continued surveillance process aimed at maintaining its purity, even after an oil spill incident has occurred in such water bodies. The ability of drones to monitor air quality and detect pollution is of great benefit to wildlife and humans by providing valuable insights into the environmental health of biodiversity (Chi et al., 2020; Nycz & Pietrucha-Urbanik, 2024).
10. *Community Engagement:* Involving our neighbours in efforts to save wildlife and making them realize why biodiversity should be kept safe for coming years, is significant to a larger extent. By utilizing AGI drones' information, this can be made lively visually through community and corporate social activities. This means one of the countries can do well in promoting eco-tourism if supported at all on this issue (Zenadrone, 2023; Coverdrone, 2024).

In the next section, we will explore real-world examples and case studies showcasing how AGI drones are actively contributing to biodiversity monitoring and conservation efforts worldwide.

9.5.1 Case Study 1: Rainforest Monitoring with AGI Drones – Amazon Rainforest, Brazil Deforestation Monitoring in Amazon

Rainforests are one of the richest ecosystems on the planet as for biodiversity and they play an essential role in supporting the balance of the global ecology. But, deforestation, global warming and illegal logging present great danger (threat) to them (Bajracharya et al., 2020; Pichler & Hartig, 2022). It is very difficult to supervise such large areas that are frequently impossible to get to. In this case study, we find out about the use of AGI drones in the Amazon rainforest which has enabled better biodiversity monitoring, preservation efforts and implementation of policies (FEDS, 2024; Zenadrone, 2023).

People say that the Amazon rainforest which is called “lungs of the planet” stretches by approximately 5.5 million square kilometres and boasts 878 species of birds and 3000 fish. It implies that this biggest tropical forest in South America has a large number of animal species including mammals like jaguars because it acts as their habitat (Koh & Wich, 2012; Christin et al., 2019).

There are limitations associated with traditional monitoring methods like ground surveys and satellite imagery. This is because ground surveys are laborious, time-consuming and frequently dangerous owing to rugged landscapes and wild animals. Whereas useful for broad observations, satellite imagery does not have sufficient detail in some cases and may be affected by clouds. AGI drones enable an overhead perspective of the ecosystem, allowing in-situ observations. This concept can be seen in Figure 5, a survey of the Amazon Rainforest, which provides data about the status of deforestation in real-time (Anderson & Gaston, 2013; Hodgson et al., 2016).

Figure 5. Aerial view of the Amazon Rainforest, Brazil.

Key objectives include developing a sophisticated approach to understanding and managing important forest ecosystems. We can see if there are patterns or trends in the changes in species and populations by following the ecology at all. This enables us to put in place specific conservation measures. At the same time, we will want to detect deforestation as a crucial tool to help us quickly identify and monitor illegal logging activities that poses danger to our forests. And also, our pledge is thorough climate impact assessments to analyse intimately how climatic changes interfere with the fragile equilibrium found in rain forests (Pichler et al., 2022; Rammer & Seidl, 2019).

Advanced data collection and analytical techniques allowed to delve deeper into the spectacular diversity, revealing previously unknown dynamics in population dynamics, ecological interactions that provide a foundation for conservation strategies with sustainable knowledge and practices (Chi et al., 2020; Wäldchen & Mäder, 2018).

The use of advanced technologies like remote sensing and machine learning algorithms has helped in establishing a reliable system that detects deforestation timely and accurately, thus lending a hand greatly to governments as well as parties that fight against it, while at the same time taking care of ecosystems with valuable forests.

In-depth study of complex climatic data and trends provided significant knowledge on how climate change affects ecosystems, species' ranges and natural habitats, guiding towards adaptive and mitigative tactics to lessen exposure and build resistance towards a transformed environment (Christin et al., 2019; FEDS, 2024).

Research has enabled key policymakers, conservationists, and stakeholders to prioritize and carry out focused conservation activities that help conserve threatened species, protect critical habitats, promote sustainable resource use, and in so doing, ensure biodiversity's brighter future, through identifying key biodiversity hotspots and areas that are highly important for conservation and restoration (Pajares, 2015; Pichler et al., 2022).

In the Amazon rainforest, deploying AGI drones has greatly increased the capability to keep track of different animals, discover the disappearance of trees and evaluate the consequences of global warming. Deforestation in the Brazilian Amazon—a vast and ecologically crucial region—has been a huge environmental concern for over three decades now, starting most notably in 1988. The urgency of this matter is reflected in its continually fluctuating rates that reflect an intricate relationship with such varied elements as government regulations vis-à-vis economy needs versus environmental protection campaigns. Clearly, such stimuli balance delicate scales, showing that deforestation in this area is multi-faceted, wherein every variable either accelerates or decelerates the clearing of forests. The fate of the Amazon Rainforest is always in debate in that governments are changing their policies, and it is expected to be dragged between development and preservation forces with financial interests. This tool is very important to environmentalists as well as leaders because there is immediate information regarding the problems that used to be very difficult to access which these drones offer (Bajracharya et al., 2020; Nycz & Pietrucha-Urbanik, 2024).

Figure 6. Drones over the Amazon

It is important to be very careful about following the complicated and difficult rules set by individual countries for flying drones across large and protected regions that are rich in living things. Equally essential is observing the rights to indigenous land by having detailed knowledge of the historical issues involved and being ready to follow any necessary laws or guidelines. In order to effectively address ethical issues, it is necessary to actively engage local indigenous communities in monitoring activities focussing on their values and inclusion of their traditional wisdom as a way of adopting a broader approach that is also sensitive culturally. It is not only in the realm of ethics that getting permission from these societies is important. During this process, which is characterized by respect, they can create powerful ties with each other where there are similarities in terms of monitoring as well as general welfare for locals (FEDS, 2024; Zenadrone, 2023).

Figure 7. Deforestation in the Brazilian Amazon since 1988, according to INPE

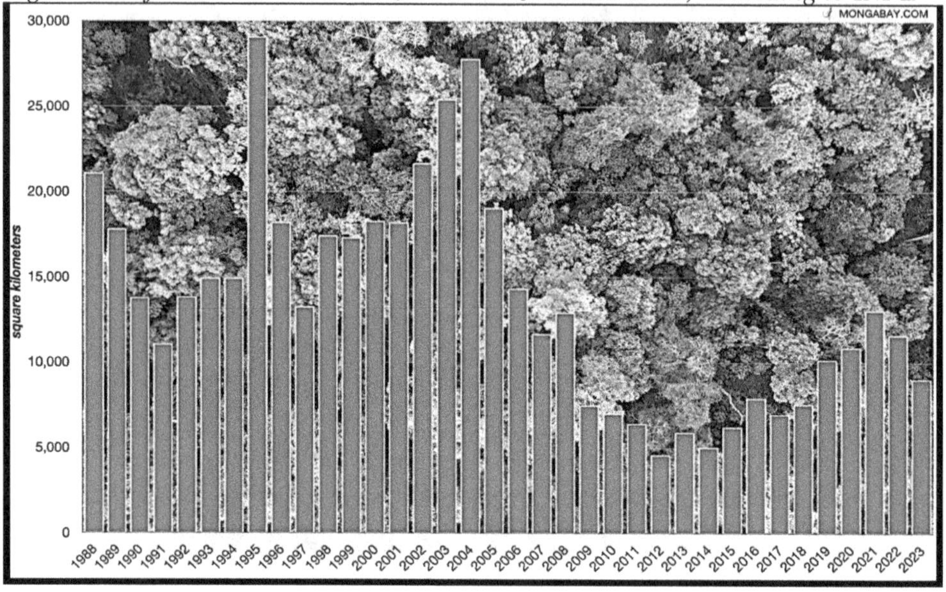

9.5.2 Case Study 2: Coastal Ecosystem

Coastal ecosystems are very varied and crucial to both the well-being of the environment and the people who depend on them for their survival. Their proper functioning supports the survival of very many species of animals and plants in the world, promoting the total biodiversity through various types of organisms (Bajracharya et al., 2020; Pichler & Hartig, 2022). These coastal habitats are home to many animals as well as provide other important functions like shielding against erosion, floods or increased water levels rising due to global warming which although less considered is still important to mention (Pajares, 2015; Rammer & Seidl, 2019).

Figure 8. (a) Aerial drone surveillance in coastal ecosystems

Figure 9. (b) Submerged drone operations in coastal ecosystems

Nevertheless, they have been confronted with several severe challenges that imperil their general condition and functional properties though they are life supporting areas along the coast. The fragility of these ecosystems implies that climate change is a great jeopardy because it leads to higher sea levels while there are more occurrences of drastic weather conditions than before. This endangers them and anyone living near or far away from them may suffer too (Pichler et al., 2022; Chi et al., 2020). Alongside this environmental hazard there comes another one which tends to deteriorate the water quality besides destroying the intricate structure of abiotic & biotic constituents prevailing in it due to human obstinacy, including plastic trash accumulation or application/release funnels for toxic substances into muddy rivers that flow through such places like this one; moreover overpopulation-all these things seem unreal! (Wäldchen & Mäder, 2018; Christin et al., 2019).

It is very necessary to maintain close monitoring and successful conservation efforts so that these ecosystems can be kept safe for future generations. Through the act of closely supervising what it is happening in coastal areas, ecologists get a chance of making out the changes happening within them thus putting up strategies aiming at lessening bad consequences. AGI drones are an advanced methodology that could provide huge advantages by ensuring that coastal ecosystems are well observed always. Their employment would help to obtain vital information concerned with these ecosystems' state, hence improved managerial decisions can be made in order to maintain them sustainably in a long-term perspective (Zenadrone, 2023; Anderson & Gaston, 2013).

9.6 Comparison of AGI Drones with Other Conservation Technologies

Historically, numerous conservation technologies have been employed for biodiversity monitoring, each with their own strengths and weaknesses. For example, satellite imaging technology is widely used to monitor large-scale environmental change, such as climate change or deforestation (Pichler & Hartig, 2022; Chi et al., 2020). A satellite however, lacks the fine spatial resolution required for monitoring species, with clouds also inhibiting real-time observation. On another note, AGI drones can provide localized high-resolution data for monitoring both species and habitat change in relation to fixed GIS data (Zenadrone, 2023; Anderson & Gaston, 2013). Another commonly used technology includes IoT sensor networks, which are sensor data networks (most commonly a fixed sensor) to monitor aspects of the environment like temperature, humidity, soil condition, etc. IoT sensors provide continuous monitoring, but they are fixed and unable to adjust to changing environmental features of an area being studied (FEDS, 2024; Wäldchen & Mäder, 2018). Although the AGI drones will not continuously monitor specific sites like the IoT sensors, AGI drones can actively survey spatio-temporal monitoring in multiple habitats in one flight, providing flexibility in biodiversity monitoring through time and habitat types (Bajracharya et al., 2020; Rammer & Seidl, 2019).

Camera traps and acoustic sensors can also be beneficial for conservation. Camera traps are particularly effective for monitoring elusive or nocturnal species, while acoustic sensors are useful for species that rely on vocal communication to study them (Hodgson et al., 2016; Christin et al., 2019). However, both of these active sensors are limited in their field of view and require human coordination for data collection, processing and returning back to any site where monitoring of the species is happening. AGI drones with AI-enabled recognition systems can produce huge forays of habitat time to monitoring species appearance and behaviour in terrestrial and aquatic environments (Pajares, 2015; Pichler et al., 2022). Although field surveys are still an important methodological tool for detailed ecological studies, they are typically time-consuming, labour-intensive and inaccessible. AGI drones mitigate the challenge of visual monitoring, in addition to a number of different potential fields, with real-time aerial monitoring, exploring dangerous or very isolated habitats and contributing to conservation research potential and efficiency (Zenadrone, 2023; Anderson & Gaston, 2013).

9.7 Drones That are Commonly Used in Biodiversity Monitoring

Biodiversity monitoring has immensely benefitted from the incorporation of AGI-enabled drones into conservation initiatives (Anderson & Gaston, 2013; Zenadrone, 2023). However, there are different types of drones – different unmanned aerial vehicles (UAV) etc – that tend to specialize more for different types of ecological uses. The next section will evaluate the most common drones used for biodiversity monitoring, including their capabilities, features, and specific use in environmental conservation (Christin et al., 2019; Wäldchen & Mäder, 2018).

9.7.1 DJI Mavic 3 Multispectral

The Mavic 3 Multispectral drone is an advanced technology tool designed for precise agriculture monitoring, natural resource assessment, and high-resolution topographic mapping (Zenadrone, 2023). It is a class-leading drone with a maximum flight time of 43 minutes that achieves a brilliant compromise of convenience and performance- its smaller footprint produces higher quality data than any other agricultural drone in its class (Christin et al., 2019). Among the high-level product features is centimetre-level RTK (real-time kinematic) positioning that allows the highest level of mapping precision available today (Anderson & Gaston, 2013). Optimized battery usage allows for greater coverage per flight on the same day for follow-up visits on previous surveys. One flight covers over 200 ha, demonstrating the utility value of this drone for environmental monitoring and surveying human-made green spaces- for example, lake water quality monitoring and the surveying of complex forest distributions (Waldchen & Mäder, 2018). Therefore, the Mavic 3 Multispectral will be a pragmatic tool for sustainable resource management and ecological preservation (Hodgson et al., 2016).

Figure 10. (a) DJI Mavic 3 multispectral drone in aerial operation

Figure 11. (b) Front view of DJI Mavic 3 multispectral drone

9.7.2 EchoLogger ECT D052S Echo Sounder

The EchoLogger ECT D052S Echo Sounder offers a specialized capability for hydrographic surveys and aquatic ecosystem studies, while the Mavic 3 Multispectral, as its name indicates, is an airborne and terrestrial monitoring device (SPH Engineering, 2023). The EchoLogger ECT D052S Echo Sounder created by SPH Engineering epitomizes flexibility in water depth surveying and aquatic habitat assessment and bottom mapping and is one of the most used instruments in this vertical (Hodgson et al., 2016). The efficacy of the EchoLogger ECT D052S Echo Sounder is buttressed by its dual-frequency sonar technology and incorporated tilt and temperature sensor that can read from 0.5m to 200m measurements in either 50kHz or 200kHz (Koh & Wich, 2012). The ability to provide high-resolution underwater topography mapping benefits plotting depth mapping analysis of aquatic ecosystems; therefore, it plays a central role in marine research initiatives, resource management plans, and environmental efficiency monitoring systems (Garcia et al., 2020). The most notable characteristic about the EchoLogger ECT D052S Echo

Sounder is the fact that it's compatible with the DJI drone systems including the M350 RTK allowing for drone-based hydrographic surveys of more remote water bodies by scientists and surveyors who want to collect data by drones without the logistical limitations of physically getting to the waterbody or simply doing the surveying at a distance (Zenadrone, 2023).

Figure 12. (a) Drone Deployment of EchoLogger ECT D052S Echo Sounder for Water Depth Measurement

Figure 13. (a) EchoLogger ECT D052S Echo Sounder Sensor for Hydrographic Surveying

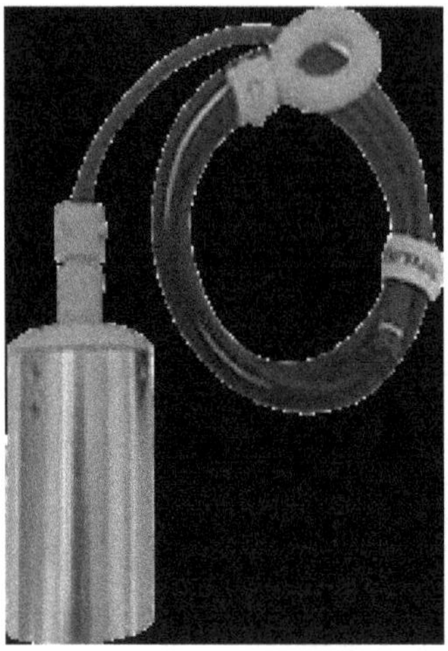

9.7.3 **YellowScan Navigator**

In addition to the EchoLogger's capabilities for hydrographic studies, we also provide the YellowScan Navigator which is specifically designed for shallow water mapping and bathymetric seafloor applications (Zenadrone, 2023). As a component of the advanced data acquisition system, the YellowScan Navigator provides very accurate topographic data acquisition in shallow water environments, such as rivers, lakes, estuaries, and coastal zones (Garcia et al., 2020). Additionally, the YellowScan Navigator provides a 44-degree field of view, therefore the data acquisition tool optimizes the area covered for each scan, all the while on board a global shutter camera is able to provide colorized imagery to enhance the data interpretation process (Koh & Wich, 2012). For mapping capabilities, the YellowScan Navigator has been prepared for use with various UAV mounting options. Efforts to streamline the data processing software also support the accuracy of the topography derived metrics, even in complex environments (Hodgson et al., 2016). Using LiDAR technology and camera equipment, processing capabilities, the YellowScan Navigator is exceptionally positioned for topographic mapping in shallow water environments (SPH Engineering, 2023).

Figure 14. (a) YellowScan Navigator LiDAR Drone Scanning an Aquatic Environment

Figure 15. (b) YellowScan Navigator LiDAR Sensor Unit

9.8 Challenges Faced During AGI Drone Monitoring

AGI-powered drones have advanced technologies, but many operational issues still impact their uses in biodiversity monitoring. From regulation to cyber risk, to limited capacity in-country, these operational issues need to be resolved to use drones effectively in conservation (Garcia et al., 2020). One of the most significant challenges with drone usage concerns airspace regulations. Each country formulates its own laws regarding the use of drones: flight altitude limits, areas that drones cannot be flown in, or licensing requirements for the drone operator (Koh & Wich, 2012). Oftentimes, obtaining either verbal or written consent to fly a drone over the borders of a national park, or through a protected reserve or government-managed area, requires quite a bit of administrative checks which can hinder conservation work or monitoring results. For example, in Kaziranga National Park in India, drone-based anti-poaching initiatives were thwarted due to drone restrictions associated with airspace regulations, which impacted the potential for real-time monitoring of illegal activities (Hodgson et al., 2016).

The other relevant concern involves cybersecurity threats because AGI drones utilize real-time data transmission and data storage in the cloud. Data can be hacked, accessed improperly, or violated which could lead the conservation data to be tampered with, as well as a hijacked drone committing illegal acts (SPH Engineering, 2023). For instance, in 2021, in Australia a wildlife drone operation was hacked by criminals who purposely disrupted the broadcast of conservation data to change land use decisions. To lower the risks, it is extremely important that data is strongly encrypted, that the communications are secured, and who has access to the data is limited (Zenadrone, 2023). There are also technical limitations that present challenges to the operationalization of AGI drones in remote locations. Drones may struggle to survey larger geographic areas, because you might have to deal with inclement weather, thick forest canopy, and limited battery life (Garcia et al., 2020). Conservationists studying the Amazon rainforest note that drones may have difficulty receiving a flight signal in areas of thick vegetation and therefore experience limited real-time data transmission. Moreover, steep or massive terrain obstacles, such as steep mountains, gullies, or water, can present challenges to AGI drones for sustaining flight paths or collecting accurate data. These limitations, particularly the challenges operationalized by inclement weather, could be solved with investment in safety and battery technology combined with AI adaptive navigation design elements (Koh & Wich, 2012).

Along with the logistical and technical challenges, there is also the aspect of public perception and community engagement. In many parts of the world, local communities may not be fully on board with or trust drone-based monitoring; they may even be resistant to the idea entirely due to concerns over land surveillance or ownership of monitoring data. For example, in Kenya's Maasai Mara National Reserve, local communities initially resisted the idea of drones being used for wildlife monitoring because of concerns about potential land access restrictions (Hodgson et al., 2016). One way of creating acceptance and support for drone technology initiatives in conservation is to achieve community engagement through education and involvement of local communities in the process.

Although these challenges pose major obstacles to drone monitoring on a large scale, improvements in policy frameworks, cybersecurity solutions, and technological advances will likely help address them. However, beyond technical and regulatory hurdles, AGI drones also present significant ethical issues that must be addressed to ensure their implementation in a manner that is consistent with responsible conservation practices (Garcia et al., 2020).

9.8.1 Case Study 1: Poaching Prevention in African Wildlife Reserves

Drones powered by Artificial General Intelligence (AGI) technology have been an incredible resource for conservationists working to protect endangered wildlife throughout Africa (Garcia et al., 2020). The complexities created by national regulations regarding airspace and wildlife protection can lead to inconsistencies in drone surveillance used to wrangle poaching across countries (Hodgson et al., 2016). The pilot of the drone must navigate an ethical line of ensuring that someone or something is protected, while simultaneously minimizing the disturbance to the wildlife (Koh & Wich, 2012). For example, in Kruger National Park in South Africa, drones equipped with thermal imaging cameras have helped to discover poaching activities, especially with rhinoceroses, which are frequently targeted for their horns (Hodgson et al., 2016). While these innovations are effective in addressing poaching, ethical dilemmas arise in regard to use—particularly regarding constant monitoring and concerns about how far the drone's surveillance extends beyond its intended conservation purpose into unauthorized surveillance (Garcia et al., 2020). Therefore, their usage must be held publicly accountable and adhere to regulatory frameworks and ethical principles so that their role as a conservation tool is maintained and public trust is fostered for responsible environmental monitoring systems.

Figure 16. Trends in rhinoceros poaching in South Africa (2013–2022)

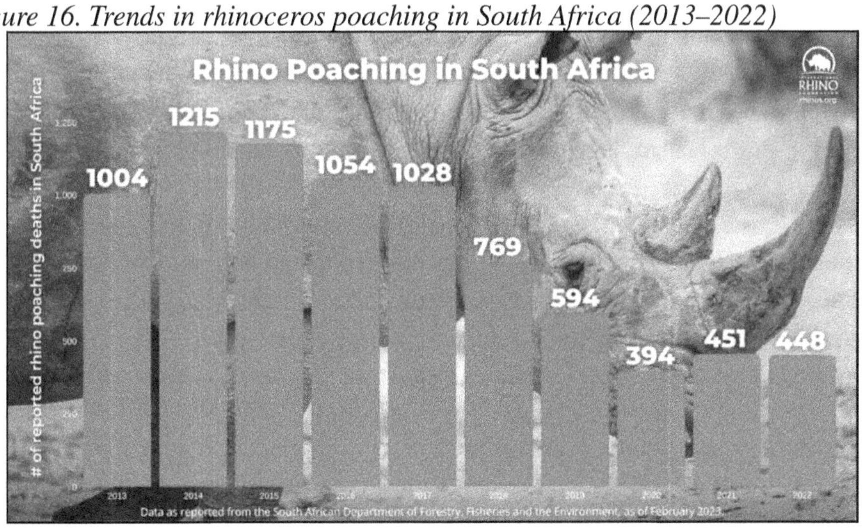

9.8.2 Ethical Concerns in AGI-Powered Biodiversity Monitoring

The uses of AGI-powered drones for biodiversity monitoring have contributed considerably to enhanced conservation. However, their deployment raises a number of ethical issues that must be thoroughly investigated and addressed to ensure their use remains appropriate and sustainable (Garcia et al., 2020). Specifically, wildlife disturbance, data privacy, AI-based decision-making, environmental consequences, and surveillance risks are among the major issues conservationists and policymakers must contend with (Koh & Wich, 2012; Sandbrook et al., 2015).

Wildlife disturbance is one of the most significant ethical concerns. When drones operate at low altitudes, they can disrupt animal behavior, causing stress and interference with natural patterns like foraging, mating, and migration (Mulero-Pázmány et al., 2017). Studies in Antarctica showed altered nesting behavior in penguin colonies due to drone presence, and research in Yellowstone National Park documented avoidance behaviors in wolves and bison when drones flew overhead (Mulero-Pázmány et al., 2017). Additionally, drones introduce noise pollution that affects sound-sensitive species such as bats, birds, and marine mammals, disrupting communication and navigation (Hodgson et al., 2016).

Another critical issue is data privacy and the rights of indigenous peoples. AGI drones collect extensive environmental data, which may inadvertently capture images of human habitation, agricultural lands, and indigenous territories (Garcia et al., 2020). For example, surveillance in the Amazon has raised concerns about the unauthorized collection of data from indigenous groups and the question of data ownership. Similar issues arose in the Canadian Arctic where AGI drones mapped Inuit territory without community consultation (Sandbrook et al., 2015). Ethical data governance must include transparency, community participation, and informed consent to prevent misuse of sensitive information.

Overreliance on AI-based decision-making is also ethically problematic. While AI-enabled drones can analyze large datasets efficiently, they are not infallible. Errors in species identification or environmental assessment can occur due to biases in training data or limitations in machine learning algorithms (Van Andel et al., 2023). For instance, AI misclassified shadows as elephants in African monitoring projects, leading to overestimated population counts. Similarly, in the Sundarbans mangrove forests, AI failed to detect habitat degradation, leading to underreporting of conservation needs. Thus, human oversight and field verification remain essential (Hodgson et al., 2016).

The environmental impact of drone use is another concern. Although drones reduce the need for resource-intensive fieldwork, they still depend on lithium-ion batteries, contributing to electronic waste and mineral extraction impacts (Sandbrook et al., 2015). Supporting equipment and transport also increase the carbon footprint.

Innovations like solar-powered drones being trialed in European wetlands offer a more sustainable alternative (Garcia et al., 2020).

Finally, surveillance risks must be considered. While AGI drones have successfully aided anti-poaching efforts using thermal imaging in areas like Kruger National Park, there are growing fears that such technologies could be repurposed for mass surveillance or non-conservation policing (Sandbrook et al., 2015). Establishing safeguards through regulatory frameworks, transparent operations, and public accountability is essential to prevent such misuse.

9.8.3 Regulatory Barriers in AGI Drone Conservation

The use of drones driven by AGI technology for biodiversity conservation is rapidly increasing, yet regulatory restrictions remain a significant obstacle. The absence of a unified global policy framework, along with diverse national regulations, hinders the full potential of drone-based environmental monitoring (Garcia et al., 2020). These challenges include restricted airspace access, complex licensing requirements, fragmented data privacy laws, and regulatory discrepancies across borders. Such legal barriers delay conservation responses, create inefficiencies, and complicate transboundary ecosystem management (Sandbrook et al., 2015).

Airspace control and flight restrictions are among the most pressing regulatory concerns. Many countries restrict or prohibit UAV operations over ecologically sensitive areas, protected zones, or international borders, due to security, environmental disruption, and surveillance concerns (Mulero-Pázmány et al., 2017). For instance, conservationists working in Australia's Great Barrier Reef region often face long delays in securing drone flight approvals for coral monitoring missions. Similarly, in Tanzania's Serengeti National Park, real-time monitoring essential for poaching prevention is often hindered by limitations on UAV use (Hodgson et al., 2016).

Complicated licensing and operational approval requirements further hinder the use of drones in conservation. Several nations mandate commercial pilot certifications, UAV registration with aviation authorities, and adherence to strict air traffic regulations. These protocols disproportionately affect NGOs and researchers, slowing deployment during ecological emergencies (Sandbrook et al., 2015). In Madagascar's rainforest reserves, for example, bureaucratic hurdles often prevent timely biodiversity assessments and responses to habitat degradation (Garcia et al., 2020).

Another emerging issue is data privacy and the ethical governance of drone-collected information. AGI drones capture high-resolution imagery, spatial data, and environmental metrics, yet many countries lack clear laws on data ownership, security, and dissemination (Koh & Wich, 2012). In Brazil's Amazon Rainforest, indigenous communities have raised concerns that drone surveillance intended for monitoring illegal deforestation could unintentionally capture data on their settlements, thereby infringing on their land rights. In Norway's Arctic regions, local fishing groups have opposed academic drone use due to fears of increased regulation stemming from drone-collected data (Sandbrook et al., 2015). These concerns highlight the need for transparent data-sharing agreements and community-centered governance frameworks.

Furthermore, regulatory divergence between countries creates barriers to conservation in shared geographies. Unlike satellite monitoring, which operates under an international legal regime, drone usage lacks harmonized cross-border policies (Garcia et al., 2020). This issue is evident in regions like the Congo Basin, where deforestation and poaching transcend national boundaries, making coordinated drone-based monitoring difficult. The Himalayas face a similar problem, where inconsistent drone laws across India, Nepal, and China hinder monitoring of glacial retreat and biodiversity change (Van Andel et al., 2023). As AGI drones grow more autonomous and sophisticated in their ecological applications, international and national policymakers must modernize legislation to uphold privacy, security, and ethical standards while enabling innovation in AI-driven environmental protection.

9.9 Strategies to Address the Challenges

To address regulatory barriers, it is crucial to outline legal structure that strikes the risght balance between accessibility and conservation and privacy considerations. Collaboration on policy development, building effective partnerships with government agencies, creating a clear licensing system for conservation drones, and managing drone data on a global level, are all strategies that can create a greater legal underpinning for AGI drone-based biodiversity monitoring. This section collaborates on a number of strategies that propose tangible approaches to addressing regulatory and ethics-based restrictions that help operationalize AGI drones in conservation stewardship across the globe (Smith et al., 2022).

9.9.1 Regulatory Compliance

Fostering strong collaboration with national and local aviation authorities is one of the main parts of regulatory compliance. So as to traverse the difficult regulation landscape and ensure that we meet all requirements, we have to make good relations with each one of them. In advance of when events happen, we need to get authorizations and deal with any form of non-compliance incident that might occur in a manner that is conscious (Johnson & Lee, 2021).

9.9.2 Enhancing Data Security

The operations are focused on data protection, which has led to strong cybersecurity measures being adopted in our company to prevent any unauthorized access or leaking of information. Our data systems need to be fortified in order to protect them, and our sensitive information must not be comprised at any point because of the potential threats present. Moreover, we make secure data storage our priority by use of encrypted databases with strict access controls which guarantees that all information will be kept safely so it can only be accessed by people who have been given permission thus improving our total level of data security (World Conservation Technology Forum, 2023).

9.9.3 Ethical Operation Practices

We are researching and developing drone technologies that are less intrusive for wildlife habitats and give priority to animal welfare as a means of minimizing the effect of our business operations on local ecosystems. This is to ensure we are true to our commitment of operating ethically. through active engagement with our communities, we show that we respect their rights and privacy by trying to act in a responsible and ethical way (UN Environmental Ethics Board, 2022).

9.9.4 Clear Ethical Guidelines

One of the essential things in our setup is clear ethical guidelines concerning getting the local communities and other stakeholders on board through informed consent before the commencement of drone operations. By providing a clear procedure on how to obtain the consent; we ensure that everyone involved in this process fully understands the importance and implications of our monitoring services. Our commitment to the autonomy and decision-making power of individuals and communities is underscored by this emphasis on informed consent. Additionally, we maintain high quality while clearly articulating our goals, approach and results in drone-based monitoring efforts. The Board's commitment to open dialogue helps build trust and accountability among stakeholders (Global Drone Policy Coalition, 2021).

9.10 Securing Sustainable Funding for AGI Drone Conservation

AI-powered drones show great potential to help protect biodiversity through habitat conservation, fighting poaching, and keeping an eye on ecosystems (Smith et al., 2022). But researchers and conservation groups face a big challenge: finding money to keep these high-tech tools running over time. The price tag for drone hardware, AI software creation, data handling, and following regulations makes it tough to use these drones (Johnson & Lee, 2021). To make AI drones a lasting success in conservation, we need to come up with different ways to fund them that are reliable and can grow as needed (World Conservation Technology Forum, 2023).

9.10.1 Challenges in Funding AGI Drone Conservation

One major hurdle in getting money for conservation projects is the big upfront cost of drone tech. Regular conservation methods don't need as much cash, but drones with AI need fancy sensors top-notch cameras smart data crunching, and systems to handle info on the fly. This drives up the price tag (Smith et al., 2022). Take drones with special cameras and LiDAR for mapping ecosystems in detail. They can set you back anywhere from $20,000 to $100,000 each. That's way too pricey for many groups trying to protect nature in poorer areas where budgets are tight (Johnson & Lee, 2021).

Also, ongoing operational costs like fixing drones, updating software teaching pilots, and storing data make money tight. Many nature projects run on short-term grants, which often don't cover long-term costs. This lack of steady funding hurts efforts to watch wildlife over time, as gaps in money can stop data collection and conservation work (World Conservation Technology Forum, 2023).

Another money problem is following rules and paying for insurance. Many countries need permits, certificates, and strict air rules, which make using drones more expensive. In places where conservation drones fly in war zones or near controlled airspace, getting approval can be costly and slow taking resources away from saving nature

9.10.2 Funding Sources and Strategies for Sustainable Investment

To remove financial obstacles, protectionists must diversify funding sources and detect new financial models that provide long -term stability. Many possible funding are included in the way:

1. ***Government Grants and Public Funding***: Several governments provide grants, subsidies, and conservation funds to assist with development of technology for environmental protection. For example, the European Green Deal, the United Nations Environment Program (UNEP) grants, and National Science Foundation (NSF) funding to develop conservation technology may all provide necessary funding to deploy AGI drones. Conservation organizations should work to

develop government partnerships that incorporate drones into their national biodiversity strategies (Nguyen et al., 2023).

2. ***Private Sector and Corporate Sponsorships***: Aligning with technology firms, AI companies, and drone manufacturers can result in system funding and equipment support. Corporations with corporate social responsibility (CSR) funding often invest in environmental initiatives, particularly climate repercussion mitigation, deforestation prevention, and wildlife preservation (for instance, removing poachers from the jungles of Africa) (Johnson & Lee, 2021). Subsequently, industry examples from tech enterprises like Google, Microsoft, and IBM have already funded and employed environmental drone monitoring with AI project funding for analytics in cloud-based AI (Smith et al., 2022).
3. ***Non-Governmental Organizations (NGOs) and Philanthropic Foundations***: Environmental non-governmental organizations and international conservation foundations are key funding sources for ecology, driven by technology. For example, the World Wildlife Fund (WWF), the Bill & Melinda Gates Foundation, and the Leonardo DiCaprio Foundation have even offered support for AI and drone projects in conservation. When pursuing smaller conservation technology grants, projects have access to diverse funding sources to cover drone acquisition, operational costs, and analysis of data with specific tools (World Conservation Technology Forum, 2023).
4. ***Crowdfunding and Public Donations***: Conservationists can use crowdfunding sites like GoFundMe, Kickstarter, and Patreon to directly engage the public and generate funds for drone-based biodiversity assessments. Fundraising efforts focused on real-world conservation success stories, and an intense need for the environmental cause his or her crowdfunding campaign is about, will be able to attract global donors dedicated to sustainability (Nguyen et al., 2023).
5. ***Eco-Tourism and Conservation Incentive Programs:*** The establishment of eco-tourism frameworks that entangle AGI drone monitoring demonstrations may create a self-sustaining funding mechanism. Visitors to protected sites may pay to experience guided drone-assisted biodiversity tours, with a portion of the funds put back into conservation. In addition, carbon credit programs may provide financial incentives for conservation programs that utilize AGI drones to monitor reforestation and climate adaptation programs to sell carbon offsets to eco-friendly companies (Smith et al., 2022).

9.10.3 Building a Sustainable Financial Model

To provide lasting funding stability, conservation organizations need to explore hybrid financial models that give rise to public funding, private investment, and community-driven funding streams. Collaborative networks with research institutions, government agencies, and technology companies can provide joint funding opportunities and shared resource availability (Johnson & Lee, 2021; Smith et al., 2022). Further, investments in cost-effective AI innovations, such as solar-powered drones and self-learning AI models, can provide cost savings in the long term, advancing AGI drone conservation affordability over time (Nguyen et al., 2023).

9.11 Future Directions for AGI Drones

It is a promising future for AGI (Artificial General Intelligence) drone application in biodiversity monitoring, as it aims at increased independent functionality and instant data processing without any human involvement (Johnson & Lee, 2021). As a result, these drones will be enabled to move through tough landscapes; identifying / tracking a wide array of animals; harmonizing with fluctuating surroundings without people's help. Improvement of brand-new battery technologies such as solid-state batteries and photovoltaic-powered systems would increase flight duration of drones to enable them cover bigger areas and operate for more extended hours (Nguyen et al., 2023). Coordinated drones using swarm intelligence will improve the efficiency and scalability of monitoring ecosystems, enabling large-scale ecological a it is large and varied. Major advances in sensor and machine teaching developments might enable collection of more sophisticated wildlife and plant runs leading to much more accurate and comprehensive biodiversity analysis (Smith et al., 2022). Improved satellite linking will ensure that data and information are available for processing or immediate response even in the remotest locations. On top of this, timely information together with collective stakes may be promoted by the employment of AGI drones on worldwide data networks which can then be seen as an increase in the capacity for observation and conservation of natural assets all over the world (Dlamini et al., 2022).

9.12 SUMMARY

1. AGI-powered drones are transforming biodiversity monitoring by providing real-time, high-resolution ecological insights.
2. Traditional biodiversity monitoring methods are expensive, time-consuming, and often inadequate for large-scale conservation efforts.
3. India, home to rich biodiversity, faces severe threats from deforestation, habitat loss, and climate change, necessitating advanced conservation strategies.
4. Biodiversity monitoring is crucial for assessing ecosystem health, detecting environmental changes, and shaping effective conservation policies.

5. While traditional methods like camera traps and field surveys are useful, emerging technologies like AI, drones, and IoT provide greater efficiency and accuracy.
6. AGI drones can autonomously collect and analyze data, track wildlife populations, assess habitat degradation, and support environmental restoration efforts.
7. These drones are widely used for applications such as wildlife population surveys, plant health monitoring, ecosystem mapping, and detecting environmental hazards.
8. In the Amazon Rainforest, AGI drones help monitor deforestation, track illegal logging, and assess the effects of climate change.
9. In coastal ecosystems, drones assist in monitoring coral reefs, mangroves, and rising sea levels, playing a vital role in marine conservation.
10. Compared to satellites, AGI drones provide localized, high-resolution monitoring, while offering more mobility than stationary IoT sensors.
11. Various drone models, such as the DJI Mavic 3 Multispectral and EchoLogger ECT D052S Echo Sounder, are extensively used in biodiversity research.
12. Strict airspace regulations, licensing challenges, and compliance requirements create major hurdles for drone-based conservation efforts.
13. Ethical concerns surrounding AGI drones include wildlife disturbance, data privacy violations, AI biases, and the risk of mass surveillance.
14. Drone-based conservation faces challenges such as battery limitations, cybersecurity threats, and resistance from local communities.
15. Anti-poaching initiatives in Africa use thermal imaging drones to track illegal hunting, though concerns about privacy and surveillance persist.
16. Addressing regulatory challenges, enhancing AI accuracy, improving cybersecurity, and engaging local communities can improve conservation drone usage.
17. The high costs of drone technology, software development, and regulatory compliance create financial barriers to widespread implementation.
18. Sustainable funding sources such as government grants, corporate sponsorships, NGOs, crowdfunding, and eco-tourism initiatives can support conservation efforts.
19. Future advancements in AI, extended battery life, and swarm intelligence will further enhance the role of drones in biodiversity monitoring.
20. AGI drones offer immense potential for conservation, but ethical, regulatory, and financial challenges must be addressed to maximize their impact.

REFERENCES

International Rhino Foundation. (2023). A New Poaching Problem in South Africa. International Rhino Foundation. https://rhinos.org/blog/a-new-poaching-problem-in-south-africa/

Measurusa. (n.d.). https://drones.measurusa.com/pages/drones-for-environmental-monitoring-and-conservation

Amazon Monitoring. (n.d.). https://amazonmonitoring.com/

Anderson, K., & Gaston, K. J. (2013). Lightweight unmanned aerial vehicles will revolutionize spatial ecology. *Frontiers in Ecology and the Environment*, *11*(3), 138–146. DOI: 10.1890/120150

Bachimanchi, H., Pinder, M. I. M., Robert, C., De Wit, P., Havenhand, J., Kinnby, A., Midtvedt, D., Selander, E., & Volpe, G. (2023). Deep-learning-powered data analysis in plankton ecology. *arXiv preprint arXiv:2309.08500.*

Bajracharya, D. M., Shrestha, A. B., & Rajbhandari, R. (2020). Endangered species and habitat loss: A regional overview. *Journal of Environmental Protection*, *11*(3), 215–223.

Chi, T., Ouyang, C., Wong, M. S., & Nichol, J. E. (2020). Smart sensing and artificial intelligence in the monitoring and management of biodiversity. *Environmental Science & Technology*, *54*(7), 3628–3640. DOI: 10.1021/acs.est.9b06279

Christin, S., Hervet, É., & Lecomte, N. (2019). Applications for deep learning in ecology. *Methods in Ecology and Evolution*, *10*(10), 1632–1644. DOI: 10.1111/2041-210X.13256

Wakefield, L. (2024). How drones monitor environmental changes. Cover Drone. https://www.coverdrone.com/how-drones-monitor-environmental-changes/

Dandois, J. P., & Ellis, E. C. (2010). Remote sensing of vegetation structure using computer vision. *Remote Sensing (Basel)*, *2*(4), 1157–1176. DOI: 10.3390/rs2041157

Weiss, S. (2024). Environmental Sensing is Here, Tracking Everything from Forest Fires to Threatened Species. Wired. https://www.wired.com/story/environmental-sensing-is-here-tracking-everything-from-forest-fires-to-threatened-species/#:~:text=7%3A30%20AM-,Environmental%20Sensing%20Is%20Here%2C%20Tracking%20Everything%20from%20Forest%20Fires%20to,listen%20out%20for%20threatened%20animals.&text=You%20are%20in%20a%20lush%20forest

European Commission. (2023). *Biodiversity policies at the global level*. https://biodiversity.europa.eu/policy/biodiversity-policies-at-the-global-level

FEDS. (2024). Drone innovation for environmental conservation. Retrieved from https://feds.ae/industry/environment-and-ecology/

Fernandez, A., Li, Q., & Banerjee, D. (2024). Long-term trends in biodiversity under climate stress. *The Science of the Total Environment*, *902*, 167221. DOI: 10.1016/j.scitotenv.2023.167221

Hart, J. K., & Martinez, K. (2006). Environmental sensor networks: A revolution in the earth system science? *Earth-Science Reviews*, *78*(3-4), 177–191. DOI: 10.1016/j.earscirev.2006.05.001

Hart, J. K., & Martinez, K. (2006). Environmental sensor networks: A revolution in the earth system science? *Earth-Science Reviews*, *78*(3–4), 177–191. DOI: 10.1016/j.earscirev.2006.05.001

Hodgson, J. C., Baylis, S. M., Mott, R., Herrod, A., & Clarke, R. H. (2016). Precision wildlife monitoring using unmanned aerial vehicles. *Scientific Reports*, *6*(1), 22574. DOI: 10.1038/srep22574 PMID: 26986721

Kellenberger, B., Marcos, D., & Tuia, D. (2018). Detecting mammals in UAV images: Best practices to address a substantially imbalanced dataset with deep learning. *Remote Sensing of Environment*, *216*, 139–153. DOI: 10.1016/j.rse.2018.06.028

Koh, L. P., & Wich, S. A. (2012). Dawn of drone ecology: Low-cost autonomous aerial vehicles for conservation. *Tropical Conservation Science*, *5*(2), 121–132. DOI: 10.1177/194008291200500202

Kumar, R., & Mehta, A. (2024). Monitoring biodiversity and ecosystem resilience: Impacts of urban expansion. *Global Ecology and Conservation*, *42*, e02543. DOI: 10.1016/j.gecco.2024.e02543

MDPI. (2023). Advancements in air quality monitoring: The role of drone technology. *Proceedings*, *105*(1), 19.

Newbold, T. (2018). Future effects of climate and land-use change on terrestrial vertebrate community diversity under different scenarios. Proceedings of the Royal Society B: Biological Sciences, 285(1881), 20180792. https://doi.org/DOI: 10.1098/rspb.2018.0792

Nycz, B., & Pietrucha-Urbanik, K. (2024). Advancements in air quality monitoring: The role of drone technology. *Proceedings*, *105*(1), 19.

Pajares, G. (2015). Overview and current status of remote sensing applications based on unmanned aerial vehicles (UAVs). *Photogrammetric Engineering and Remote Sensing*, *81*(4), 281–329. DOI: 10.14358/PERS.81.4.281

Pettorelli, N., Laurance, W. F., O'Brien, T. G., Wegmann, M., Nagendra, H., & Turner, W. (2014). Satellite remote sensing for applied ecologists: Opportunities and challenges. *Journal of Applied Ecology*, *51*(4), 839–848. DOI: 10.1111/1365-2664.12261

Pichler, P., & Hartig, F. (2022). Deep learning as a tool for ecology and evolution. *Methods in Ecology and Evolution*, *13*(9), 1903–1920.

Pichler, P., Hartig, F., & Dormann, C. F. (2022). Machine learning and deep learning—A review for ecologists. *Methods in Ecology and Evolution*, *13*(7), 1640–1656.

Rammer, W., & Seidl, R. (2019). Harnessing deep learning in ecology: An example predicting bark beetle outbreaks. *Frontiers in Plant Science*, *10*, 1327. DOI: 10.3389/fpls.2019.01327 PMID: 31719829

Rault, T., Bouabdallah, A., & Challal, Y. (2014). Energy efficiency in wireless sensor networks: A top-down survey. *Computer Networks*, *67*, 104–122. DOI: 10.1016/j.comnet.2014.03.027

Rault, T., Bouabdallah, A., & Challal, Y. (2023). Advances and applications of machine learning and deep learning in environmental ecology and health: A review. *Environmental Pollution*, *306*, 119347.

Reiche, J., Lucas, R., Mitchell, A. L., Verbesselt, J., Hoekman, D., Haarpaintner, J., Kellndorfer, J. M., Rosenqvist, A., Lehmann, E. A., Woodcock, C. E., Seifert, F. M., & Herold, M. (2016). Combining satellite data for better tropical forest monitoring. *Nature Climate Change*, *6*(2), 120–122. DOI: 10.1038/nclimate2919

Sachs, J. D., Schmidt-Traub, G., & Williams, J. (2023). *Biodiversity and ecosystem services: Monitoring for sustainability*. Springer.

Secretariat of the Convention on Biological Diversity. (2022). *Biodiversity-related conventions*. https://www.cbd.int/brc

Singh, S. (2020). *Biodiversity hotspots in India and the challenges of conservation*. Oxford Research Encyclopedia of Environmental Science., DOI: 10.1093/acrefore/9780199389414.013.95

Turner, W. (2014). Sensing biodiversity. *Science*, *346*(6207), 301–302. DOI: 10.1126/science.1256014 PMID: 25324372

Turner, W., Spector, S., Gardiner, N., Fladeland, M., Sterling, E., & Steininger, M. (2003). Remote sensing for biodiversity science and conservation. *Trends in Ecology & Evolution, 18*(6), 306–314. DOI: 10.1016/S0169-5347(03)00070-3

UNESCO. (2023). World Network of Biosphere Reserves. Retrieved from https://www.unesco.org

Wäldchen, J., & Mäder, P. (2018). Plant species identification using computer vision techniques: A systematic literature review. *Archives of Computational Methods in Engineering, 25*(2), 507–543. DOI: 10.1007/s11831-016-9206-z PMID: 29962832

Wilson, R. J., Thomas, C. D., & Maclean, I. M. D. (2020). Ecological monitoring for biodiversity conservation. *Biological Conservation, 248*, 108658. DOI: 10.1016/j.biocon.2020.108658

World Health Organization. (2020). *Biodiversity and health.* https://www.who.int/news-room/fact-sheets/detail/biodiversity

Zenadrone. (2023). Drones for a greener future: Innovations in environmental monitoring. Retrieved from https://www.zenadrone.com/drones-for-a-greener-future-innovations-in-environmental-monitoring/

Chapter 9
Revolutionizing Sea Level Rise Monitoring in India Leveraging AGI-Powered Drones for Climate Adaptation

Gabriela Michael
https://orcid.org/0000-0002-8173-4692
Symbiosis Law School Nagpur, Symbiosis International University, India

Anurag Sarthi
https://orcid.org/0009-0002-0913-8789
Faculty of Law, Delhi University, India

ABSTRACT

The Indian subcontinental peninsula has a large coastline with land borders extending more than 15,000 km and a coastline stretching over 7000 km supporting coastal communities, diverse ecosystems, and several economic activities making it one of the countries highly susceptible to the detrimental effects of sea level rise making it the need of the hour for a formidable solution. This chapter will highlight how AGI powered drones monitors sea level rise in India. The current need is for the AGI drone system which uses cutting-edge computer vision and ML algorithms to collect high-resolution data on flooding, coastal erosion and sea level. Instant detections of rapid anomalies and trend analysis by using real-time data and pattern recognition, speeds up the response and gives time for mitigation plans for the coastal regions. The adaptability and autonomy of the drone system allow it to function in various coastal environments which would immensely help the Indian

DOI: 10.4018/979-8-3693-6457-4.ch009

subcontinent. The chapter will also include international best practices for monitoring and assessing sea level rise.

1. INTRODUCTION

Since 1880, the global mean rise in sea level is about 21-24 centimetres (8-9 inches) primarily due to the combination of melting waters from glaciers due to global warming and the thermal expansion of seawater. However, in 2022, the global mean sea level rise was 4 inches above the one recorded in 1993, making it the highest annual average, according to satellite records (Lindsey, 2022). Due to oceanographic dynamics and uneven distribution of gravitational field the sea level rise does not occur equally in all parts of the ocean and thus does not affect every country in the same manner (National Aeronautics and Space Administration, n.d.).

Rising sea levels directly impact coastal and island regions where a substantial percentage of the world's population lives (Anthoff et al. 2006). Even if greenhouse gas (GHG) emissions are reduced, the atmospheric concentrations of these gases stabilize, and sea levels are predicted to rise for centuries to come. Forecasting climate change and its effects in future requires an accurate estimation of historical sea-level rise and representation of global ocean heat uptake in climate models. The largest uncertainty in projections of sea-level rise up to 2100 is the uncertainty in global mean sea level (GMSL) and thus improving estimates of GMSL rise (as well as regional variations in sea level) remains a high priority (Lambeck et al. 2002).

Approximately 250 million people live in India's coastal areas, where they engage in fishing, tourism, maritime activities, and industry that make significant contributions to the national economy. Nevertheless, these areas that are densely populated and economically vibrant also happen to be quite susceptible to the effects of sea level rise. As a result, increased coastal erosion accompanying rising sea levels leads to higher amounts of floods and their intensities, intrusion into freshwater sources, and risks posed towards critical infrastructure, habitats, and livelihoods. To combat and address the complicated nature of vulnerabilities posed by the coastal regions, countries are trying to understand and develop effective and efficient mitigation strategies (Ruckelshaus, M., et al, 2020).

Utilizing AGI-powered drones in maritime surveillance will expedite and facilitate the monitoring of sea level rise on Indian coasts. AGI-powered drones have huge potential to harness all kinds of information including sea level rise with both aerial and undersea unmanned vehicles as they are well equipped with all kinds of sensors, cameras and other instruments. Such drones not only have uses in military surveillance but are equipped enough to expand beyond traditional use by monitoring marine environments. AGI-powered drones help in comprehensive monitoring as it uses advanced AI called AGI which has human-like intelligence and can collect real-time data and images for analysing the changes over time. These drones are extremely useful in forecasting and developing early warning systems for almost all types of disasters including floods, tsunamis etc. and can also be useful in identifying soil erosion, monitoring vegetation health, and assessing the shifts in shorelines while evaluating the maritime infrastructures.

Data collected and obtained on a regular basis could assist stakeholders make informed decisions about mitigation techniques, plan resource allocation, and manage coastal areas. The study conducted in the Netherlands demonstrated how drones equipped with advanced AI may be used to map and trace coastline erosion and the effects of increasing sea levels. These drones, which incorporate AI-driven algorithms, image recognition, and high-tech cameras, can identify and measure changes to the coastline over time. This essential data is then utilised to build policies and prioritise coastal management adaptation actions.

Drones with artificial general intelligence have the potential to address various maritime challenges associated with coastal areas, going beyond mere monitoring of sea level rise. It is able to identify the quality of ocean water for example it can provide data relating to dissolved oxygen, pH levels and nutrient concentrations. The obtained data is crucial for pinpointing and mitigating sources of pollution, thereby protecting the health of marine ecosystems. Besides that, AI-powered drones can be helpful in search and rescue, offering swift aerial assistance during disasters or emergencies occurring in coasts of India. However, there are various constraints like technological developments, moral issues, and other obstacles such as lack of regulatory frameworks and governance structures which need to be addressed in order to revolutionize the tracking and monitoring system to harness India's full potential. Collaborative efforts among all the stakeholders like researchers, scientists, local coastal communities, government authorities and policymakers becomes quintessential ensuring sustainable partnership based on mutual trust and understanding. Lastly, as AGI powered drones collect important information and data which are confidential as it can compromise India's national security, putting policies which can reduce the data breach becomes a priority.

2. WHAT ARE AGI POWERED DRONES?

Artificial General Intelligence (AGI) is a type of machine or software that can carry out intellectual tasks normally performed by humans, including knowledge acquisition, reasoning, abstract thought, common sense, transfer learning, and the capacity to distinguish between cause and effect. Since Alan Turing first presented the renowned Turing test in the 20th century, the idea of artificial general intelligence

(AGI) has been around. This test assesses whether a machine exhibits human-like intelligence if it can converse with a human without being recognized as a machine.

Unlike other drones, AGI-powered drones can perform tasks emulating humans and aim to replicate human cognitive abilities which allows them to tackle unfamiliar tasks, learn from new experiences, and apply that knowledge in creative ways. Enhanced flexibility and learning capacities are demonstrated by these drones that allows for higher learning capacities and gaining knowledge and skills from various sources, mastering new tasks with minimal supervision, and to apply comprehension to unfamiliar situations. These drones possess great levels of autonomy, enabling them to take decisions and problem-solving assessments in a wide range of scenarios.

Several initiatives, pilot projects, and case studies have been conducted worldwide demonstrating the significance and need for AGI-equipped drones to monitor and surveill the coastal regions for enhanced resilience.

As a Worldwide Heritage site the Sundarbans Delta exists as an intricate delta system of mangrove forests and estuarine waterways. Significant threatening forces such as tidal inundations along with climate change and anthropogenic activities affect this region (Ghosh et al., 2022). AGI drones have been employed for region-wide mapping of tidal inundation areas while performing condition assessments on the health of mangrove forests. Through advanced sensors along with imaging technology the drones efficiently obtain detailed environmental data across large inaccessible areas according to Sharma & Chatterjee (2023). Together academic institutions work alongside government bodies and local communities to carry out the deployment of AGI-enabled drone operations in Sundarbans. Through their collaborative effort the obtained data becomes both inclusive and practical. The use of drones enables the detection of areas with mangrove degradation so proper reforestation procedures can be implemented in a timely manner. The identification of areas prone to tidal inundations enables authorities to put in place specific protection measures that safeguard communities and the environment (Mitra et al., 2021).

The archipelagic country of Indonesia faces high vulnerability to sea-level rise and coastal erosion because of its vast coastline. The implementation of projects using drones with AGI technology enables coastal communities to evaluate their areas at risk due to sea-level changes. Drone-collected data when integrated with geographical information systems (GIS) and predictive models enables researchers to establish which areas face the highest risk of flooding (Rahman et al., 2022). Using AGI technology drones obtain the capacity to automatically review collected data while generating real-time choices through analysis. The capability enables officials to prepare disaster readiness by allowing quick deployment of evacuation plans and infrastructure strengthening (Siregar et al., 2023). With drone applications, data collection becomes semi-automated so human beings remain safe from dangerous environments while conducting regular detailed assessment operations.

3. TECHNOLOGIES USED BY AGI-POWERED DRONES FOR MONITORING THE SEA LEVEL RISE

The technological advancements in Machine Learning and Artificial intelligence have upgraded the performance of drones, particularly in the areas of monitoring and surveillance. Lately, the AGI powered drones have developed the potential to use various technologies to collect data with a level of accuracy very similar to humans which have been mentioned below:

3.1. Photogrammetry

Photogrammetry is yet another technological advancement used to monitor sea level rise. However, this phenomenon is nearly as old as photography itself. As stated by Wolf and Dewitt (2000) and McGlone (2004) photogrammetry is the *"art, science, and technology of obtaining reliable information about physical objects and the environment through processes of recording, measuring, and interpreting photographic images and patterns of recorded radiant electromagnetic energy and other phenomena"*.

The main goal of photogrammetry is to use bundle adjustment, a computer vision technique, to accurately measure three-dimensional objects and terrain characteristics from two-dimensional photographs. A drone is used in photogrammetry to take many high-resolution pictures across a region. The identical spot on the ground appears in multiple pictures taken from various angles because of this image overlap. This can be performed using AGI drones while capturing aerial images. (Aber, Marzolff, & Ries, 2010)

Using aerial triangulation techniques such as bundle-block adjustment, complete blocks of overlapping photos can be oriented simultaneously with relatively few ground control points when used with appropriate digital photogrammetry software. The produced orientated stereo models can be used to gather measurements and manually map 3D objects in a stereo viewing environment. Alternatively, automatic height point extraction using picture-matching algorithms enables for the creation of digital terrain representations such as raster or vector elevation models.

A case study conducted by researchers from the University of Exeter has combined drones and the technique of photogrammetry fixed with GPS satellite location to create a detailed digital model of the coastal habitat. This study was undertaken on the beaches of Alagadi in Cyprus which is a key nesting site for marine species known as loggerhead and green turtles. It was found that a sea level rise of 1.2 metres would result in the loss of up to 67% of loggerhead nesting sites and 59% of those used by green turtles (Varela et al., 2018). The lead author of the Centre of Ecology and Conservation on the University of Exeter's Penryn Campus in Cornwall,

Miguel Varela, has said "Being able to accurately map out terrains in such a fine and local scale is crucial if we want to develop effective conservation strategies. Most of the studies on climate change impacts are done on a global scale, but it is also crucial to understand what will happen in local areas. This method allows us to create realistic digital models that are highly accurate (less than 10cm error). This is a fine scale when compared to global projections. It's ideal for surveying coastal sites, but could also be applied to other habitats and species, and be used to help identify wildlife protection areas."

3.2. Digital Twin Technology

The concept of the digital twin was envisioned by and was first documented in 2002 (Grieves, 2019); it refers to a virtual model that accurately reflects a physical system or object. The digital twin technology is used for many purposes such as design optimization, quality control, monitoring, decision-making and policy formulation. The application of this technology can be achieved in the maritime field as well, where it is considered as a tool for understanding the difficult problems which are not easy to discover in reality.

Digital twin is a virtual representation or a prototype of an actual real-world physical product, process or service designed to provide real-time data which are sent from sensors. The concept of digital twin is not new and has been in use in various sectors such as the manufacturing industry, urban planning & construction industry, health industry, automotive industry etc., for simulation, integration, testing, monitoring, and maintenance. (Moi, Cibicik, & Rølvåg, 2020).

AGI-powered drones equipped with digital twin technology can be utilized to monitor and assess the sea level conditions. When sensors gather data from a connected item, the collected data can be utilised to promptly update a "digital twin" replica of the equipment's current condition. These drones utilize a model digital twin which can monitor, and provide diagnostics and prognostics to ensure enhanced performance and utilization. The integration of collected data, historical data, and human expertise in this domain would be capable of enhancing the accuracy of prognostics. The ability of these drones can also be used to ensure coastal restoration such as seagrass meadows and to monitor and model the effects of sea level rise as well as anthropogenic effects on coastal ecosystems.

This technology can also be used further to support coastal restoration (e.g. of seagrass meadows), and to monitor and model the climate (e.g. sea level rise) and anthropogenic effects on coastal environment. Nature-based solutions and collaborative efforts for coastline protection and preservation can be seen in several coastal areas such as the North-Atlantic, Wadden sea, Black Sea, Mediterranean sea, and Ghana's eastern coast. In this context, the digital twin can help in identifying the locations where the seagrasses are most vulnerable to sea level rise and provide multiple mitigation strategies to promote marine resilience (Staneva et al., 2024).

3.3. Bathymetric LiDAR (Light Detection and Ranging)

Through the utilization of laser light, the remote sensing technology called LiDAR (Light Detection and Ranging) is able to accurately measure distances and generate high-resolution three-dimensional models of both objects and environments. It does this by sending out laser pulses and then determining the time-of-flight and wavelength of the signals that are reflected back from the laser.

Bathymetric LiDAR is a specialized technology designed for measuring water depth and mapping underwater terrain. It utilizes a laser sensor that emits light pulses that penetrate the water and reflect on the seabed before returning to the sensor. The time delay between the emitted and returned pulses of light is used to calculate the water depth.

Using Bathymetric LiDAR for monitoring sea level rise is advantageous as it has the ability to accurately measure the depth of water with detailed information of the shape and contours of the seabed and efficiently measures large landscapes. It not only collects real-time data, but does it quickly, saving valuable time and resources, which further helps in prompt decision making. It has the ability to penetrate through the ocean surface, without any invasive or destructive techniques, ensuring protection of the marine environment. Even though such LiDAR has high take off weight and is large in size, it is still considered better when compared to the outcome of photogrammetric methods (Glira, Pfeifer, & Mandlburger, 2019).

The combination of bathymetric LiDAR with Unmanned Aerial Vehicle technology enhances coastal observation processes through sea level measurements and coastal resilience management operations. Current drone-based LiDAR technology exists in several coastal communities across the globe even though AGI advancement remains under development.

- **Coastal Mapping in the Philippines**

The Philippine coast underwent improvements in mapping with the use of UAV-based bathymetric LiDAR technology that brought TOPODRONE and RASA Surveying together in 2023. The AQUAMAPPER operated by TOPODRONE functions as a drone-mounted LiDAR system that gathers highly accurate elevation and depth data when used for coastal management and disaster readiness (TOPODRONE, 2023). The technology helps spot dangerous coastal areas whereas sustainable land-use planning helps reduce potential dangers of increasing sea levels.

- **An Investigation Tracks Coastal Erosion Patterns Throughout the Southern French Region**

LiDAR collected by UAVs successfully tracked shoreline erosion patterns according to research performed at the Mediterranean coast of southern France. Operators obtained detailed terrain models of a 900-meter area through the YellowScan Surveyor LiDAR installed on UAVs within short flight durations (GIM International, 2023). The high-resolution collected data enabled researchers to conduct morphological analyses for better understanding coastal transitions while assisting with extended environmental planning.

- **Coastal Embankment Mapping in Bangladesh**

The long coastal region of Bangladesh makes it highly exposed to elevated sea levels and storm surges. The SURF-IT project utilized drones with LiDAR sensors to achieve coastal embankment mapping because they needed to deal with these security threats. The project executed terrain data acquisition by DJI Matrice 350 RTK drones equipped with Zenmuse L2 LiDAR to identify weak embankments which need reinforcement prior to disaster events (CLARE Programme, 2023). The disaster preparedness strategy uses early warning systems along with a swift response capability to safeguard coastal population safety.

- **The South Carolina Coastal Area Includes an Aerial Mapping Operation Focused on Intertidal Estuaries in USA.**

Aerial LiDAR technology operated by UAVs tracks intertidal estuaries particularly the North Inlet estuary located in South Carolina within the United States. Drones operated by DJI Matrice 300 RTK provided precise elevation models for marshland areas using the ROCK Robotic R2A LiDAR system in 2022 (Tucker et al., 2023). Research teams utilize the collected data to determine how increasing sea level heights affect coastal habitats and direct protection and resilience development plans.

4. NEED FOR AGI-POWERED DRONES IN INDIA

The Indian peninsula, with a vast coastline and a rich marine life with millions of species of flora, fauna and other diversified ecosystems is being threatened due to anthropogenic activities in the region. Sea level rise modelling and prediction is complicated by several variables, including ice melt rate uncertainty and complex interactions between ocean-atmosphere-ice sheet. Sea level rise has various consequences, making it harder to anticipate its implications. However, the loss of marine habitats and ecosystem imbalance can be seen as one of the certain implications of sea level rise.

One of India's initial cases witnessed in Odisha showcases the managed retreat due to several decades of gradual changes in receding shoreline and coastal erosion that accelerated the forced migration and displacement in multiple villages around the Kendrapada district. When sea level rise and cyclones caused the Satabhaya villages to disappear one after another in the 1980s and 1990s, with the last village disappearing in 2011, the state government initiated the planned relocation process at that same time. The government built a resettlement colony 12km away in 2016, as the situation worsened. The state government's Revenue and Disaster Management Department announced a resettlement and rehabilitation policy in 2011, which resulted in the relocation of 571 families through the use of a relief package. The policy cited the "threat perception on account of their proximity to the sea." The plan took into account renovating 247 homes at the Bagapatia resettlement colony in addition to these 571 families. These families are considered as India's first "climate refugees", by few people and were provided agricultural plots, housing, and other facilities as compensation.

Similarly, the Sundarbans delta region of India, where 54 out of 104 islands are inhabited, has long been threatened by sea-level rise and coastal erosion. Data shows that the average annual sea-level rise in the Sundarbans is more than double the global average, at 8 millimetres compared to 3 millimetres. Additionally, the Indian part of the Sundarbans and its delta are sinking at 2 to 4 millimetres per year. Between 1973 and 2010, the Sundarbans' coastline lost 170 square kilometres of land due to climate change, rising sea levels, and other factors. Given the threats of sea-level rise and coastal flooding, it is projected that by 2050, 1 million people will need to be relocated from the most vulnerable regions of the Indian Sundarbans, necessitating extensive managed relocation efforts in the future. (Danda et al. 2019).

The growing concern of coastal erosion in India was officially recognized in 1977 when the Central Government identified the problem. Rising sea levels pose several hazards to the livelihood of people living in the coastal regions eventually leading to displacement. (Dhara, 2019). It is estimated that even a one-meter rise in sea level could approximately displace around 7.1 million people from the coastal regions and lead to the loss of around 5,764 square kilometres of land area, along with 4,200 kilometres of road networks (Sanyal & Shuvo, 2020). Several strides are being made in technological advancements to address and combat these cropping issues. States and organizations are continuously forecasting weather and environmental data from satellites, aircraft (including drones), ground/water-based sensors, and other instruments using technology such as artificial intelligence (AI), machine learning (ML), data visualization, the Internet of Things (IoT), and digital twins.

This can increase public awareness of climate change issues and give government officials and other stakeholders the information they need to make policies that are based on scientific facts to mitigate such threats.

4.1 The Islands of Andaman & Nicobar and Lakshwadeep

India's 2 island territories: the Andaman and Nicobar Islands and the Lakshadweep are the Union Territory administered by the Union Government of India. Lakshadweep (UTL) is a group of islands in the Laccadive Sea located off India's southwest coast. The Lakshadweep archipelago is home to the world's largest atoll and the most extensive coral reef in the Indian Ocean. Coral reefs, in addition to containing significant biodiversity and serving as a breeding ground for fisheries, act as a 'natural defence mechanism' against sea surges and storms. Several islands comprise the Union Territory of Lakshadweep and due to its low levels, it makes it very sensitive to the changes in the sea level. Both the islands of Lakshadweep and Andaman & Nicobar are of strategic importance for India both in terms of security concerns as well as economic gains. The Andaman Islands have been providing with meteorological data for the shipping industry in the Bay of Bengal highlighting the importance in the region since 1868. The islands are also home to lush vegetation and tropical flora and fauna, corals including several endangered species, and forests covering almost 80% of the land region.

The rise in the mean sea level would be devastating for an island with a diverse ecosystem, long coastlines and low-lying land. However, there is no uniformity in the rise of sea level and considering the geographical location of these islands, it is predicted to only worsen along with local climatic changes and disturbances. The Andamans have witnessed a decade-long of coastal development, sand mining, and reclamation of wetlands which has further increased the effects of sea level rise in the region (Anujan, n.d.). The effects of sea level rise on coastal countries is very different as compared to island regions since the entire island thrives on lush vegetation, sea level could swallow coastal property, coastal forests, and adjoining low-lying evergreen and moist deciduous forests which are at sea level which could threaten the very existence of the islands. Landmasses that are now connected by low-lying bridges would get disconnected which would impede the movement of small reptiles and amphibians. The 2004 tsunami was a landmark event that left lasting impacts on the Andaman archipelago in terms of human and habitat losses. The IPCC report also states that rising sea levels increase the probability of tsunamis (Anujan, n.d.).

Mangroves are considered as the lifeline of coastal regions, protecting it from the adverse changes in the tropical climate and the impact of sea level rise has particularly become worrisome for these ecosystems. It has been indicated by research that due to severe and frequent changes in climatic conditions, the mangroves are not able to keep pace leading to a loss of mangrove patches. The islands of Lakshadweep and Andaman and Nicobar Islands have fallen victim of these effects making it crucial to establish effective monitoring and adaptation strategies. Thus, drones that are equipped with AGI technology could have immense potential benefits in continuously monitoring and mapping the changes in the health of the mangrove forests enabling policymakers to make informed decisions while undertaking strategies for coastal management and resilience.

When AI technology such as remote sensing and geospatial tools are combined, it becomes effective and efficient to comprehensively understand the dynamic changes in the coastal environment of these islands which becomes imperative in identifying and implementing adaptive strategies for sustainability and leverage nature based solutions to protect the delicate and fragile diverse ecosystems.

4.2 Sir Creek

For several years India and Pakistan disputed control over the Sir Creek region. Although the disagreement over Sir Creek remains moderate in nature it becomes more intricate because the creek water has gradually shifted from its original location. The coastal boundary definition determines the maritime border and EEZ space claims between the two countries within their ocean region. Analyzing changing border placement two km from the coastline potentially results in losing several hundred square kilometers of EEZ space where gas and oil resources are anticipated (Rajen, 1999).

Figure 1. Proposed area for a cooperative coastal monitoring project (Rajen, 1999)

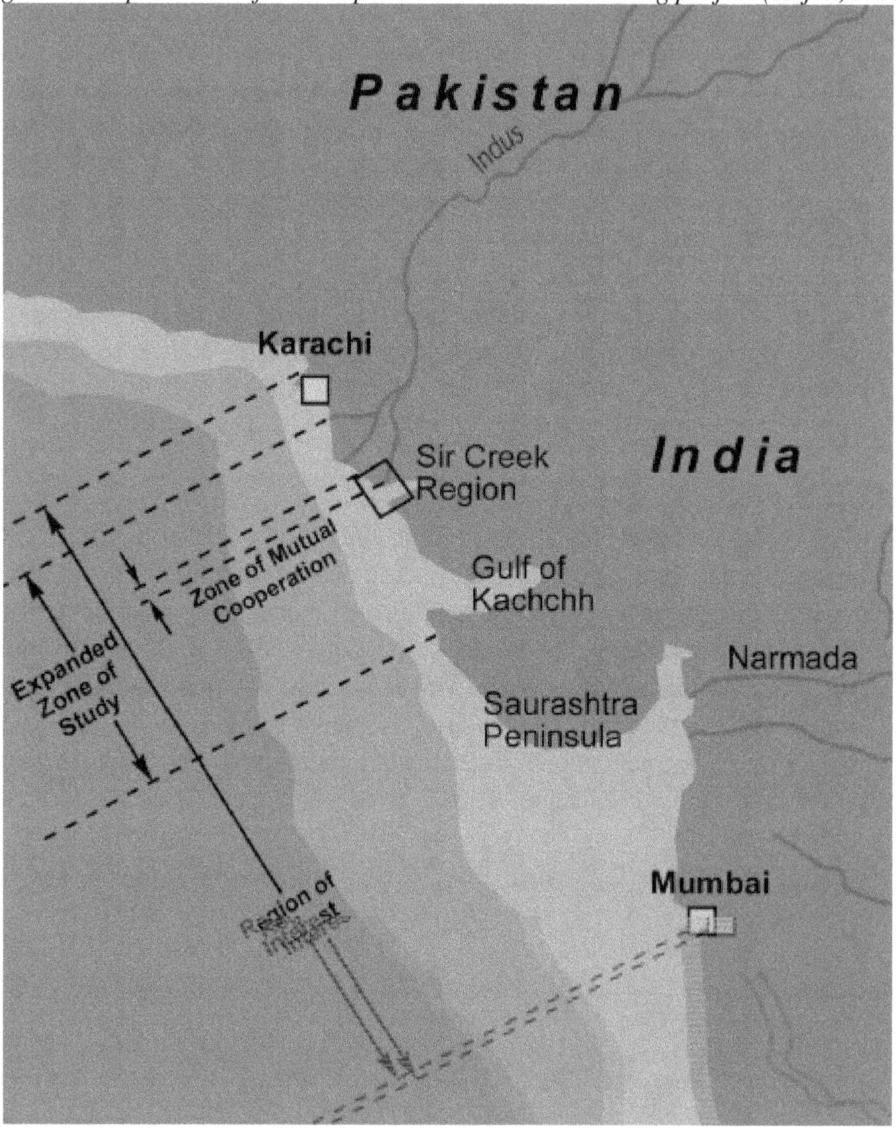

Research into sediment transport and water movement at the Indus Delta and Gulf of Kachchh demands both historical records analysis and sediment core assessment as well as variable data measurement which makes AGI-powered drones a revolutionary platform for exploration. Research on the natural development of this complex environment will reveal how human actions affect its evolution and also clarify how sediments and particles move across the area.

Successful implementation of monitoring experiments depends on the sensitivity of the data that researchers decide to gather. The bathymetric and ocean current and salinity information will enhance submarine navigation capabilities using drones operated by AGI technology. The development of efficient non-military monitoring technologies will enable India to verify the terms of agreements made with Pakistan. Terrestrial and maritime monitoring systems need documentation with verifiable evidence as it forms a critical element for building national agreements when ratifying agreements. The two countries should begin with joint environmental monitoring studies of their coastal regions while ignoring their current border conflicts because this approach will establish a deeper knowledge of the areas' essential ecosystems and their fundamental physical structures (Rajen, 1999).

4.3 Sundarbans Mangroves

The Sundarbans, which covers about 10,000 km^2, is the world's largest contiguous mangrove forest. The Sundarbans stretches across areas of India and Bangladesh while maintaining its position as the main ecosystem for the critically endangered Royal Bengal Tiger. It was estimated by a school of scientists that the Bangladesh Sundarbans would be completely inundated and lost under a 1.0 m sea level rise, while another school of scientists predicted that most parts would be inundated for a sea level rise of 0.28 m using 2000 as a base year. Mukul et al. (2019) [S.A. Mukul *et al.* Combined effects of climate change and sea-level rise project dramatic habitat loss of the globally endangered Bengal tiger in the Bangladesh Sundarbans] predicted that the entire Bangladesh Sundarbans will be submerged by 2070 due to the combined effect of climate change and sea-level rise. In contrast to these rather pessimistic findings, argued that the Sundarbans mangroves could survive a much more rapid Sea level rise scenario (1.48 m by 2100). Similarly, it was projected only around 10% loss in the mangrove area of Bangladesh Sundarbans by 2100 under the same 1.48 m Sea level rise scenario. The majority of Sea level rise impact assessments conducted to date have focused on the Bangladesh Sundarbans region while showing different prediction results. The diverse outcome of studies about future changes in the Sundarbans coupled with its worldwide significance emphasizes the necessity for additional research to understand the management implications of such transformations.

AGI-powered drones can be helpful for better analysis of real-time data. The AGI-powered drones could successfully be used for other such shoreline change analysis for similar difficult terrains of the World. The study recommends the employment of such surveillance for the monitoring of the shoreline experiencing frequent changes caused due to Climate change.

5. OTHER USES OF DRONES

There are several instances of drone technology being used in protecting and safeguarding marine life to enhance resilience. One such benefit of this technology was seen in research which was conducted in the US using mapping technology to protect one of the endangered whale species. Researchers benefit from using satellite ocean monitoring together with physical ocean models similar to the newest released model to improve their population estimates of whales throughout the East Coast of America. NOAA Fisheries as a part of the National Oceanic and Atmospheric Administration through its National Marine Fisheries Service branch uses this model to evaluate threats against large whale populations from activities including trap and pot fishing and vessel traffic and naval testing as well as offshore energy exploration. The feeding behaviour of the right whales regulates the health of ocean ecosystems together with all marine species in the food chain. The modification of whale migration patterns due to climate change reduces their prey population which leads to a higher risk for human activities to affect their reproductive capacity.

Another nature-based solution could be gathered from research conducted around Lord Howe Island, Australia, using object-based image analysis. Coral reefs and their associated marine communities are increasingly threatened by anthropogenic climate change. A key step in the management of climate threats is an efficient and accurate end-to-end system of coral monitoring that can be generally applied to shallow water reefs. The researchers used drone-based imagery and a deep learning algorithm to develop a system of classifying bleached and unbleached corals. The imagery was collected five times across one year, between November 2018 and November 2019, to assess coral bleaching and potential recovery. One of the scientists went on to explain "We take those two, the current temperature and the climatology, to generate products focused specifically on corals that allow us to predict when corals are likely to be bleaching,"

IUCN launched the PWFI project in the Caribbean, Mediterranean, and Oceania in 2019, with support from NORAD (the Norwegian Agency for Development Cooperation), with the goal of promoting the island circular economy and demonstrating effective, quantifiable solutions to addressing plastic leakage from Small Island Developing States. AnimaMundi created software that uses artificial intelligence, computer vision, and object identification to automatically record, count, and register the existence and amount of plastic bottles on our coastlines in real time. This data may be collected utilising an app that includes both still photography and drone imagery. The data is automatically uploaded, processed, and saved on the company's servers. This information will allow for results-driven, demonstrable influence on project investments.

Researchers employed unmanned aerial vehicles (UAVs) and thermal sensor technologies to develop a new survey approach for burrow-nesting seabirds that are active at night. Seabird surveys are used to monitor population demographics and distribution, as well as to better understand the anthropogenic stresses on seabird species. Burrow-nesting seabirds are difficult to observe. The present ground survey procedures cause damage to colony health and occupy considerable time while being destructively invasive. Short survey transects obtained from ground activities need extension to calculate entire population numbers but the process creates sampling biases because of spatial irregularities in burrow locations and land elevation differences.

The researchers inspected the short-tailed shearwater (Ardenna tenuirostris) colony within three hectares of land located in Tasmania Australia. Every occupied burrow within the colony elicited thermal data through this survey method. The research proved that UAV aerial thermal sensors integrated with automated count systems enable better estimation of seabird populations that live in burrows.

6. ADVANTAGES OF USING AGI-POWERED DRONES IN INDIA

Before moving to the advantages of AGI-powered drones, it is imperative to understand the base technology that is involved to distinguish it from the regular drones that are currently being used. The advanced technology heavily relied on machine learning and deep learning algorithms that helps in improving the accuracy of predictions and automation of navigating systems to avoid obstacles. Its pivotal role in visual and sensor data, enables it to allow for real time decision making. Precision in performance levels, detecting objects and advanced image processing combined together make it all the more advantageous for countries to adopt this technology for enhancing coastal resilience. AI can be powerful tool in climate change adaptation in the Indian coasts especially, by analyzing large amounts of data to identify vulnerabilities and develop targeted adaptation plans. One example could be the constant threat of erosions affecting the coastal regions which could be predicted and avoided with the help of analyzing satellite imagery to identify areas at risk of flooding or erosion due to sea-level rise. By incorporating and integrating machine learning algorithms to large datasets of satellite imagery, pattern could be identified and predictions about how various areas could be affected by sea level rise could be made which might be missed by human analysts.

Real-time intelligence collection coupled with situational awareness is the most powerful benefit that AGI-powered drones offer. These advanced capabilities of AGI drones would serve the Indian Navy and Coast Guard by creating more effective maritime security operations and Illegal, Unregulated and Unreported fishing barriers and disaster response capabilities. The current surveillance methods through traditional methods require costly resources while demanding manned flying vehicles combined with numerous human personnel according to Sharma (2021). Advanced sensor systems, high-definition cameras, and live video capabilities of AGI drones enable them to enhance current methods of maritime monitoring resulting in improved security proficiency.

AGI-powered drones possess Vertical Take-Off and Landing (VTOL) functionality that enables them to function efficiently in confined operational areas. The VTOL features of AGI drones allow them to take off and land from small decks so they suit deployment operations on naval vessels and offshore platforms as well as patrol boats. The Sir Creek maritime area between India and Pakistan needs maneuverable forces for quick deployments because these capabilities ensure territorial security in this disputed territory according to Chatterjee (2020). AGI drones cut down on maintenance costs by dispelling the requirement for extensive vessel navigation prior to starting patrol operations.

The monitoring of sea level changes represents a critical function for which Indian officials utilize AGI drones to uphold coastal resilience strategies. The combination of global warming and ice melting poses significant threats to Indian coastal areas through both coastal drowning and shoreline destruction. Accurate shoreline and tidal pattern and coastal flooding observations can be obtained frequently through AGI drones which utilize LiDAR technology and high-resolution imaging sensors. The immediate availability of this data enables policy-makers and environmental agencies and disaster management authorities to make adaptive strategies which reduce sea level rise impacts (Ramesh & Kumar, 2023). The integration of drone inspection data with satellite photos alongside oceanic analytic models enables India to better detect coastal hazard potential hence improving their planning of targeted protective measures supporting vulnerable areas.

AGI drones enable environmental surveillance as well as marine protection work. These drones serve a dual purpose to track marine pollution and observe illegal waste disposal and measure the environmental effects of human ocean activities. The Indian government seeks active promotion of blue economy programs and sustainable maritime practices through which AGI drones deliver critical environmental compliance features and ecological conservation capabilities (Indian Maritime Foundation, 2023).

The CoastalAI project is one such initiative undertaken in the US employed to asses the consequences of sea level rise on the coastal structure through the use of AI. These models can help recreate the possible effects by projecting the same with the data collected and combining the vast database with machine learning algorithms, thus forecasting the consequences and helping in chartering a mitigation strategy. This model could be used in India as well to ensure that the coastal populations are sufficiently aware about the impending effects of sea level rise in the region. Another example is the CoastalDEM project (Kulp & Strauss, 2018), which employs artificial intelligence to generate high-resolution elevation maps of coastal regions. The technology uses satellite images and machine learning algorithms to detect and rectify problems in elevation data, resulting in more accurate and comprehensive maps of coastal zones. This information may be used to identify locations at danger of flooding or erosion as sea levels rise, as well as to build targeted adaptation plans to safeguard vulnerable communities and infrastructure. Overall, using AI to analyse satellite data is a valuable tool for detecting places at risk of flooding or erosion as sea levels rise and generating effective adaptation methods to alleviate such risks.

Nevertheless, it is also important to note the challenges and limitations posed by such high-quality technology and the potential for bias and errors which could occur, specifically the possibility of malfunctioning and also the need for experts to constantly monitor and over see the use of the technology at every stage.

7. BEST PRACTICES

The inclusion of AGI-powered drones for monitoring sea level rise is not new as it is already in use by various nations globally in multiple ways that cannot be overlooked.

7.1 Ocean Digital Twin: Innovative Breakthrough by Japan

Japanese Fujitsu Limited empowered its research and development division by detailing technology which utilizes Light Detection And Ranging (LiDAR) with AI to obtain three-dimensional high-resolution biological and natural ocean data such as coral reefs and manmade underwater wind turbine infrastructure through autonomous underwater vehicles (AUV). This technology enables researchers to construct digital representations of underwater ecosystems for accurate environmental prediction as well as conservation effect simulation.

The technology applies a real-time scanning method originally developed by Fujitsu for the "Judging Support System" that the International Gymnastics Federation produced together with Fujitsu for gymnastics judging purposes to achieve AUV scanning despite adverse conditions from rough waves and currents. Through its AI image enhancement technology, Fujitsu achieves bright and clear image results to measure targets accurately down to several centimeters even when operating in turbid underwater conditions.

1) Image enhancement AI technology that restores the color and contour of underwater subjects

Fujitsu achieved the development of deep learning AI technology designed for underwater subjects which generates high-definition 3D images from underwater photos taken in murky environments with colour distortion and unclear object outlines. There are two types of AI that operate within this technology to first restore object colors and sharpen blurred object borders while improving images before conversion and a separate AI which eliminates water scum and enables contour recovery. The measurement of each object's shape takes place without subject recognition or 3D processing errors thanks to this approach.

2) Underwater 3D measurement technology that allows real-time 3D measurement from a moving AUV

A high-speed sampling technology is being used that combines high-speed scanning and short-cycle laser emission for its "Judging Support System," which was created in partnership with the International Gymnastics Federation to help with gymnastics judging, to measure in 3D in real time underwater. Moreover, Fujitsu unveiled an underwater LiDAR gadget that lets users choose from three laser wavelengths the ideal measurement wavelength based on the state of the sea. This will allow for the measurement of three dimensions from an autonomous unmanned submarine in motion as well as the development of technologies that will track the movement of objects and measure them.

7.2 Mozambique

Many developing nations in the Southwest Indian Ocean regions are susceptible to the changes in sea levels particularly due to the intense tropical cyclone activities especially from October to March each year. (GFDRR, 2019). Mozambique is one of the vulnerable coastal countries of Africa that is susceptible to such challenges and a recent study shows that 23% of the coastline has high or very high levels of exposure to coastal hazards and rising sea levels would worsen the already dynamic condition of the region. (Ballesteros & Esteves, 2021)

The usage of drones for mapping coastal topography and changes in sea level is currently being implemented in Mozambique, which is most prone to flooding. To provide high-accuracy three-dimensional maps of the shoreline, these drones make use of the Light Detection and Ranging (LiDAR) technology. This comprehensive data makes it possible to forecast future sea-level changes and identify regions that are most susceptible to floods. This provides policymakers with assistance in developing climate adaptation policies that are both effective and efficient. (africanDRONE Organization, 2023)

7.3 Tonga

A panel at the last UN Conference on Climate Change (COP28) discussed the benefits of utilizing AI to analyze and predict sea level rise in Tonga and simulate climate change scenarios and adaptation methods. Nvidia and RSS-Hydro collaborated on this demonstration. Local stakeholders including government officials, community leaders, and youth groups can use the app to brainstorm and explore solutions and outcomes.

Tonga island, like other low-lying nations, faces serious sea level rise. RSS-Hydro, the Commonwealth, and Nvidia created a 3D visualization of Tonga's climate change impact using Nvidia's Omniverse technology. The application's geospatial methods and models can inform on present and future sea level rise, climate change, natural disasters, and extreme weather events and their projected impacts on local communities, infrastructure, and biodiversity linked to habitat loss.

RSS-Hydro transposed current climate model projections onto land elevations to analyze and visualize the impact of permanently submerged Tongan lands under intermediate and high GHG emissions scenarios. Projection visualizations begin with the current sea level. They rise by 2030, the United Nation's Sustainable Development Goals (SDG) deadline, and by 2050, most net zero targets deadline. Finally, the visualizations illustrate sea level increase by 2100 and 2150, representing future generations. The tool may also assess the effects of possible loss on urban infrastructure, agriculture, and critical habitats including forests, wetlands, and mangroves using satellite imagery of land cover.

8. CONCLUSION & WAY FORWARD

This chapter is conceptual research that has explored the mechanisms for monitoring sea level rise using AGI-powered drones in order to revolutionize sea level monitoring for climate adaptation. On exploration, it is found that the inclusion of AGI-powered drones in maritime/coastal surveillance is extremely advantageous for India as compared to other monitoring mechanisms as it brings out a clear and nuanced understanding of the sea level rise. As AGI-powered drones are an amalgamation of Artificial Intelligence as well as human-like intelligence, the efficiency and accuracy of data collection increase manifold. For a developing country like India using AGI-powered drones is more beneficial as the Indian peninsular terrain is rough, difficult, and inaccessible and it does not have big pockets for such monitoring, as it still isn't the priority. The inclusion of AGI-powered drones will not only be helpful for cutting cost or reducing errors in sea level monitoring but would also boost the confidence. Developing such technologies will not only make India a pioneer in this field but would also enhance its soft power diplomacy besides selling such technologies globally.

India has already taken the first step by coming up with 'Sagar Manthan' which is the "Real-time Performance Monitoring Dashboard" of the Ministry of Ports, Shipping and Waterways, introduced in the year 2023. Its main characteristics include the representation of data in visual form, continuous monitoring of information in real-time, enhanced communication capabilities, making decisions based on data analysis, and a higher level of responsibility and transparency. In the future, this dashboard will be enhanced by integrating information from surveillance cameras, live streaming from drones, an Artificial Intelligence-based algorithm to map the actual progress, a digital twin feature on board, and a mobile application for simple access and utilization by all stakeholders, to increase efficiency. The introduction of the 'Sagar Manthan' Dashboard signifies progress in the digitalization and transparency of the maritime industry which is a welcome step in the correct direction.

REFERENCES

Aber, J. S., Marzolff, I., & Ries, J. B. (2010). Photogrammetry. In Aber, J. S., Marzolff, I., & Ries, J. B. (Eds.), *Small-format aerial photography* (pp. 23–39). Elsevier., DOI: 10.1016/B978-0-444-53260-2.10003-1

Anthoff, D., Nicholls, R. J., Tol, R. S. J., & Vafeidis, A. T. (2006). *Global and regional exposure to large rises in sea-level: A sensitivity analysis.* Tyndall Centre for Climate Change Research Working Paper 96.

Bogucki, R., Cygan, M., Khan, C B., Klimek, M., Milczek, J K., & Mucha, M. (2018, November 28). Applying deep learning to right whale photo identification. Wiley-Blackwell, 33(3), 676-684. DOI: 10.1111/cobi.13226

Boogaard, F., Lima, R. L. P. D., & Graaf, R. D. (2020, April 23). Innovative Water Quality and Ecology Monitoring Using Underwater Unmanned Vehicles: Field Applications, Challenges and Feedback from Water Managers. *Water (Basel)*, *12*(4), 1196–1196. DOI: 10.3390/w12041196

Chatterjee, S. (2020). *Maritime Security Challenges in India: The Role of Emerging Technologies*. Maritime Institute.

CLARE Programme. (2023). *From sky to safety: How Uttaran is introducing UAV technology in coastal embankment mapping*. Retrieved from https://clareprogramme.org/update/from-sky-to-safety-how-uttaran-is-introducing-uav-technology-in-coastal-embankment-mapping/

Danda, A. A., Ghosh, N., Bandyopadhyay, J., & Hazra, S. (2019). Managed retreat: Adaptation to climate change in the Sundarbans ecoregion in the Bengal Delta. *Journal of the Indian Ocean Region*, *15*(3), 317–335. DOI: 10.1080/19480881.2019.1652974

Dhara, C. (2019, March 21). West Bengal's climate change conundrum part III: Extraordinarily rapid sea-level rise in Sundarbans turns families into refugees. *Acclimatise News*.

Ghosh, A., Mukherjee, S., & Dutta, S. (2022). Tidal inundation and mangrove health assessment in the Sundarbans using remote sensing technologies. *Environmental Monitoring and Assessment*, *194*(7), 4682. DOI: 10.1007/s10661-022-09876-2

Giles, A., Ren, K., Davies, J., Abrego, D., & Kelaher, B. P. (2023, April 23). Combining Drones and Deep Learning to Automate Coral Reef Assessment with RGB Imagery. *Remote Sensing (Basel)*, *15*(9), 2238–2238. DOI: 10.3390/rs15092238

GIM International. (2023). *Monitoring coastal erosion with UAV LiDAR*. Retrieved from https://www.gim-international.com/content/article/monitoring-coastal-erosion-with-uav-lidar

Glira, P., Pfeifer, N., & Mandlburger, G. (2019). Hybrid orientation of airborne LiDAR point clouds and aerial images. *ISPRS Annals of the Photogrammetry, Remote Sensing and Spatial Information Sciences*, *IV-2*(W5), 567–574. DOI: 10.5194/isprs-annals-IV-2-W5-567-2019

Grieves, M. (2019). Virtually intelligent product systems: Digital and physical twins. In Flumerfelt, S., Schwartz, K. G., Mavris, D., & Briceno, S. (Eds.), *Complex systems engineering: Theory and practice* (pp. 175–200). American Institute of Aeronautics and Astronautics. DOI: 10.2514/5.9781624105654.0175.0200

Hensel, E., Wenclawski, S., & Layman, C. A. (2018, November 10). Using a small, consumer grade drone to identify and count marine megafauna in shallow habitats. Escuela de Ciencias del Mar, Facultad de Recursos Naturales. *Latin American Journal of Aquatic Research*, *46*(5), 1025–1033. DOI: 10.3856/vol46-issue5-fulltext-15

Indian Maritime Foundation. (2023). *Blue Economy and Environmental Sustainability: The Indian Perspective*. IM Foundation.

Kandrot, S., Hayes, S., & Holloway, P. (2021, October 18). Applications of Uncrewed Aerial Vehicles (UAV) Technology to Support Integrated Coastal Zone Management and the UN Sustainable Development Goals at the Coast. Springer Science+Business Media, 45(5), 1230-1249. DOI: 10.1007/s12237-021-01001-5

Kesavan, P C., & Swaminathan, M S. (2006, June 27). Managing extreme natural disasters in coastal areas. Royal Society, 364(1845), 2191-2216. DOI: 10.1098/rsta.2006.1822

Lambeck, K., Esat, T., & Potter, E. K. (2002). Links between climate and sea levels for the past three million years. *Nature*, *419*(6903), 199–206. DOI: 10.1038/nature01089 PMID: 12226674

Lindsey, R. (2022, April 19). Climate change: Global sea level. *Climate.gov*. Reviewed by Lumpkin, R., Johnson, G., Thompson, P., & Sweet, W. Retrieved from https://www.climate.gov/news-features/understanding-climate/climate-change-global-sea-level

López, J. J., & Mulero-Pázmány, M. (2019, January 9). Drones for Conservation in Protected Areas: Present and Future. *Drones (Basel)*, *3*(1), 10–10. DOI: 10.3390/drones3010010

Mitra, D., Bhattacharya, A., & Banerjee, K. (2021). Mapping coastal vulnerability in the Indian Sundarbans using geospatial techniques. *Journal of Coastal Research, 37*(2), 377–389. DOI: 10.2112/JCOASTRES-D-20-00085.1

Modern Technology For Enhancing Agricultural Productivity. (2020, August 29). https://practicalagrculture1214.blogspot.com/search/label/%23Agriculture

Moi, T., Cibicik, A., & Rølvåg, T. (2020). Digital twin based condition monitoring of a knuckle boom crane: An experimental study. *Engineering Failure Analysis, 112*, 104517. DOI: 10.1016/j.engfailanal.2020.104517

Myagmar-Ochir, Y., & Kim, W. (2023, August 23). A Survey of Video Surveillance Systems in Smart City. *Electronics (Basel), 12*(17), 3567–3567. DOI: 10.3390/electronics12173567

National Aeronautics and Space Administration. (n.d.). Are sea levels rising the same all over the world, as if we're filling a giant bathtub? *Sea Level Change: Observations from Space.* Retrieved from https://sealevel.nasa.gov/faq/9/are-sea-levels-rising-the-same-all-over-the-world-as-if-were-filling-a-giant-bathtub/

Rahman, H., Siregar, V. P., & Putra, R. S. (2022). Coastal inundation risk assessment in Indonesia using AGI-powered drones. *Marine Geodesy, 45*(1), 29–47. DOI: 10.1080/01490419.2022.2045678

Rajen, G. (1999). *Cooperative environmental monitoring in the coastal regions of India and Pakistan.* Cooperative Monitoring Center Occasional Paper/11, Sandia National Laboratories. Retrieved from http://www.cmc.sandia.gov/issues/papers/coastal/index.html

Ramesh, K., & Kumar, V. (2023). *Sea Level Rise Monitoring in India: The Role of UAVs and Remote Sensing Technologies.* Oceanographic Research Institute.

McClelland, G. T. W., Bond, A. L., Sardana, A., & Glass, T. (2016). Rapid population estimate of a surface-nesting seabird on a remote island using a low-cost unmanned aerial vehicle. *Marine Ornithology, 44*, 215–220.

Ruckelshaus, M., Reguero, B. G., Arkema, K., Compeán, R. G., Weekes, K., Bailey, A., & Silver, J. (2020). Harnessing new data technologies for nature-based solutions in assessing and managing risk in coastal zones. *International Journal of Disaster Risk Reduction, 51*, 101795. DOI: 10.1016/j.ijdrr.2020.101795

Ryan, M., Nuhoff-Isakhanyan, G., & Tekinerdoğan, B. (2023, January 30). An interdisciplinary approach to artificial intelligence in agriculture. Taylor & Francis, 95(1). DOI: 10.1080/27685241.2023.2168568

Samanta, S., Hazra, S., Mondal, P. P., Chanda, A., Giri, S., French, J. R., & Nicholls, R. J. (2021). Assessment and attribution of mangrove forest changes in the Indian Sundarbans from 2000 to 2020. *Remote Sensing (Basel), 13*(24), 4957. DOI: 10.3390/rs13244957

Sanyal, N., & Shuvo, S. R. S. (2020, October 29). Coastal Hazards in Asian Countries in the Context of Climate Change., 20. http://www.rericjournal.ait.ac.th/index.php/reric/article/download/2531/pdf

Sharma, R. (2021). *The Future of Maritime Surveillance: Integrating UAVs into India's Coastal Defense*. National Defense Review.

Sharma, R., & Chatterjee, A. (2023). Advancements in drone technology for environmental monitoring in India: A case study of the Sundarbans. *Sustainability, 15*(4), 2341. DOI: 10.3390/su15042341

Siregar, V. P., Rahman, H., & Lubis, R. F. (2023). AI-driven disaster management using drones: Case studies from Indonesia. *International Journal of Disaster Risk Reduction, 84*, 103456. DOI: 10.1016/j.ijdrr.2023.103456

Staneva, J., Pinardi, N., Coppini, G., Jacob, B., Chen, W., Jayson-Quashigah, P.-N., Alessandri, J., Mentaschi, L., & Drillet, Y. (2024). Advancing marine sustainability through digital twin what-if scenarios in nature-based solutions. *EGU General Assembly 2024*, Vienna, Austria, 14–19 Apr 2024, EGU24-14266. DOI: 10.5194/egusphere-egu24-14266

TOPODRONE. (2023). *TOPODRONE-RASA Surveying partnership augments coastal mapping, serves resilient communities*. LIDAR Magazine. Retrieved from https://lidarmag.com/2023/04/04/topodrone-rasa-surveying-partnership-augments-coastal-mapping-serves-resilient-communities/

Tucker, C., Smith, J., & Wilson, R. (2023). Mapping intertidal estuaries using UAV LiDAR: A case study from South Carolina. *Sustainability, 15*(22), 15823. https://www.mdpi.com/2071-1050/15/22/15823

Varela, M. R., Patrício, A. R., Anderson, K., Broderick, A. C., DeBell, L., Hawkes, L. A., Tilley, D., Snape, R. T. E., Westoby, M. J., & Godley, B. J. (2018). Assessing climate change associated sea-level rise impacts on sea turtle nesting beaches using drones, photogrammetry and a novel GPS system. *Global Change Biology*. Advance online publication. DOI: 10.1111/gcb.14526 PMID: 30430701

Young, J C., Arthur, R., Spruce, M., & Williams, H T P. (2022, May 1). Social sensing of flood impacts in India: A case study of Kerala 2018. Elsevier BV, 74, 102908-102908. https://doi.org/DOI: 10.1016/j.ijdrr.2022.102908

Chapter 10
Autonomous Eyes on the Earth:
AGI Drones for Climate Change Solutions

Muhammad Hammad Nawaz
https://orcid.org/0009-0002-0264-4152
University of Mianwali, Pakistan

Faisal Rehman
University of Mianwali, Pakistan

ABSTRACT

The purpose of this chapter is to review the dispersed uses of drone technologies within the built environment and as an instrument of climate change study. Unmanned aerial vehicles (or drones) have become very useful for environmental scientists to collect data, monitor, and analyse in the urban environment. The first section provides extensive details about drones operating. The systems showcase complex capabilities because they incorporate high-quality cameras together with LiDAR sensors and thermal imaging equipment. Modern tools developed for drones allow researchers to obtain precise measurements of space and environment information needed for urban climate analysis. Furthermore, the chapter investigates the integration of drone technology into environmental research and urban analysis. The paper describes essential applications of drone systems beginning with urban heat island mapping followed by building energy efficiency assessment and air quality management and finishing with greenhouse gas emission location work.

DOI: 10.4018/979-8-3693-6457-4.ch010

1.INTRODUCTION

Presented with the widespread use of drone technologies, our organization leads innovation in climate change analysis focusing on urban areas. Secure data collection about spatial and temporal built environment details through these advancements allows researchers to conduct extensive evaluations of climate change effects on urban areas. Present-day drones produce precise observational and cartographic information at multiple levels and viewpoints and satellite imagery delivers broad spatial data that was previously highly significant. Various high-resolution imaging equipment and LiDAR scanners and specialized instruments can be installed on drones, enabling advanced data collection and monitoring capabilities. To make exact and accurate 3D models that provide temperature measurements and quantity of global warming with high resolution. But drones are absolutely key to the work being done to prepare for and address the effects of climate change in rapidly urbanizing areas prone to being knocked out by urban heat island and infrastructure imperils (Anderson & Gaston, 2013). UAVs are being used in the built environment to research climate change, both to map the extent of urban heat islands, to understand the effects of buildings on energy efficiency. UAV's can take pictures of thermal leakage or atmospheric changes in a city with higher details. In addition, UAVs can perform routine inspections of building insulation and identify weak points in the building envelope, thereby supporting efforts to enhance energy efficiency. The high mobility of the drones allows for small and inexpensive sampling of air quality in different altitudes and in different places in a metropolitan area. Cities will need granular quantitative data like these to count the vulnerability of the climate, as well as to apply risk profiles on top of them. Given high density of buildings in dense urban area (Anderson & Gaston, 2013; Nex & Remondino, 2014). UAVs still have advantages over conventional UAVs and can routinely inspect thermally images greater success in climate change adaptation through green infrastructure occurs when urban areas improve their predictive climate models and scenario models. The integration of UAV-generated extensive data with artificial intelligence and machine learning techniques leads to this achievement. Neural networks enable researchers to discover patterns between drone sensor data and imagery in order to conduct simulations which evaluate different adaptation methods. The growing importance of UAVs in environmental studies makes them perfect for delivering detailed information needed by cities for defeating the worsening effects of climate change.

This chapter presents a review of the use of drones today in cities around the world for measuring and tackling issues associated with climate change. The chapter elaborates the kinds of sensing payloads and drones empowered with data analytics that resulted to the insights and present the current research which combines the use of UAVs in urban climate change strategies. the deployment of larger UAVs for environmental research is examined in the context of regulatory and ethical issues. This chapter aims to inspire further innovation in the application of UAVs and artificial intelligence (AI), emphasizing their role as a crucial factor in urban efforts to adapt to or mitigate the impacts of climate change.

1.1. Inspiration and Objective

Indeed, UAVs are starting to be recognized as a crucial element for minimizing completion time, enhancing performance, and improving flexibility across various tasks. This is because as climate change changes risk to the built environment and brings UAVs into such systems, they must be dealt with. The pertinent requirements must be met, but UAVs are to be cared against so as to be used for bolstering resilience. The use of UAVs demonstrates significant potential in accelerating structural assessments, emergency responses, damage evaluations, and mitigating climate-related risks. Their capabilities extend to evaluating the condition of buildings, transportation infrastructure, utilities, and communication networks to be higher in adaptability and durability to the increasing impacts of climate is due to the increased adaptability and durability of strategic agencies capability under the UAV coordination. With risks on the rise, UAVs can be fully exploited for protecting the built environment from climate risks, but further research and planning is needed. Since UAVs' application in a rapidly changing urban infrastructure can support these missions, but experiencing successes in doing so is difficult. For specific applications, (Pajares, 2015) UAV capability tailoring is necessary in order to address specific application problems. For example, we must prepare in advance the deployment of UAVs to survey damage to urban areas and aid in emergency response to a climate disaster, in such a way as to best realize the benefits of resilience. When combined with intensified climate change threats, further research on integrated UAV in built environment may enable them to interact and fulfil the assigned tasks on ground users. Since extremely direct application of AI enabled UAV is to disclose real time hazard assessment as well as to precisely monitor the constructed infrastructure and facilitate the coordination of emergency response following climate-related disasters there is a need for continued development of current UAV technology. However, to exploit all new UAV and AI potential, such UAV and AI capabilities should be properly integrated into existing systems.

1.2. Evolution and Innovations of UAVs

Over the past years, there have been great developments in UAVs, and they are intensively applied in a variety of areas. One section of the master thesis is devoted to the analysis of UAV technology, with an emphasis on its development and prevailing factors that attributed to approval. The advancements in UAV technology were fuelled by growing capability requirement in the field of military operations as well as new arenas of application in civilian areas like agriculture, environment monitoring and infrastructure management. Furthermore, UAVs also used the advanced technologies like the (AI), (IoT), and (ML) to increase abilities and possibilities.

1.3. Impact of Military Developments on UAV Technology

Even in the first two decades of the 20th century, when the attempt to perform unmanned flights started, technology of unmanned flights is already available. Beginning of the early UAV technology has turned out to be applicable to civilian applications like agriculture and aerial photography. In the development of UAVs, development of miniaturization, sensor development, as well as communication system development were important. UAVs were originally developed for military applications solely for the sake of lieu of human lives but UAVs did that for surveillance and reconnaissance. As time passes, the missions that were considered too complex became taken up by UAVs that had progressed and became more sophisticated (Pajares, 2015; Tmušić et al., 2020). The development of UAV technology has been boosted by the military, which has really had a good kick off into this technology. From the early stages, the predominant deployment of UAVs was for military surveillance and survey. Research and development were funded in the large investments with autonomous systems need these operations at the same time as material advancements and propulsion system upgrades and avionic system capabilities and payload abilities because they are becoming essential. On the other hand, these technologies also have aided in miniaturization of UAVs, becoming more agile as well as versatile for use in multiple environments.

The use of UAVs in the military field is one of the greatest advantages of using such drones, due to the capacity to mitigate potential dangers posed to military operatives. UAV's can be employed for reconnaissance missions' way ahead by the military forces without placing the lives of the soldiers at risk to obtain important intelligence. Therefore, military operations can be ethical and legal, as well as more accurate for targeting activity and less proportional for collateral damage.

1.4. Drones and Their Civilian Uses

UAV's have been widely used in a lot of ways in civilian sectors and they indeed have been a revolutionary device in many industries, and have answers to many problems. UAVs have also been valuable tools for use in environmental monitoring and research. What they allow is to collect very high-resolution information of ecosystems, biodiversity, and environmental changes. An example of uses of such

UAVs is the employment to monitor the vegetative health, making assessments on land cover changes and mapping the natural habit. Additionally, UAVs have been used to monitor wildlife, such that views on animal behaviour, population dynamics and habitat assessment are gained. states that the study was based on the great advancement that unmanned aerial vehicles (UAVs) have realized in environmental research, the benefits of modern techniques outweigh those of traditional data collection methods.

The work was for monitoring wildlife population where there is a need for accurate and precise population count. This study proves that UAVs can be used practically in wildlife population monitoring. Improving accuracy is just one of the applications that this technology has potential to enhance the precision with which we can estimate populations, but it is also able to help us better understand the dynamics of ecological processes more generally, and, thereby, to be better able to conserve them. UAVs are being used in environmental research and will at least provide an improved understanding, but better ecosystem management.

1.5. Cutting-Edge Technologies and UAV Innovations

Various UAVs exist at different sizes and designs with varying capabilities however, UAV platforms equip multiple sensor components alongside independent navigation systems which link to remote networks and share flight control systems for automated flight operations and data retrieval. UAV technology has progressed because of important technological developments and modifications including high prices and reduced flight duration and diminished payload capacity. The technical advancements allowed both improvement of battery technology and production of light-weight materials and integration of satellite navigation systems and boosted analytical computation capabilities in real time. UAVs at consumer price points now exist in the market and this affordability results in more use scenarios which allows people to develop bespoke airborne environmental sensor systems. The lack of bounds on the electrical charge has become a hobbling force behind the push for drone adoption: the electrical charge is required for rotors, sensors, and navigation systems and also remains lightweight. Additionally, they offer range storage to payload ratios needed for sustained flight time in excess of 30 min for several types of UAV. At the same time charging technologies have also evolved to the point where we can indeed recharge very rapidly between missions. The creation of strong yet light-weight composite materials exists because of material development advances that produced carbon-fiber which led to drone airframe and component manufacture. Microprocessor and sensor enhancements allow drones to operate effectively in demanding conditions since they can tolerate environmental elements such as temperature changes and humidity and mechanical vibrations.

1.6. (UAVs) and Advanced Data Capabilities

By most accounts, climate change is heavily impacted by the urban environment and it in turn, is heavily impacted by climate change. To develop such strategies, it is important to understand the climate change in urban areas (Villa et al., 2016; Yigitcanlar et al., 2020). The research and understanding of the relationships between the urbanization, climate and human activities is initiated with the urban environment data collection, monitoring and analysis related to the climate change. Satellite imagery has a long history as one of a very few important means to monitor climate change in an urban context. However, satellite imagery has a limited advance and it is using existing gaps in satellite imagery valid monitoring of dynamic urban environments as well as the delivery of detailed fine-scale information requires assessing and monitoring climate-related phenomena in urban areas becomes difficult because spatial resolution, temporal coverage and cloud cover interference factor into the process. The factors create difficulties when scientists attempt to collect exact data measurements. The major UAV benefit emerges from their capability to generate detailed imagery which achieves broader inspection and precise evaluation results. The scale of detail which drones and other systems provide becomes clear when observing Figure 1 for easy recognition of UAV management capabilities. The technology of unmanned aerial vehicles also serves different purposes in the agricultural domain. Satellite imagery provides worldwide perspective which analyses large areas that exceed kilo-meters in distance. UAVs possess the ability to fly extended distances while achieving imaging precision at or below 1 cm and reaching beyond several kilo-meters in range. UAVs precisely recognize and identify numerous terrain characteristics during their extensive range of several kilo-meters. The capturing capabilities of UAVs surpass traditional geomatics techniques for high-level three-dimensional data collection of study sites illustrated in Figure 1.

Equip LiDAR technology enables UAVs to exceed TLS systems by allowing surveys of tough terrain areas along with tall structures. There are two major differences between the operations of TLS and UAV systems; TLS needs static scanning positions and generates occlusion gaps yet UAVs provide superior data acquisition through multiple flight paths. In addition, UAV LiDAR also captures obscured geometries that cannot be seen from the ground, for example, rooftops. Concurrently, UAVs have lower altitude measurement resolution compared to crewed aircraft LiDAR throughout the 0–20 m range, with resolution as low as centimetres, and up to 80% cost saving.

Figure 1. (a) UAVs and different geomatics technologies; (b) UAV coverage scale compared to other data collection technologies (Piekkoontod et al., 2020; Zohdi, 2018)

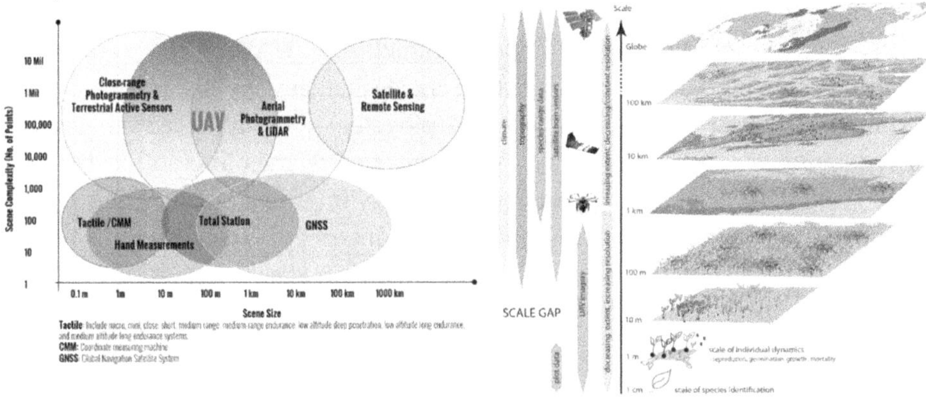

(a) (b)

UAVs typically operate at altitudes below 7 kilometers, offering imagery resolution at the centimeter scale over areas spanning several square kilometers. This makes UAVs particularly useful for capturing high-resolution data over relatively small to medium-sized regions, ideal for tasks like mapping, surveillance, and environmental monitoring. Their growing popularity can be attributed to their flexibility and relatively low cost compared to other aerial platforms. UAVs are capable of performing (Piekkoontod et al., 2020; Zohdi, 2018) a variety of missions, ranging from short-range inspections to large-area imaging, making them versatile tools in many industries.

However, UAVs have notable limitations, particularly in terms of flight endurance. With battery life typically lasting no more than 30 minutes, they are suited for relatively short-duration flights. This short flight time can be restrictive, especially for applications requiring extended operation periods or covering larger areas. In contrast, manned helicopters, particularly larger models, offer much greater flight endurance and can carry heavy payloads, making them more suited for long-duration flights or tasks requiring substantial sensor equipment. Helicopters are able to remain airborne for several hours, which is ideal for missions like continuous surveillance or data collection over extended periods.

Moreover, helicopters, while offering greater endurance and payload capacity, are typically more expensive to operate than UAVs. However (Piekkoontod et al., 2020; Zohdi, 2018) they provide significant advantages in terms of the ability to carry larger and heavier sensors, making them suitable for more complex operations. Despite this, there are fewer economical helicopter options available that can offer both extended flight times and the capacity to carry large payloads at a cost-effective price. In comparison, airplanes—especially fixed-wing UAVs—can provide extended flight times and have the added advantage of being able to hover and operate at low altitudes, though they still fall short in terms of maneuverability and close-proximity inspections compared to multirotor UAVs.

Ultimately, while UAVs are an appealing choice for many applications due to their flexibility, cost-effectiveness, and ease of deployment, they still face limitations in flight duration and payload capacity when compared to larger aircraft like helicopters or fixed-wing airplanes. As battery technology improves and new innovations emerge, UAVs may become more capable of addressing these constraints, but for now, they are best suited for shorter missions or areas where high resolution and flexibility are more critical than long flight times.

Figure 2. Comparison between UAVs and conventional aerial technologies regarding resolution, payload capacity, and operation costs. Based on (Colomina & Molina, 2014).

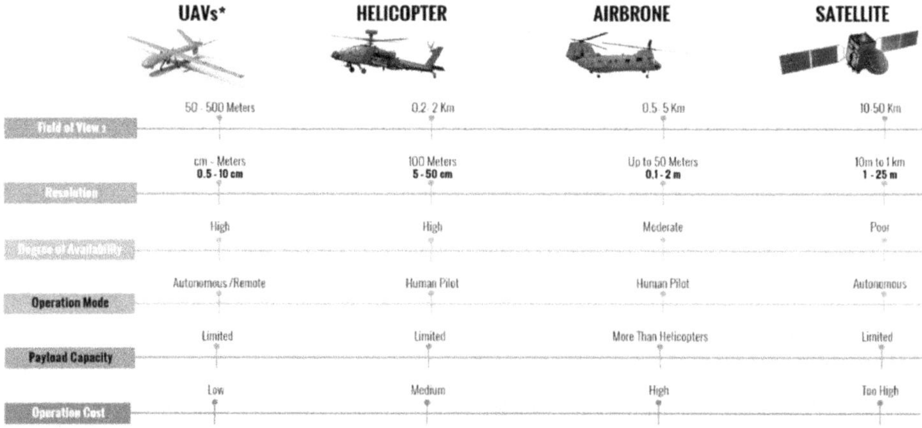

Satellites achieve complete data mapping but UAVs specifically target resolution capabilities to capture imagery at levels higher than what most commercial satellites can achieve. The assessed imagery enables computer vision technology to produce 3D point clouds that achieve accuracy measurements down to centimeter precision. UAVs function for real-time monitoring combined with environmental analysis in urban areas. UAVs accept the attachment of different sensors which allows them to collect real-time climate-related data by measuring temperature and humidity while also monitoring air quality and solar radiation. Through this kind of real time monitoring, there is near to real time feedback and decisions can be made in a timely manner in response to climate related events and aid in the process of urban planning and management (Qi et al., 2021; Ventura et al., 2023). UAV platforms are flexible, automated and miniaturized that enable nimble, customizable 3D data capture for project sites. To achieve comparable perspectives and positional flexibility. The efficient capture of real time, close ranges 3D spatial data through UAVs makes this yet unique and valuable geomatics tool, a uniquely valuable tool in support of 3D spatial data acquisition, with high quality and high scalability.

2. UAV PLATFORM EVOLUTION

During operation in urban areas, UAVs have been employed to provide localized data collection of multiple types of aircraft equipment serve the purpose of urban climate research. Multiple UAV platforms with their distinctive configurations serve research needs in climate change investigation and environmental assessment of urban areas. Before UAV technology gained popularity throughout the last few years several distinct UAV platforms started to emerge and their capabilities began to specialize in urban environment data gathering and built infrastructure examination and site monitoring operations. The following segment presents an overview of multirotor drones together with fixed-wing drones and hybrid VTOL drones along with their properties, benefits, constraints, and suitable applications for climate research in dense urban areas.

2.1. UAV Platforms Categorized by Aerodynamic Characteristics

The use of the UAV for many roles and applications and the diversity of aerodynamic characteristics make them the most diverse aerodynamic vehicles. Perhaps the most numerous drones to create a revolution in the world, as they are aptly termed as the most manoeuvrable and the drone of choice the vehicles possess the ability to perform take offs and landings together with hovering capabilities inside urban zones. The vertical thrust needed for hovering and low-altitude flight can be produced through the four lifting rotors positioned horizontally on quadcopters (Naqvi et al., 2020). Unlike helicopters and their main rotors, which have only a single lift source, quadopters consist of four more or less small (mean smaller than main rotors but greater payload for given size and redundancy and safeness), less polluted rotors. In 3D space, orimance allows the speed of each rotor to be adjusted to stabilize omnidirectionally, and also to tight urban areas inaccessible by fixed wing

UAVs. A strict summarization of different UAV platforms in relation to aerodynamic characteristics, application, technical limitations is carried out.

UAVs with ducted fans merge characteristics between multirotor and fixed-wing platforms through ducted rotor system integration. The safety precautions implemented by these UAVs depend on circular duct-controlled fans that provide vertical lift functions. The ducting system directs the airflow to improve rotor efficiency which produces better results than open rotor configurations. A ducted fan UAV system can sustain flight operations between 45 and 60 minutes thus outperforming quadcopters in flight duration. Ducted fan UAVs often come with fixed wings for flight propulsion just like conventional UAVs as per references Minkina (2020) and Jiang et al. (2021). UAVs with ducted fans present flexible flight operations that combine hovering with high-speed and low-speed flight modes thus permitting efficient area scanning over large areas at quadcopter hover speeds. The contraposition to fixed-wing UAV airframes shows the construction of their streamlined structures that use forward airspeed to produce lift exactly like regular airplanes. The aerodynamic shape that uses non-rotating propulsion enables fixed-wing drones to stay airborne for longer periods and travel farther distances compared to multirotor drones which allows them to perform wide-area mapping operations (Naqvi et al., 2020). Fixed-wing (Nawaz et al., 2021) UAVs possess limitations because they cannot perform vertical takeoff or landing actions. Aircraft launch or catapult with human assistance powers initial flight until touch down occurs through skids or by landing on the belly. Fixed-wing UAVs possess restricted capabilities when it comes to inspecting and scanning local urban features because hovering and close proximity operations remain essential. The inability to maneuver as well as multirotor UAVs constitutes one disadvantage of fixed-wing UAVs. Fixed-wing UAVs function excellently for acquiring aerial images covering extensive urban regions notwithstanding their technical restrictions. The current advancements in battery performance enable UAV aircraft to operate for 30 to 60 minutes across different UAV types. Books and other materials are now distributed autonomously because satellite navigation systems brought improvements to stability control. Fixed-wing UAVs require human assistance for launch, typically using a catapult or manual throw to initiate flight. Once airborne, these UAVs rely on forward momentum to maintain lift, and they typically land either by skidding or by belly landing. This method of launch and landing is a significant constraint for fixed-wing UAVs, as they cannot take off, land vertically, or hover like multirotor UAVs. These limitations make fixed-wing UAVs less suitable for tasks requiring precise, close-proximity inspections or detailed scanning of localized urban structures, where hovering is essential. Despite these constraints, fixed-wing UAVs excel in applications that require extensive coverage of large urban areas. Their aerodynamic design, which relies on forward flight to generate lift, allows them to cover wide areas efficiently, making them ideal for tasks like large-scale aerial imaging or mapping of urban environments. However, their inability to hover or maneuver as freely as multirotor UAVs limits their suitability for detailed or localized inspections, where closer proximity and the ability to stay stationary in the air are critical.

Recent advancements in battery technology have extended the flight times of UAVs across various types, with current capabilities allowing for flight durations ranging from 30 to 60 minutes, depending on the specific UAV model and operational conditions. These advancements have made UAVs more versatile and capable of performing longer missions, increasing their potential for use in a wide range of applications.

Additionally, the integration of satellite navigation systems has greatly improved the stability and control of UAVs. These systems enable more accurate navigation and better autonomous operation, which has enhanced the UAVs' ability to perform complex tasks without direct human intervention. This development has also facilitated autonomous operations, such as the distribution of materials, allowing UAVs to deliver goods or collect data with minimal human involvement.

Figure 3. UAV classifications based on their aerodynamic features, level of autonomy, sensors, and power sources. Based on (Zohdi, 2018).

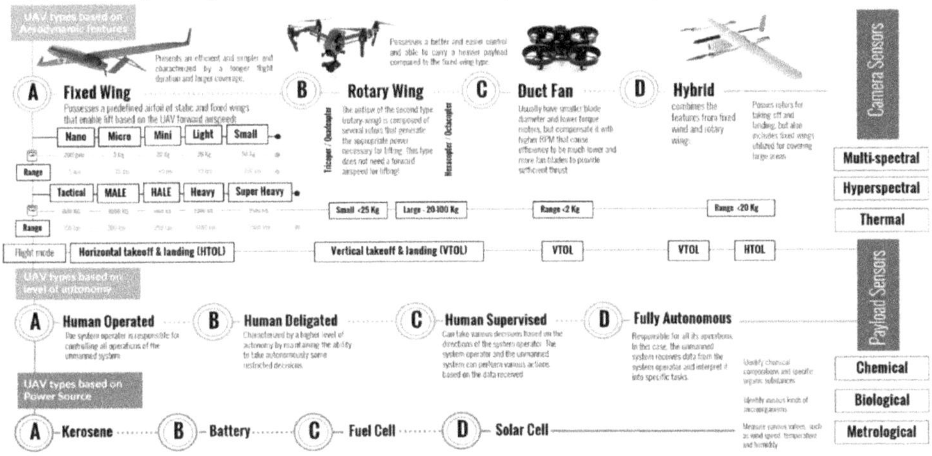

Modern UAV technologies can fly between 30–60 minutes while using various battery types which make these features applicable in present times. The optimal mix for prolonged UAV flight times comes from lithium-ion and lithium-polymer battery systems. The combination of satellite navigation systems allows autonomous control and stability through which airframe design and avionic and battery technology improvements increase UAV capabilities for resilient climate research in urban environments affected by turbulence. UAV designs adopt existing cost-performance levels from conventional aircraft models while providing customizable sensor payload systems which enable the collection of multiple climate-related data. With 3-hour flight durations visible spectrum cameras can produce detailed 2D and 3D urban mapping data of infrastructures alongside vegetation and morphologies. Thermal infrared sensors help detect microclimate patterns in addition to their ability to locate heat density zones and evaluate building wall insulation effectiveness. Miniaturized gas sensors have the capability to identify pollutants including carbon dioxide, particulate matter together with volatile organic compounds such as benzene. The technology of LiDAR scanners allows users to obtain solid 3D point cloud data for buildings and terrain alongside other structures. The exchange of sensors between missions remains possible and UAVs undergo aerodynamic enhancements to full-fill particular data acquisition requirements. The combination of onboard processing along with AI capabilities enables UAVs to conduct real-time analytics which makes sensors and data processing systems work efficiently while in flight.

2.2. UAV Sensors

The ability to deploy advanced active remote sensing of the evolving and complex knowledge requires to climate change in urban environments through previously unattempt specialized airborne data capture has been demonstrated to be more effective than the conventional way. At present, the multiple elements shaping the response of urban built environments to natural disasters while affecting their levels of sustainability and gas emissions are currently presented in terms of the diversity in UAV payload. This section introduces relevant UAV sensor types for climate research and outlines the importance of UAV sensors thereof.

A. HRVS Imaging Cameras

Extremely high-resolution-visible-spectrum (cameras used to generate highly detailed aerial photos covering urban metric has become very common on the UAV payloads). Low UAV altitudes can be used to miniaturized cameras with CMOS sensors to resolve in the sub centimeters per pixel range in still images or video. It enables the 3D modelling and digital generation of the precision of urban form factors that correlate to a climate effect such as heat retention have, for instance, correlated the urban geometry and the heat island intensity using 2 cm resolution land surface temperature variations derived from Ortho-mosaics were used using UAV derived ortho-mosaics. By remapping often, the system captures slow transformations in city and countryside vegetation areas as well as land usage patterns and physical urban planning which are needed to create adaptation solutions. Now existing with many leading commercial UAV platforms in the form of remote sensing and mapping or imaging applications suitable visual band cameras with high resolution. The most widespread and common platform of UAVs in the commercial market is DJI because their technical capabilities and the variety available on this platform. Camera line and these include payloads that are small down to large formats to suit resolution requirements. However, the UAV cameras most commonly use nadir perspectives, but obliques can supplement them to get more immersive urban structural data. We then compare the most common commercial UAVs in Fig 4.

Figure 4. Comparison between common commercial UAV platforms and their high-resolution sensor capabilities (Radmanesh et al., 2015; Sarim et al., 2019; Manfreda et al., 2018)

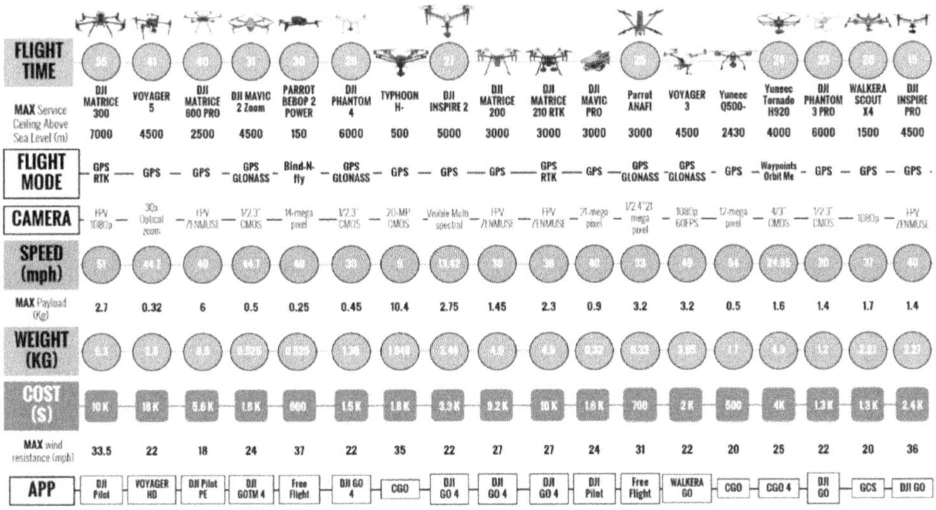

B. Multispectral Sensors

Collecting distinct informational data regarding vegetative health, water stress, biomass, and material characteristics that is not available from conventional RGB cameras. Because many UHI mitigation strategies entail urban greenery expansion, multispectral UAV data can record growth, canopy size and condition of these heat regulating plants over time. Near-infrared bands also allow easy urban ground surface and materials surfaces to be differentiated that are important for the urban energy model. In figure 5 multispectral sensors that are UAV manufacturers compare their products by evaluating resolution quality along with price and system compatibility. A wide range of cities have undergone climate vulnerability assessments through the use of multispectral UAVs. Climate change research benefits strongly from the multispectral sensor integration on UAVs because it helps track urban vegetation health alongside its evolutionary patterns. Urban vegetation evaluations benefit from these sensors which measure chlorophyll content together with leaf area index and vegetation stress information for providing valuable insights about vegetation health. Green spaces are able to cool, quantifying this is enabled by combined thermal and visible imagery. On the other hand, UAV spectral data are also used to build 3D urban models based on the rooftop materials with intent to target solar panel deployment.

Figure 5. Comparison between standard multispectral UAV sensors, resolution, costs, and compatibility (Merkert & Bushell, 2020; Bansod et al., 2017).

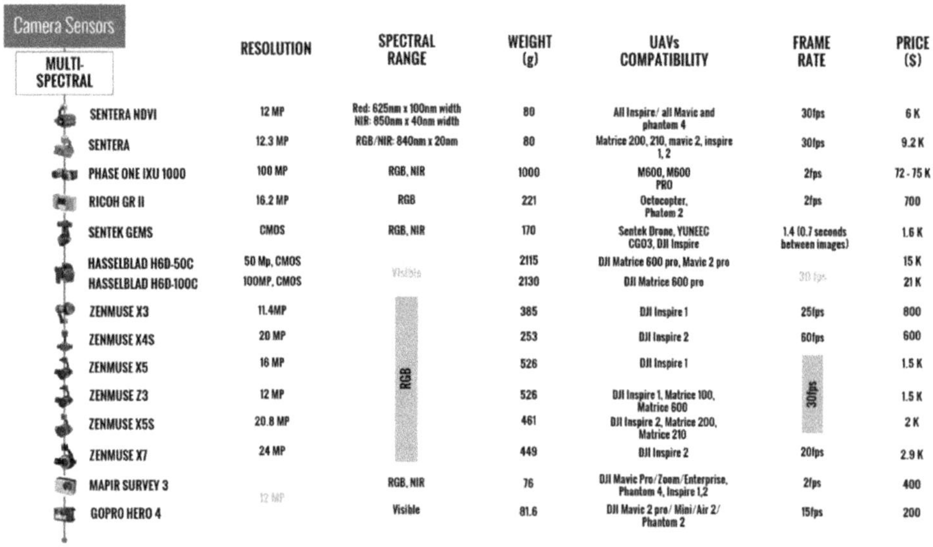

Camera Sensors – MULTI-SPECTRAL	RESOLUTION	SPECTRAL RANGE	WEIGHT (g)	UAVs COMPATIBILITY	FRAME RATE	PRICE ($)
SENTERA NDVI	12 MP	Red: 625nm x 100nm width NIR: 850nm x 40nm width	80	All Inspire/ all Mavic and phantom 4	30fps	6 K
SENTERA	12.3 MP	RGB/NIR: 840nm x 20nm	80	Matrice 200, 210, mavic 2, inspire 1, 2	30fps	9.2 K
PHASE ONE IXU 1000	100 MP	RGB, NIR	1000	M600, M600 PRO	2fps	72 - 75 K
RICOH GR II	16.2 MP	RGB	221	Octocopter, Phatom 2	2fps	700
SENTEK GEMS	CMOS	RGB, NIR	170	Sentek Drone, YUNEEC CGO3, DJI Inspire	1.4 (0.7 seconds between images)	1.6 K
HASSELBLAD H6D-50C	50 Mp, CMOS	Visible	2115	DJI Matrice 600 pro, Mavic 2 pro	30 fps	15 K
HASSELBLAD H6D-100C	100MP, CMOS		2130	DJI Matrice 600 pro		21 K
ZENMUSE X3	11.4MP	RGB	385	DJI Inspire 1	25fps	800
ZENMUSE X4S	20 MP		253	DJI Inspire 2	60fps	600
ZENMUSE X5	16 MP		526	DJI Inspire 1	30fps	1.5 K
ZENMUSE Z3	12 MP		526	DJI Inspire 1, Matrice 100, Matrice 600		1.5 K
ZENMUSE X5S	20.8 MP		461	DJI Inspire 2, Matrice 200, Matrice 210		2 K
ZENMUSE X7	24 MP		449	DJI Inspire 2	20fps	2.9 K
MAPIR SURVEY 3	12 MP	RGB, NIR	76	DJI Mavic Pro/Zoom/Enterprise, Phantom 4, Inspire 1,2	2fps	400
GOPRO HERO 4		Visible	81.6	DJI Mavic 2 pro/ Mini/Air 2/ Phantom 2	15fps	200

C. Hyperspectral Sensors

UAVs have hyperspectral sensors that are able to acquire hundreds of contiguous spectral bands covering wide range of wavelengths. Hyperspectral sensors are typical which cover the range of 400 to 250nm with a resolution less than 10nm. By so doing, they can realize a very broad light spectrum band to form rich data set to detect features that sensors with limited bandwidth can't detect. A very small number, as compared to the number of spectral bands, can be identified. Many applications use increasingly more sensing lines, as shown in Figure 6, the developers of small, rugged hyperspectral systems suitable for UAV platforms include companies such as Headwall Photonics. Engineering challenges associated with integration of hyperspectral imaging into the suites of standard UAV electro-optical and infrared sensors have been met as recently demonstrated with high-resolution UAV based hyperspectral analysis. Such an exploitation of the spectral signature of hyperspectral data facilitates UAV remote sensing tasks. Hyperspectral data allow superior identification and discrimination of targets compared to the data obtained in multispectral mode because the information is acquired in narrow bandwidths. Realization of the benefits of the rich spectral information by carefully selecting and calibrating hyperspectral sensors and processing their data is enabled. High spectral resolution allows the identification of specific materials, and urban composition such as vegetation type, water quality, and infrastructure materials, within the built environment. Being hyperspectral sensors, they also assist to construct 3D models to classify building materials based on their energy efficiency opportunities, the test environment helps us find energy problems associated with roof defects or insulation issues.

Figure 6. Different hyperspectral sensors and their spatial and spectral resolutions (Nawaz et al., 2021).

Camera Sensors (HYPERSPECTRAL)	SENSOR	SPECTRAL RANGE	SPECTRAL RESOLUTION (nm)	ACQUISITION MODE	NO. OF BANDS	SPATIAL RESOLUTION (px)	WEIGHT (g)
HEADWALL PHOTONICS	Nano HyperSpec	400 - 1000	6a	PUSHBROOM	270 - 775	640	1200
	Micro Hyperspec VNIR	380 - 1000	2.5a		837 923	1004 1600	≤ 3900
NOVA SOL	vis-NIR microHSI	400 - 800 400 - 1000 380 - 880	3.3c		120 180 150	680	< 450
	Alpha-vis micro HSI	400 - 800 350 - 1000	10c		40 60	1280	< 2100
	SWIR 640 micro HSI	850 - 1700 600 - 1700	5c		170 200	640	3500
	Alpha-SWIR micro HSI	900 - 1700	5c		160	640	1200
	Extra-SWIR micro HSI	964 - 2500	6c		256	320	2600
Cubert GmbH	S 185–FIREFLEYE SE	450 - 950 355 - 750	4c	SNAPSHOT	125	50 X 50	490
	S 485–FIREFLEYE XL	450 - 950 550 - 1000	4.5c		125	70 X 70	1200
	Q 285–FIREFLEYE QE	450 - 950	4c		125	50 X 50	3000
RESONON	Pika - L	400 - 1000	2.1c	PUSHROOM	281	900	600
	Pika - XC2	400 - 1000	1.3c		447	1600	2200
	Pika - NIR	900 - 1700	4.9c		164	320	2700
	Pika - NUV	350 - 800	2.3c		196	1600	2100
BRANDYWINE PHOTONICS	CHAI S-640	825 - 2125	2.5c - 5c		260	640 X 512	5000
	CHAI V-640	350 - 1080	5c - 10c		256		480
PHOTONFOCUS	MV1-D2048x1088-HS05-96-G2	470 - 900	10 - 12b		150	2048 X 1088	265
MOSAICMILL	Rikola	500 - 900	10b	SNAPSHOT	50a	1010 X 1010	720
SENOP	VIS-VNIR Snapshot	400 - 900	10b		380		720
BAYSPEC	OCI-UAV-1000	500 - 900	<5b	PUSHROOM	100	2048	272
XIMEA	MQ022HG-IM-LS100-NIR	600 - 975	4c		100-	2048 X 8	32
	MQ022HG-IM-LS150-VISNIR	470 - 900	3c		150-	2048 X 5	300

a: 380 in laboratory - b: at FWHM (full width at half maximum) - c: by Sampling

D. Meteorological, Chemical, &LiDAR Sensors

Technical specification and gather necessary data of meteorological sensors for UAVs. Generally, these sensors are instruments for measuring some parameter include air pressure, temperature, relative humidity, atmospheric pressure, wind speed and wind direction. For these measurements to have an accuracy and precision enough for us to understand the dynamics of the atmosphere and what is going on between it and the environment, they must be very accurate and exactly measure what they should be measuring. It can be sampled at low altitudes as opposed to the remote weather balloons or remote weather stations (for instance). As a result, it permits measurement of urban heat island and air pollution layers and urban microclimates in the near surface. For examples, thermal comfort level and wind velocity distribution can be detected at neighborhood level by thermometers, hygrometers and anemometers erected on UAVs. Additionally, UAV meteorological sensors enable it to help assess the urban area air quality and pollution. In the case of these sensors, the data they are capable of emitting can include particulate matter, carbon dioxide and so on, that can be measured to help us know how climate change is impacting the air quality and human health.

3. APPLICATIONS OF UAV'S IN CCR

Urban communities now use UAV technology to study and solve climate change problems because they realize the growing ecological dangers. The following part explains how UAV technology helps researchers do better climate research for urban areas and develop better policies to build resilience against future risks.

3.1. Unmanned Aerial Vehicles (UAVs) in CCR

Early 2000s, the invention of micro-sized sensors, UAVs light weight remote sensing platforms suitable for climate change research. Small fixed wing UAVs have been demonstrated as useful for atmospheric dynamics, composition and thermodynamics; this work was undertaken early in the first decade of the new century. It had the effect of confirming that where UAVs were a polyvalent (or 'transpose') resource. Originally in the early 2000s, when manned aircraft flights were not permitted during disasters, UAVs were used by nonprofit organizations for the purpose of aerial damage assessment on humanitarian grounds. In 2004, small UAVs were flown in the humanitarian sector to assess the extent of damage from Haiti earthquake and following post hurricane flooding to survey disaster outcomes. Much of the first work in this area was to point at the possibility of equipping UAVs so as to facilitate disaster relief by providing fast emergency mapping even in situations of inaccessibility.

Figure 7. UAV innovation timeline and integration with humanitarian action research.

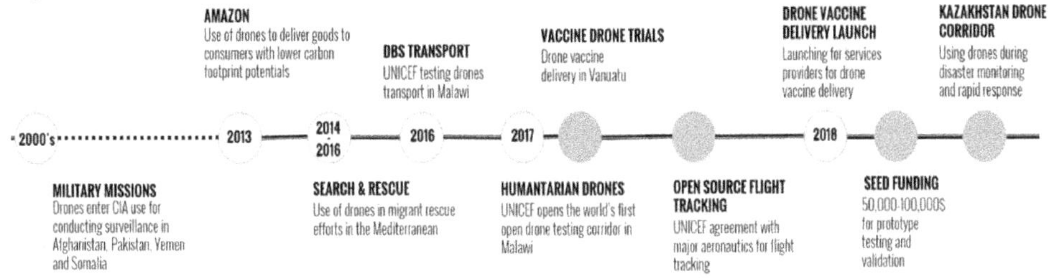

In recent urban expansion and influence of climate change, urban monitoring and sustainably planning are sought to be improved by Djimantoro and Suhardjanto. Researchers have developed UAVs for surveying building infrastructures and making 3D surface models with accuracy down to one meter without human assistance. UAVs collected geospatial data faster and safer while saving money and making the process more flexible than manned aircraft systems. City planning departments can benefit more from the ongoing use of UAVs according to the research results, and with the immediacy of information necessary, gain 3D mapping of a city to improve management of the growth of the city in the face of fixed budgets and the limited availability of personnel.

4. OBSTACLES AND FUTURE PERSPECTIVES

The quick development of UAV systems enabled revolutionary opportunities throughout different business sectors, scientific areas, and public service operations. UAVs encounter multiple barriers in achieving integration and evolution because of technical matters, regulatory barriers and ethical considerations. Current laws together with regulatory guidelines restrict what areas UAV operations can access along with restricting operational methods. Network capabilities and battery management affect data quality because technical constraints including flight time and data resolution and signal blockage exist during operations. Current constraints prevent the advancement of UAV technology because they stem from battery limitations and autonomous operational restrictions and payload capacity issues and resistance to weather elements and operational interaction complexities with human operators. Standards that safeguard safety and privacy conditions and determine specific usage contexts need to be developed into formal legal and policy structures. The exciting prospects of UAVs demand that technological improvements traverse barriers to guarantee new developments match what society accepts. The following part describes essential problems and introduces methods to advance UAV technology securely while fighting climate change and applying it to build environment uses.

4.1. Technical Challenges

Multiple restrictions exist when UAVs are employed for climate studies in complicated urban areas. The short flight duration of UAVs creates difficulties when performing prolonged environmental assessment of heat variation together with air quality measurements throughout metropolitan zones. The operational duration of commercial UAVs reaches less than half an hour yet improvements in battery cells and energy management systems can enhance this period. Specific urban climate measurements along with payload-weight specifications restrict sensor selection, sensor weight limits and measurement factor requirements. The battery-powered limits affect how long rotary-wing UAV aircraft can stay airborne. Extended remote operations place demands on power supply that exceeds what a single battery capacity can sustain.

4.2. Regulations and Security

Yet and still, although the number of UAVs multiplies rapidly, the regulatory bodies still need to create the standardised regulations for operating drones across countries. But one prominent obstacle to the use of UAVs on a broader scale is the cloudiness or lack of standards and rules to designate what the airspace permitted to UAVs is, what its size should be, what is the allowable height, what are privacy considerations, cost, what safety requirements, what some operating characteristics are. Secondly, the government rules regarding UAV deployment are not standardized. For instance, the drone operations were certified and air traffic rules were issued by the Federal Aviation Administration (FAA) in the US.

4.3. The Future of UAVs

The implementation of machine learning along with neural net solutions has previously occurred in UAVs (Nawaz et al., 2021; Jiang et al., 2021) but deep learning and reinforcement learning applications have experienced limited adoption. Droning power systems and processing power stand as the main limiting factors. Researchers should establish deep learning strategies particularly for UAV SAR missions which represents an open research area for deep learning application development. These schemes have the ability to understand their environment and develop trajectory-based learning methods. Small UAV dimensions create obstacles that prevent researchers from adopting deep learning and reinforcement learning methods. Scientists keep working to develop computer models for deep learning systems which operate effectively within drone power capacity and processing capabilities. AI researchers possess a chance to establish autonomous functionality which enables UAVs to learn independently for various applications such as search and rescue operations.

5. CONCLUSIONS

UAV technology has become more and more common within many fields of research including climate change, which has resulted in a slew of patents and scientific articles associated with drone technology. This is driven by the further expanding R&D for UAVs that allows more mobility, increased autonomy and increased range. These advancements have enhanced drone application in many places. Today patents, publications, and R&D are not only accelerating in pace faster and faster, but drones with better and better performances are rapidly taking on critical civilian, commercial, and governmental jobs. However, the many prospects which UAVs can offer to be utilized by cities in all places to achieve dedicated tailored, UAVs used with multispectral sensors can identify where an urban greenery intervention can most of all cool the communities likely to be affected. They also enable quick assessment of the state of infrastructure integrity and the possibility for renewable energy on toppers to help guide up-grades and decentralization of energy. Because UAVs allow for real time mapping of damage that permit rapid service on key services, UAVs are used for disaster responses to enable damage mapping.

Additionally, persistent monitoring of UAVs gives an early warning of new threats, from the increased temperatures of heat islands to polluting agents following infrastructure failure. The timely triggered alerts and protective measures could be triggered by continuous streamed drone sensor data. But the issues are still in processing the data, endurance of flight, citizen concerns of privacy, and airspace management. With all these concerns, cities need to address data security, regulation, privacy of the citizen and air traffic management. It has to adequately balance innovation with safety and public interests. But UAVs have a bright outlook to offer unique, spatiotemporal detail of urban climate change to cities across the world, enabling new ideas of how to target and neighbourhood scale mitigation and adaptation strategies.

REFERENCES

Abdelkader, M., & Koubaa, A. (2023). *Unmanned aerial vehicles applications: Challenges and trends*. Springer. DOI: 10.1007/978-3-031-32037-8

Ader, M., & Axelsson, D. (2017). *Drones in Arctic environments* (Master's thesis). KTH School of Industrial Engineering and Management, Stockholm, Sweden.

Ahmed, F., Mohanta, J. C., Keshari, A., & Yadav, P. S. (2022). Recent advances in unmanned aerial vehicles: A review. *Arabian Journal for Science and Engineering*, *47*(7), 7963–7984. DOI: 10.1007/s13369-022-06738-0 PMID: 35492958

Alanezi, M. A., Shahriar, M. S., Hasan, B., Ahmed, S., Sha'aban, Y. A., & Bouchekara, H. R. E. H. (2022). Livestock management with unmanned aerial vehicles: A review. *IEEE Access : Practical Innovations, Open Solutions*, *10*, 45001–45028. DOI: 10.1109/ACCESS.2022.3168295

Anderson, K., & Gaston, K. J. (2013). Lightweight unmanned aerial vehicles will revolutionize spatial ecology. *Frontiers in Ecology and the Environment*, *11*(3), 138–146. DOI: 10.1890/120150

Bansod, B., Singh, R., Thakur, R., & Singhal, G. (2017). A comparison between satellite-based and drone-based remote sensing technology to achieve sustainable development: A review. *Journal of Agriculture and Environment for International Development*, *111*, 383–407. DOI: 10.12895/jaeid.20171.657

Bouvry, P., Chaumette, S., Danoy, G., Guerrini, G., Jurquet, G., Kuwertz, A., Muller, W., Rosalie, M., & Sander, J. (2016, September 19–21). Using heterogeneous multilevel swarms of UAVs and high-level data fusion to support situation management in surveillance scenarios. In *Proceedings of the 2016 IEEE International Conference on Multisensor Fusion and Integration for Intelligent Systems (MFI)* (pp. 424–429). IEEE. DOI: 10.1109/MFI.2016.7849525

Cai, H., & Geng, Q. (2015). Research on the development process and trend of unmanned aerial vehicle. In *Proceedings of the 2015 International Industrial Informatics and Computer Engineering Conference* (pp. 37–40). Atlantis Press. DOI: 10.2991/iiicec-15.2015.418

Chisholm, R. A., Cui, J., Lum, S. K. Y., & Chen, B. M. (2013). UAV LiDAR for below-canopy forest surveys. *Journal of Unmanned Vehicle Systems*, *1*(1), 61–68. DOI: 10.1139/juvs-2013-0017

Colomina, I., & Molina, P. (2014). Unmanned aerial systems for photogrammetry and remote sensing: A review. *ISPRS Journal of Photogrammetry and Remote Sensing*, *92*, 79–97. DOI: 10.1016/j.isprsjprs.2014.02.013

Dario, J., & Millian, R. (2019). *Towards the application of UAS for road maintenance at the Norvik Port* (Master's thesis). KTH Royal Institute of Technology, Stockholm, Sweden.

Deane, S., Avdelidis, N. P., Ibarra-Castanedo, C., Williamson, A. A., Withers, S., Zolotas, A., Maldague, X. P. V., Ahmadi, M., Pant, S., & Genest, M.. (2022). Development of a thermal excitation source used in an active thermographic UAV platform. *Quantitative Infrared Thermography Journal*, •••, 1–32. DOI: 10.1080/17686733.2022.2102523

Hobbs, A., & Lyall, B. (2016). Human factors guidelines for unmanned aircraft systems. *Ergonomics in Design*, *24*(4), 23–28. DOI: 10.1177/1064804616640632

Hodgson, J. C., Baylis, S. M., Mott, R., Herrod, A., & Clarke, R. H. (2016). Precision wildlife monitoring using unmanned aerial vehicles. *Scientific Reports*, *6*(1), 22574. DOI: 10.1038/srep22574 PMID: 26986721

Isibue, E. W., & Pingel, T. J. (2019). Unmanned aerial vehicle based measurement of urban forests. *Urban Forestry & Urban Greening*, *48*, 126574. DOI: 10.1016/j.ufug.2019.126574

Jiang, W., Liu, L., Xiao, H., Zhu, S., Li, W., & Liu, Y. (2021). Composition and distribution of vegetation in the water level fluctuating zone of the Lantsang cascade reservoir system using UAV multispectral imagery. *PLoS One*, *16*(2), e0247682. DOI: 10.1371/journal.pone.0247682 PMID: 33780473

Khosiawan, Y., Park, Y., Moon, I., Nilakantan, J. M., & Nielsen, I. (2018). Task scheduling system for UAV operations in indoor environment. *Neural Computing & Applications*, *31*(9), 5431–5459. DOI: 10.1007/s00521-018-3373-9

Koubaa, A., Ammar, A., Abdelkader, M., Alhabashi, Y., & Ghouti, L. (2023). AERO: AI-enabled remote sensing observation with onboard edge computing in UAVs. *Remote Sensing (Basel)*, *15*(7), 1873. DOI: 10.3390/rs15071873

Lan, Y., & Chen, S. (2018). Current status and trends of plant protection UAV and its spraying technology in China. *International Journal of Precision Agricultural Aviation*, *1*(1), 1–9. DOI: 10.33440/j.ijpaa.20180101.0002

Mader, D., Blaskow, R., Westfeld, P., & Maas, H.-G. (2015). UAV-based acquisition of 3D point cloud—A comparison of a low-cost laser scanner and SfM-tools. *ISPRS International Archives of the Photogrammetry. Remote Sensing and Spatial Information Sciences, XL-3*(W3), 335–341. DOI: 10.5194/isprsarchives-XL-3-W3-335-2015

Manfreda, S., McCabe, M. F., Miller, P. E., Lucas, R., Madrigal, V. P., Mallinis, G., Ben Dor, E., Helman, D., Estes, L., & Ciraolo, G.. (2018). On the use of unmanned aerial systems for environmental monitoring. *Remote Sensing (Basel), 10*(5), 641. DOI: 10.3390/rs10040641

Mangewa, L. J., Ndakidemi, P. A., & Munishi, L. K. (2019). Integrating UAV technology in an ecological monitoring system for community wildlife management areas in Tanzania. *Sustainability (Basel), 11*(21), 6116. DOI: 10.3390/su11216116

Merkert, R., & Bushell, J. (2020). Managing the drone revolution: A systematic literature review into the current use of airborne drones and future strategic directions for their effective control. *Journal of Air Transport Management, 89*, 101929. DOI: 10.1016/j.jairtraman.2020.101929 PMID: 32952321

Minkina, W. (2020). Theoretical basics of radiant heat transfer—Practical examples of calculation for the infrared (IR) used in infrared thermography measurements. *Quantitative Infrared Thermography Journal, 18*(3), 269–282. DOI: 10.1080/17686733.2020.1739260

Mohsan, S. A. H., Othman, N. Q. H., Li, Y., Alsharif, M. H., & Khan, M. A. (2023). Unmanned aerial vehicles (UAVs): Practical aspects, applications, open challenges, security issues, and future trends. *Intelligent Service Robotics, 16*(1), 1–29. DOI: 10.1007/s11370-022-00452-4 PMID: 36687780

Myburgh, A., Botha, H., Downs, C. T., & Woodborne, S. M. (2021). The application and limitations of a low-cost UAV platform and open-source software combination for ecological mapping and monitoring. *African Journal of Wildlife Research, 51*(1), 166–177. DOI: 10.3957/056.051.0166

Naqvi, R. A., Hussain, D., & Loh, W.-K. (2020). Artificial intelligence-based semantic segmentation of ocular regions for biometrics and healthcare applications. *Computer Modeling in Engineering & Sciences, 66*, 715–732. DOI: 10.32604/cmc.2020.011016

Nawaz, M., Mehmood, Z., Nazir, T., Naqvi, R. A., Rehman, A., Iqbal, M., & Saba, T. (2021). Skin cancer detection from dermoscopic images using deep learning and fuzzy k-means clustering. *Microscopy Research and Technique, 85*(2), 339–351. DOI: 10.1002/jemt.23846 PMID: 34448519

Nawaz, M., Mehmood, Z., Nazir, T., Naqvi, R. A., Rehman, A., Iqbal, M., & Saba, T. (2021). Skin cancer detection from dermoscopic images using deep learning and fuzzy k-means clustering. *Microscopy Research and Technique*, *85*(1), 339–351. DOI: 10.1002/jemt.23908 PMID: 34448519

Nex, F., & Remondino, F. (2014). UAV for 3D mapping applications: A review. *Applied Geomatics*, *6*(1), 1–15. DOI: 10.1007/s12518-013-0120-x

Olson, D., & Anderson, J. (2021). Review on unmanned aerial vehicles, remote sensors, imagery processing, and their applications in agriculture. *Agronomy Journal*, *113*(2), 971–992. DOI: 10.1002/agj2.20595

Pajares, G. (2015). Overview and current status of remote sensing applications based on unmanned aerial vehicles (UAVs). *Photogrammetric Engineering and Remote Sensing*, *81*(4), 281–330. DOI: 10.14358/PERS.81.4.281

Piekkoontod, T., Pachana, B., Hrimpeng, K., & Charoenjit, K. (2020). Assessments of nipa forest using Landsat imagery enhanced with unmanned aerial vehicle photography. *Applied Environmental Research*, *42*(1), 49–59. DOI: 10.35762/AER.2020.42.3.5

Qi, N., Wang, W., Ye, D., Wang, M., Tsiftsis, T. A., & Yao, R. (2021). Energy-efficient full-duplex UAV relaying networks: Trajectory design for channel-model-free scenarios. *ETRI Journal*, *43*(3), 436–446. DOI: 10.4218/etrij.2020-0060

Radmanesh, M., Kumar, M., Nemati, A., & Sarim, M. (2015). Dynamic optimal UAV trajectory planning in the national airspace system via mixed integer linear programming. *Proceedings of the Institution of Mechanical Engineers. Part G, Journal of Aerospace Engineering*, *230*(9), 1668–1682. DOI: 10.1177/0954410015609361

Radoglou-Grammatikis, P., Sarigiannidis, P., Lagkas, T., & Moscholios, I. (2020). A compilation of UAV applications for precision agriculture. *Computer Networks*, *172*, 107148. DOI: 10.1016/j.comnet.2020.107148

Rakha, T., Liberty, A., Gorodetsky, A., Kakillioglu, B., & Velipasalar, S. (2018). Heat mapping drones: An autonomous computer-vision-based procedure for building envelope inspection using unmanned aerial systems (UAS). *Technology | Architecture + Design, 2*(1), 30–44. https://doi.org/DOI: 10.1080/24751448.2018.1428137

Ranyal, E., & Jain, K. (2020). Unmanned aerial vehicle's vulnerability to GPS spoofing: A review. *Photonirvachak (Dehra Dun)*, *49*(3), 585–591. DOI: 10.1007/s12524-020-01225-1

Roldán, J. J., Joossen, G., Sanz, D., Del Cerro, J., & Barrientos, A. (2015). Mini-UAV based sensory system for measuring environmental variables in greenhouses. *Sensors (Basel)*, *15*(2), 3334–3350. DOI: 10.3390/s150203334 PMID: 25648713

Sarim, M., Radmanesh, M., Dechering, M., Kumar, M., Pragada, R., & Cohen, K. (2019). Distributed detect-and-avoid for multiple unmanned aerial vehicles in national air space. *Journal of Dynamic Systems, Measurement, and Control*, *141*(7), 071014. DOI: 10.1115/1.4043190

Serafinelli, E. (2022). Imagining the social future of drones. *Convergence (London)*, *28*(6), 1376–1391. DOI: 10.1177/13548565211054904

Shakhatreh, H., Sawalmeh, A. H., Al-Fuqaha, A., Dou, Z., Almaita, E., Khalil, I., Othman, N. S., Khreishah, A., & Guizani, M. (2019). Unmanned aerial vehicles (UAVs): A survey on civil applications and key research challenges. *IEEE Access : Practical Innovations, Open Solutions*, *7*, 48572–48634. DOI: 10.1109/ACCESS.2019.2909530

Suran, N. A., Shafri, H. Z. M., Shaharum, N. S. N., Radzali, N. A. W. M., & Kumar, V. (2019). UAV-based hyperspectral data analysis for urban area mapping. *ISPRS International Archives of the Photogrammetry. The International Archives of the Photogrammetry, Remote Sensing and Spatial Information Sciences*, *XLII-4*(W16), 621–626. DOI: 10.5194/isprs-archives-XLII-4-W16-621-2019

Szalanczi-Orban, V., & Vaczi, D. (2022). Use of drones in logistics: Options in inventory control systems. *Interdisciplinary Description of Complex Systems*, *20*(3), 295–303. DOI: 10.7906/indecs.20.3.9

Tmušić, G., Manfreda, S., Aasen, H., James, M. R., Gonçalves, G., Ben-Dor, E., Brook, A., Polinova, M., Arranz, J. J., Mészáros, J., Zhuang, R., Johansen, K., Malbeteau, Y., de Lima, I. P., Davids, C., Herban, S., & McCabe, M. F. (2020). Current practices in UAS-based environmental monitoring. *Remote Sensing (Basel)*, *12*(6), 1001. DOI: 10.3390/rs12061001

Ventura, D., Grosso, L., Pensa, D., Casoli, E., Mancini, G., Valente, T., Scardi, M., & Rakaj, A. (2023). Coastal benthic habitat mapping and monitoring by integrating aerial and water surface low-cost drones. *Frontiers in Marine Science*, *9*, 1096594. DOI: 10.3389/fmars.2022.1096594

Villa, T. F., Gonzalez, F., Miljievic, B., Ristovski, Z. D., & Morawska, L. (2016). An overview of small unmanned aerial vehicles for air quality measurements: Present applications and future prospectives. *Sensors (Basel)*, *16*(7), 1072. DOI: 10.3390/s16071072 PMID: 27420065

Wagner, B., & Egerer, M. (2022). Application of UAV remote sensing and machine learning to model and map land use in urban gardens. *Journal of Urban Ecology*, *8*(1), juac008. Advance online publication. DOI: 10.1093/jue/juac008

Wang, Q., Mao, B., Stoliarov, S. I., & Sun, J. (2019). A review of lithium-ion battery failure mechanisms and fire prevention strategies. *Progress in Energy and Combustion Science*, *73*, 95–131. DOI: 10.1016/j.pecs.2019.03.002

Yigitcanlar, T., Desouza, K. C., Butler, L., & Roozkhosh, F. (2020). Contributions and risks of artificial intelligence (AI) in building smarter cities: Insights from a systematic review of the literature. *Energies*, *13*(6), 1473. DOI: 10.3390/en13061473

Zhang, J., Jung, J., Sohn, G., & Cohen, M. (2015). Thermal infrared inspection of roof insulation using unmanned aerial vehicles. *ISPRS International Archives of the Photogrammetry. The International Archives of the Photogrammetry, Remote Sensing and Spatial Information Sciences*, *XL-1*(W4), 381–386. DOI: 10.5194/isprsarchives-XL-1-W4-381-2015

Zohdi, T. I. (2018). Multiple UAVs for mapping: A review of basic modeling, simulation, and applications. *Annual Review of Environment and Resources*, *43*(1), 523–543. DOI: 10.1146/annurev-environ-102017-025912

Chapter 11 Cutting-Edge Innovations: The Future of AGI Drones in Climate Change

Bodhisattwa Baidya
https://orcid.org/0009-0002-3663-4497
Ramakrishna Mission Vidyamandira, India

Atanu Mondal
Ramakrishna Mission Vidyamandira, India

ABSTRACT

This chapter explores the cutting-edge integration of Artificial General Intelligence (AGI) powered drones with Software-Defined Networking (SDN), quantum computing, edge computing, LIDAR technology, and satellite systems for advanced climate change monitoring and mitigation. We present a comprehensive framework for long-term autonomous monitoring that leverages these technologies to revolutionize environmental data collection, analysis, and response strategies. The chapter discusses the synergistic potential of AGI drones equipped with LIDAR, connected through SDN architectures, and supported by quantum and edge computing capabilities. We examine how these systems can interact with Low Earth Orbit (LEO) and Geostationary Earth Orbit (GEO) satellites to create a global, real-time environmental monitoring network. Through detailed case studies, methodologies, and algorithms, we illustrate the transformative impact of this integrated approach on climate change research and mitigation efforts.

DOI: 10.4018/979-8-3693-6457-4.ch011

INTRODUCTION

As the global community grapples with the escalating challenges of climate change, the need for innovative, efficient, and comprehensive monitoring and response systems has never been more critical. The integration of Artificial General Intelligence (AGI) with drone technology has already shown immense promise in environmental monitoring and climate change mitigation. However, the future of AGI drones lies in their seamless integration with other cutting-edge technologies to create a holistic, autonomous, and highly efficient climate monitoring and response system.

This chapter delves into the future of AGI drones in climate change, focusing on their integration with Software-Defined Networking (SDN), quantum computing, edge computing, LIDAR technology, and satellite systems. We explore how these technologies can work together to enable long-term autonomous monitoring, providing unprecedented insights into climate patterns, environmental changes, and potential mitigation strategies.

1. SDN-AGI ENABLED DRONE ARCHITECTURE

The proposed SDN-AGI Drone Architecture for Climate Change Monitoring is designed to create a flexible, scalable, and intelligent network that can adapt to climate monitoring missions' diverse and changing needs. This architecture leverages SDN's programmability and centralized control to manage a complex network of AGI-powered drones, edge computing nodes, satellite links, and ground control stations.

Key Components:-

AGI Drones: These are the primary data collection units, equipped with various sensors (LIDAR, multispectral cameras, atmospheric sensors) and dual communication interfaces (long-range/low-data-rate and short-range/high-data-rate) (Alharthi et al., 2019).

SDN Controller: The brain of the network, implementing a blockchain-based distributed control plane for enhanced resilience and security (Kasantikul et al., 2018). This approach addresses the challenge of maintaining controller connectivity in dynamic drone networks.

Edge Computing Nodes: Mobile or fixed units that provide local data processing capabilities, reducing latency and bandwidth requirements for data transmission (Ateya et al., 2022).

Satellite Integration: The incorporation of LEO and GEO satellites for global connectivity and data relay, is essential for comprehensive climate monitoring (Ayan et al., 2022).
Ground Control Station: Provides human oversight, data analysis, and integration with global climate research networks.

1.2 Architecture Diagram

Figure 1. SDN enabled AGI drone architecture

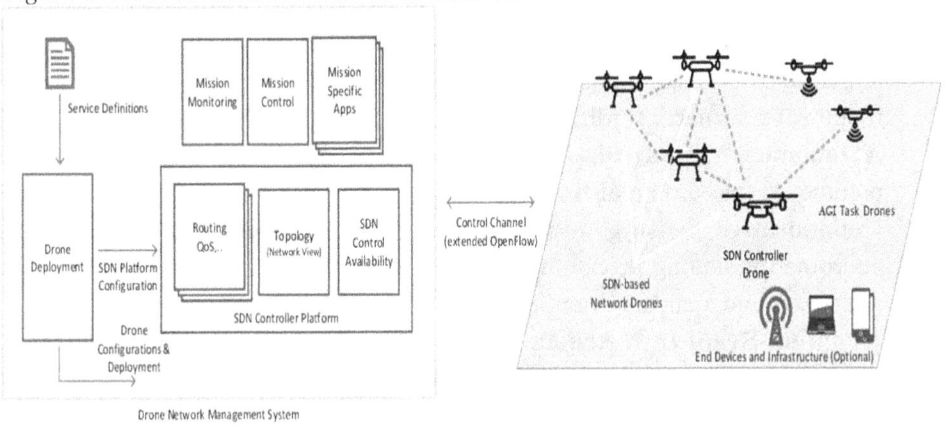

Fig 1 :- SDN Enabled AGI Drone Architecture

1.3 Key Features

Dynamic Network Slicing: The SDN controller creates virtual network slices for different types of climate data (e.g., urgent alerts vs. routine measurements), ensuring efficient resource allocation and prioritization (Alharthi et al., 2019).
Adaptive Routing and Topology Management: The network can dynamically adjust data routes and drone positions based on network conditions, data priorities, and emerging climate events (Kasantikul et al., 2018).
Intelligent Data Mule Operations: The SDN controller uses machine learning algorithms to forecast optimal positions for data collection drones, enhancing the efficiency of data collection in areas of particular interest (Ateya et al., 2022).

Enhanced Security Framework: Incorporates a multi-layered security approach, including blockchain-based control, deep-learning-based Intrusion Detection Systems, and secure communication protocols (Kasantikul et al., 2018).

Cognitive Networking and Context-Aware Communication: The SDN controller leverages AGI capabilities to implement cognitive networking principles, allowing the network to learn from past experiences and predict future network states (Alharthi et al., 2019).

Cross-Layer Optimization: Facilitates joint optimization of network, sensing, and data processing parameters, enhancing overall system efficiency (Ateya et al., 2022).

Seamless Integration of Heterogeneous Technologies: Allows for easy integration of different communication technologies to adapt to various climate monitoring scenarios (Alharthi et al., 2019).

Autonomous Mission Adaptation: The system can autonomously adapt mission parameters based on real-time climate data analysis (Mendoza et al., 2021).

Collaborative Sensing and Distributed Intelligence: Enables complex coordination among drone swarms, facilitating comprehensive climate data collection and analysis (Mendoza et al., 2021).

Quantum-Ready Infrastructure: Designed to accommodate future integration of quantum communication and computing technologies.

1.4 Benefits of SDN in AGI Drone Networks for Climate Change Monitoring

1.4.1 Enhanced Flexibility and Adaptability: SDN's programmable nature allows the network to adapt quickly to changing environmental conditions and monitoring requirements. This is particularly crucial in climate change monitoring, where the focus of observation may need to shift rapidly based on emerging phenomena or unexpected events.

1.4.2 Improved Resource Utilization: Through dynamic network slicing and adaptive routing, SDN enables more efficient use of network resources. This is essential in drone networks where energy and bandwidth are often constrained, allowing for longer mission durations and more comprehensive data collection.

1.4.3 Real-time Data Prioritization: The ability to prioritize data streams in real-time ensures that critical climate information reaches researchers and decision-makers without delay. This can be crucial in scenarios such as early warning systems for extreme weather events or rapid environmental changes.

1.4.4 Enhanced Security and Resilience: The centralized control and programmability of SDN allow for rapid implementation of security policies across the entire network. This is vital for protecting sensitive climate data and ensuring the integrity of the monitoring system.

1.4.5 Scalability: SDN architectures can easily accommodate the addition of new drones, sensors, or edge computing nodes, allowing the monitoring network to expand or contract as needed without significant reconfiguration.

1.4.6 Interoperability: SDN can facilitate the integration of diverse technologies and protocols, enabling AGI drone networks to interact seamlessly with existing climate monitoring systems, satellites, and ground-based sensors.

1.4.7 Cognitive Networking Capabilities: By leveraging AGI, SDN controllers can implement advanced cognitive networking principles, allowing the network to learn from past experiences and optimize its performance over time.

1.4.8 Energy Efficiency: Through intelligent routing and resource allocation, SDN can help optimize the energy consumption of the drone network, extending mission durations and reducing the environmental impact of the monitoring system itself.

1.5 Case Studies

Case Study 1: SDN-Enabled AGI Drone Swarm for Arctic Sea Ice Monitoring

Background

Arctic sea ice plays a crucial role in global climate regulation, but its rapid decline poses significant challenges for traditional monitoring methods. This case study explores the deployment of an SDN-enabled AGI drone swarm for comprehensive, real-time monitoring of Arctic sea ice.

System Setup

(i) 50 AGI drones equipped with LIDAR, multispectral cameras, and ice-penetrating radar.
(ii) 5 mobile edge computing nodes deployed on research vessels.
(iii) SDN controller with distributed control plane using blockchain technology.
(iv) Satellite uplink for data relay and global connectivity.

Key SDN Features Utilized

(i)**Dynamic Network Slicing:** Prioritizing critical ice thickness data over routine atmospheric measurements.

(ii)**Adaptive Routing:** Adjusting data paths based on ice conditions and drone positions.

(iii)**Cognitive Networking:** Learning and predicting optimal drone positioning for data collection.

Outcomes

(i) Real-time mapping of sea ice extent and thickness with unprecedented resolution.
(ii) Early detection of rapid melt events through AI-driven anomaly detection.
(iii) 30% increase in energy efficiency through SDN-optimized routing and task allocation.
(iv) Seamless integration of drone-collected data with satellite observations and ground-based measurements.

Challenges Encountered

(i)Maintaining controller connectivity in harsh Arctic conditions
(ii)Balancing data quality with transmission bandwidth limitations

Case Study 2: Urban Heat Island Mitigation using SDN-AGI Drone Network

Background

Urban heat islands contribute significantly to local climate change impacts. This case study examines the use of an SDN-enabled AGI drone network for monitoring and mitigating urban heat island effects in a major metropolitan area.

System Setup

(i)100 AGI drones equipped with thermal cameras and atmospheric sensors.

(ii) 20 fixed-edge computing nodes distributed across the city.
(iii) Centralized SDN controller with machine learning capabilities.
(iv) Integration with the city's IoT infrastructure and weather stations.

Key SDN Features Utilized

(i)**Intelligent Data Mule Operations:** Optimizing drone positions for data collection in heat-prone areas.

(ii)**Cross-Layer Optimization:** Jointly optimizing sensing parameters and network resource allocation.

(iii)**Autonomous Mission Adaptation:** Dynamically adjusting monitoring patterns based on detected heat anomalies.

Outcomes

(i)Creation of high-resolution, real-time urban heat maps.
(ii)Identification of key contributors to heat island effects (e.g., lack of green spaces, high-emission areas).
(iii)Guided deployment of mitigation strategies (e.g., green roof initiatives, reflective surface applications).
(iv)25% reduction in peak urban temperatures over two years.

Challenges Encountered

(i)Managing network congestion during peak data collection periods.
(ii)Ensuring data privacy in densely populated urban environments.

Case Study 3: Global Forest Carbon Monitoring with SDN-AGI Drone-Satellite Hybrid Network

Background

Accurate monitoring of forest carbon stocks is crucial for understanding and mitigating climate change. This case study explores a global-scale SDN-AGI drone-satellite hybrid network for comprehensive forest carbon monitoring.

System Setup

(i)1000 AGI drones deployed across major forest regions globally.

(ii) Integration with LEO and GEO satellites for data relay and broad-scale imaging.
(iii) Distributed SDN control plane with regional sub-controllers.
(iv) Global network of ground stations and edge computing nodes.

Key SDN Features Utilized

(i) **Seamless Integration of Heterogeneous Technologies:** Coordinating drone, satellite, and ground-based data collection.
(ii) **Dynamic Network Slicing:** Allocating network resources based on forest type and monitoring priorities.

(iii) **Collaborative Sensing and Distributed Intelligence:** Enabling coordinated data collection among drone swarms and satellites.

Outcomes

(i) High-resolution, global-scale forest carbon stock estimates updated in near-real-time.
(ii)Early detection of deforestation and forest degradation events.
(iii)Improved accuracy of climate models through enhanced forest data inputs.
(iv)Support for international carbon credit and REDD+ programs.

Challenges Encountered

(i)Managing the vast scale of the network and ensuring consistent performance across diverse geographical regions.
(ii)Developing standardized data formats and protocols for integrating multi-source data.

1.6. Future Research Directions and Challenges for SDN-AGI Drones in Climate Change Monitoring

1.6.1 Dynamic Controller Placement and Network Resilience

Future Research:

(i) Developing adaptive algorithms for optimal controller placement in highly dynamic drone networks.
(ii) Exploring distributed and hierarchical control architectures to enhance network resilience and scalability.
(iii) Investigating the use of machine learning techniques to predict optimal controller positions based on mission parameters and environmental conditions.

Challenges:

(i)Maintaining consistent connectivity between controllers and drones in diverse and challenging environments (e.g., Arctic regions, dense forests).
(ii)Balancing the trade-off between controller proximity and network overhead.
(iii)Ensuring seamless handover between controllers as drones move across different coverage areas.

1.6.2 Security and Privacy in SDN-AGI Drone Networks

Future Research:

(i)Developing advanced Intrusion Detection Systems (IDS) specifically designed for SDN-enabled drone networks in climate monitoring scenarios.
(ii)Exploring blockchain-based solutions for secure data sharing and controller consensus in distributed SDN architectures.
(iii)Investigating privacy-preserving techniques for sensitive environmental data collected by drones.

Challenges:

(i)Implementing robust security measures without compromising network performance and latency.
(ii)Addressing the unique security vulnerabilities introduced by the SDN architecture in drone networks.
(iii)Ensuring compliance with evolving data protection regulations across different jurisdictions.

1.6.3 Control Plane Visibility and Forensics

Future Research:

(i)Developing integrated logging and monitoring solutions for SDN controllers in drone networks, similar to the POXIEM approach.
(ii)Exploring AI-driven anomaly detection techniques for identifying potential threats or malfunctions in the control plane.
(iii)Investigating methods for secure and efficient storage of control plane logs for long-term climate data analysis.

Challenges:-

(i)Balancing the need for detailed logging with the resource constraints of drone-based systems.
(ii)Ensuring the integrity and authenticity of logged data in distributed drone networks.
(iii)Developing standardized forensic procedures for SDN-AGI drone networks in climate monitoring applications.

1.6.4 Autonomous Connectivity Maintenance

Future Research:

(i)Developing algorithms for safe connectivity maintenance in underactuated multi-agent networks, adapting techniques from oceanic environments to aerial drone networks.
(ii)Exploring swarm intelligence approaches for maintaining network connectivity in dynamic and unpredictable climate monitoring scenarios.
(iii)Investigating the use of reinforcement learning for adaptive network topology management.

Challenges:

(i)Ensuring robust connectivity in the face of extreme weather conditions or equipment failures.
(ii)Balancing connectivity maintenance with other mission objectives (e.g., data collection, energy efficiency).
(iii)Developing solutions that can scale to large numbers of drones across vast geographical areas.

1.6.5 Integration with Emerging Technologies

Future Research:

(i)Exploring the integration of SDN-AGI drone networks with 5G/6G technologies for enhanced communication capabilities.
(ii)Investigating the potential of quantum computing and quantum communication in SDN control planes for drone networks.

(iii) Developing frameworks for seamless integration of SDN-AGI drones with other climate monitoring systems (e.g., ground sensors, satellites).

Challenges:

(i)Ensuring interoperability between diverse technologies and protocols.
(ii)Managing the increased complexity introduced by integrating multiple advanced technologies.
(iii)Addressing the potential impact of emerging technologies on energy consumption and environmental footprint.

1.6.6 Energy Efficiency and Sustainability

Future Research:

(i)Developing energy-aware SDN control algorithms that optimize drone flight patterns and network resource allocation.
(ii)Exploring the use of renewable energy sources (e.g., solar, wind) for powering drones and network infrastructure in long-term climate monitoring missions.
(iii)Investigating the potential of energy harvesting techniques to extend drone operation times in remote areas.

Challenges:

(i)Balancing energy efficiency with network performance and data collection requirements.
(ii)Developing lightweight SDN protocols and control mechanisms suitable for energy-constrained drone networks.
(iii)Minimizing the environmental impact of large-scale drone deployments for climate monitoring.

1.6.7 Scalability and Heterogeneity

Future Research:

(i)Developing SDN architectures capable of managing heterogeneous drone networks with varying capabilities and sensor types.
(ii)Exploring hierarchical and federated SDN control structures for managing global-scale climate monitoring networks.
(iii)Investigating dynamic resource allocation techniques for adapting to changing mission requirements and environmental conditions.

Challenges:

(i)Ensuring consistent performance across diverse drone types and communication technologies.
(ii)Managing the complexity of large-scale, heterogeneous networks spanning multiple climate zones and geographical regions.
(iii)Developing standardized interfaces and protocols for integrating diverse drone platforms and sensors.

1.6.8 Ethical and Regulatory Compliance

Future Research:

(i)Developing frameworks for ethical decision-making in autonomous SDN-AGI drone networks.
(ii)Exploring adaptive compliance mechanisms that can adjust to varying regulatory requirements across different regions.
(iii)Investigating transparent and explainable AI techniques for SDN control decisions in climate monitoring applications.

Challenges:

(i)Navigating the complex and evolving regulatory landscape governing drone operations and data collection.
(ii)Addressing ethical concerns related to autonomous decision-making in environmentally sensitive areas.
(iii)Ensuring transparency and accountability in AI-driven network management decisions.

By addressing these research directions and overcoming the associated challenges, SDN-AGI drone networks can realize their full potential in revolutionizing climate change monitoring and mitigation efforts. The integration of advanced networking paradigms with artificial general intelligence and drone technology promises to provide unprecedented capabilities for understanding and responding to the complex dynamics of our changing climate.

2. QUANTUM AND EDGE COMPUTING FOR AGI DRONES

The urgency of addressing climate change cannot be overstated. As global temperatures rise and extreme weather events become more frequent, the need for sophisticated, real-time monitoring and analysis tools has never been greater. Traditional methods of climate data collection and analysis, while valuable, are often limited in their scope, resolution, and ability to provide timely insights. Enter the era of AGI-enabled drones, powered by quantum computing and supported by edge computing infrastructure – a technological triad that promises to transform our capacity to monitor and respond to environmental changes on a global scale.

We will explore how quantum computing can revolutionize data processing and analysis, how edge computing can enable real-time decision-making at the point of data collection, and how AGI can tie these elements together to create a truly intelligent and adaptive climate monitoring network. Through case studies, we will illustrate the potential real-world applications of this technology, from tracking glacial retreats to monitoring urban heat islands and managing forest ecosystems.

As we look to the future, we will also consider the challenges that lie ahead and the exciting directions for further research and development. The integration of quantum computing, edge computing, and AGI in drone-based climate monitoring is not just a technological advancement – it represents a paradigm shift in our approach to understanding and protecting our planet.

2.1Architecture of Quantum and Edge Computing-Enabled AGI Drones for Climate Change Monitoring

2.1.1 Overview

The architecture of this futuristic system integrates quantum computing, edge computing, and AGI to create a highly efficient, adaptive, and intelligent climate monitoring network. At its core, this architecture aims to leverage the strengths of each technology to overcome the limitations of traditional monitoring systems.

2.2 Key Components

2.2.1 AGI-Enabled Drones

(i)Advanced sensor arrays (LIDAR, multispectral cameras, atmospheric sensors) (Qin & Pournaras, 2023).
(ii)Onboard edge computing units.
(iii)Quantum-resistant communication modules.
(iv)AI-driven flight control and mission planning systems.

2.2.2 Edge Computing Nodes

(i)Distributed processing units on drones and ground stations.
(ii)Local data storage and analysis capabilities.
(iii)AI-powered decision-making algorithms.
(iv)Low-latency communication interfaces.

2.2.3 Quantum Computing Core

(i)Central quantum processing unit for complex calculations.
(ii)Quantum machine learning algorithms (Rajagopal et al., 2021).
(iii)Quantum-secure communication protocols.
(iv)Quantum-enhanced optimization for network management.

2.2.4 Global Control and Analysis Center

(i)AGI-powered central control system.
(ii)Quantum-classical hybrid computing infrastructure.
(iii)Advanced visualization and simulation capabilities.
(iv)Interface with global climate research networks.

2.2.5 Satellite Integration

(i)Quantum key distribution via satellite.
(ii)Global data relay and communication network.
(iii)Complementary broad-scale Earth observation.

2.3 Architectural Diagram

Figure 2. Quantum and edge computing enabled AGI drone architecture

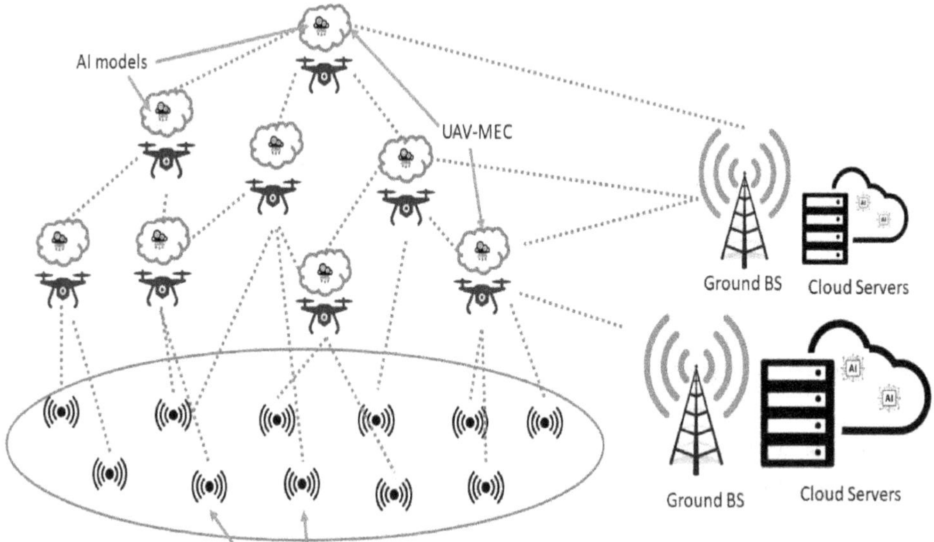

Fig 2:- Quantum and Edge Computing Enabled AGI Drone Architecture

2.4 Data Flow and Processing

(i)AGI drones collect environmental data using advanced sensors.
(ii)Onboard edge computing units perform initial data processing and analysis (Mertens et al., 2020)
(iii)Critical data and insights are transmitted to nearby edge nodes or directly to the global control center.
(iv)The quantum computing core processes complex calculations and runs advanced climate models (Rajagopal et al., 2021).
(v)The AGI-powered global control center analyzes results, makes decisions, and updates mission parameters.
(vi)Updated instructions are sent back to drones and edge nodes, adapting the monitoring strategy in real time.

2.5 Key Architectural Features

2.5.1 Distributed Intelligence

The architecture distributes intelligence across multiple levels, from individual drones to edge nodes and the central quantum core. This allows for rapid local decision-making while leveraging the power of quantum computing for complex analyses.

2.5.2 Quantum-Classical Hybrid Processing

By combining quantum computing with classical edge computing, the system can handle a wide range of computational tasks, from real-time sensor data processing to complex climate model simulations (Resop et al., 2019).

2.5.3 Adaptive Network Topology

The AGI-powered control system dynamically adjusts the network topology, optimizing drone positions and data flow paths based on current environmental conditions and monitoring priorities (Rajagopal et al., 2021).

2.5.4 Quantum-Secure Communication

Utilizing quantum key distribution via satellite, the architecture ensures secure communication across the entire network, protecting sensitive climate data from interception or tampering (Mendoza et al., 2021).

2.5.5 Scalable and Modular Design

The architecture is designed to be scalable, allowing for the seamless integration of additional drones, edge nodes, or quantum computing resources as needed.

2.3Key Features and Benefits

2.3.1 Quantum-Enhanced Data Processing and Analysis

2.3.1.1 Feature: Quantum Machine Learning Algorithms

(i)Utilizes quantum algorithms for pattern recognition and anomaly detection in climate data.

(ii)Enables processing of high-dimensional datasets that are intractable for classical computers.

Benefit: This feature allows for the identification of subtle climate patterns and trends that might be missed by traditional analysis methods, leading to more accurate predictions and deeper insights into climate change dynamics.

2.3.1.2 Feature: Quantum-Inspired Optimization

(i)Applies quantum-inspired algorithms to optimize drone swarm behavior and resource allocation.

(ii)Enables real-time optimization of data collection strategies

Benefit: This optimization leads to more efficient use of resources, extended mission durations, and more comprehensive data collection, ultimately providing a more complete picture of climate change impacts.

2.3.1.3 Feature: Quantum Simulation of Climate Models

(i)Leverages quantum computing to simulate complex climate systems at unprecedented scales

(ii)Enables the exploration of climate scenarios that are computationally infeasible with classical systems

Benefit: This capability allows for more accurate long-term climate predictions and a better understanding of the potential impacts of various mitigation strategies.

2.3.2 Edge Computing for Real-Time Processing and Decision-Making

2.3.2.1 Feature: On-Board Data Processing

(i)Performs initial data analysis and compression on the drone itself.

(ii)Enables real-time decision-making at the point of data collection.

Benefit: This feature reduces data transmission requirements, enables faster response to critical environmental changes, and extends the operational range of drones.

2.3.2.2 Feature: Distributed AI Algorithms

(i)Runs AI algorithms across a network of edge nodes.

(ii)Enables collaborative data analysis and decision-making among drone swarms.

Benefit: This distributed approach enhances the system's resilience, allows for more comprehensive environmental monitoring, and enables adaptive sampling strategies based on real-time insights.

2.3.2.3 Feature: Local Caching and Data Management

(i)Stores and manages data locally on edge nodes.

(ii)Enables continuous operation in areas with limited connectivity

Benefit: This feature ensures uninterrupted monitoring in remote or challenging environments, improving the completeness and reliability of climate data collection.

2.3.3 AGI-Driven Adaptive Monitoring and Response

2.3.3.1 Feature: Autonomous Mission Planning and Adaptation

(i)Uses AGI to dynamically adjust monitoring strategies based on collected data and environmental conditions.

(ii)Enables proactive repositioning of drones to areas of interest or emerging environmental phenomena.

Benefit: This adaptability ensures that monitoring resources are always focused on the most critical areas, maximizing the value of collected data and improving the system's ability to detect and respond to rapid environmental changes.

2.3.3.2 Feature: Multi-Objective Optimization

(i)Balances multiple monitoring objectives (e.g., data quality, energy efficiency, coverage) using AGI.
(ii)Enables sophisticated trade-off analysis in real-time

Benefit: This feature ensures that the monitoring system can adapt to changing priorities and constraints, optimizing its performance across a range of criteria simultaneously.

2.3.3.3 Feature: Predictive Maintenance and Self-Healing

(i)Uses AGI to predict maintenance needs and optimize resource allocation.
(ii)Enables autonomous fault detection and recovery.

Benefit: This capability increases the reliability and longevity of the monitoring system, reducing downtime and maintenance costs while ensuring continuous data collection.

2.3.4 Quantum-Secure Communication and Data Protection

2.3.4.1 Feature: Quantum Key Distribution

(i)Utilizes quantum principles to establish secure communication keys.
(ii)Enables tamper-evident data transmission across the network.

Benefit: This feature ensures the integrity and confidentiality of collected climate data, protecting against potential cyber threats and ensuring the trustworthiness of the monitoring system.

2.3.4.2 Feature: Post-Quantum Cryptography

(i)Implements cryptographic algorithms resistant to quantum computer attacks.
(ii)Ensures long-term security of stored climate data.

Benefit: This forward-looking approach protects the system against future quantum computing threats, ensuring the long-term value and integrity of collected climate data.

2.3.5 Integrated Multi-Scale Monitoring

2.3.5.1 Feature: Seamless Integration of Drone and Satellite Data

(i)Combines high-resolution drone data with broad-scale satellite observations.
(ii)Enables multi-scale analysis of climate phenomena.

Benefit: This integrated approach provides a more comprehensive view of climate change impacts, from local microclimates to global patterns, enhancing our understanding of climate dynamics across different scales.

2.3.5.2 Feature: Cross-Domain Data Fusion

(i)Integrates data from multiple sensor types and domains (atmospheric, terrestrial, oceanic).
(ii)Enables holistic analysis of climate system interactions.

Benefit: This feature allows for a more nuanced understanding of climate change impacts, revealing complex interactions between different environmental systems and improving the accuracy of climate models.

2.4 Case Studies

2.4.1 Case Study 1: Quantum-Enhanced Glacier Monitoring in the Himalayas

Background:

Glaciers in the Himalayas serve as crucial water sources for millions of people and are highly sensitive to climate change. Traditional monitoring methods struggle to capture the complex dynamics of these glaciers across vast and often inaccessible terrain.

System Setup:

(i)100 AGI-enabled drones equipped with LIDAR, multispectral cameras, and ground-penetrating radar.
(ii)Edge computing nodes distributed across base camps and research stations.
(iii)Quantum computing core for complex glacial flow modelling.
(iv)Integration with satellite data for broad-scale observations

Key Features Utilized:

(i)Quantum Machine Learning for Ice Flow Prediction
(ii)Edge Computing for Real-Time Hazard Detection
(iii)AGI-Driven Adaptive Sampling Strategies

Outcomes:

(i)High-resolution 3D mapping of glacier volumes with unprecedented accuracy
(ii)Real-time detection and prediction of glacial lake outburst floods
(iii)Quantum-enhanced modeling of long-term glacial retreat scenarios
(iv)40% improvement in water resource forecasting for downstream communities

Challenges Encountered:

(i)Maintaining drone operations in extreme high-altitude environments
(ii)Balancing high-resolution data collection with battery life constraints
(iii)Ensuring reliable quantum-secure communication in remote areas

This case study demonstrates the power of integrating quantum computing, edge processing, and AGI in tackling complex climate monitoring challenges in difficult terrains. The system's ability to provide both detailed, local-scale data and long-term predictions showcases its potential for enhancing our understanding of climate change impacts on critical water resources.

2.4.2 Case Study 2: Urban Heat Island Mitigation with Quantum-AGI Drone Swarms

Background:
Urban heat islands significantly contribute to energy consumption and health risks in cities. Effective mitigation requires detailed, real-time monitoring of temperature patterns and the impacts of various interventions.
System Setup:

(i)500 AGI drones equipped with thermal cameras and atmospheric sensors
(ii)Edge computing nodes integrated into city infrastructure (buildings, street lights)
(iii)Quantum computing core for complex urban climate modeling
(iv)Integration with city IoT networks and weather stations

Key Features Utilized:

(i)Quantum-Inspired Optimization for Drone Swarm Coordination
(ii)Edge-Enabled Real-Time Temperature Mapping
(iii)AGI-Driven Predictive Modeling of Intervention Impacts

Outcomes:

(i) Creation of high-resolution, real-time 3D urban heat maps
(ii) Quantum-enhanced simulation of the impacts of various heat mitigation strategies
(iii) AGI-optimized deployment of temporary cooling measures during heat waves
(iv)30% reduction in urban heat-related energy consumption over three years

Challenges Encountered:

(i)Managing privacy concerns in densely populated urban environments
(ii)Coordinating large drone swarms in complex urban airspaces

(iii) Integrating diverse data sources (drones, IoT, satellites) for comprehensive analysis

This case study illustrates the potential of quantum-AGI drone systems in addressing complex urban climate challenges. The system's ability to provide real-time insights and predictive modeling capabilities demonstrates its value in supporting data-driven urban planning and climate adaptation strategies.

2.4.3 Case Study 3: Global Forest Carbon Monitoring and Wildfire Prevention

Background:
Forests play a crucial role in carbon sequestration and climate regulation. Accurate monitoring of forest health, carbon stocks, and fire risks is essential for effective climate change mitigation strategies.
System Setup:

(i)1000 AGI drones deployed across major forest regions globally.
(ii)Edge computing nodes in forest monitoring stations and fire lookout towers.
(iii)Quantum computing core for global carbon cycle modelling.
(iv)Integration with satellite imagery and ground-based sensor networks.

Key Features Utilized:

(i)Quantum Machine Learning for Forest Health Assessment
(ii)Edge-Enabled Early Fire Detection
(iii)AGI-Driven Predictive Modeling of Forest Carbon Dynamics

Outcomes:

(i)High-resolution, global-scale forest carbon stock estimates updated in near-real-time.
(ii)Quantum-enhanced modeling of long-term forest-climate interactions.
(iii)50% improvement in early wildfire detection and prediction accuracy.
(iv)AGI-optimized forest management strategies for maximizing carbon sequestration.

Challenges Encountered:

(i)Ensuring consistent drone coverage across vast and diverse forest ecosystems.
(ii)Managing the massive data volumes generated by global-scale monitoring.
(iii)Adapting monitoring strategies to different forest types and climate zones.

This case study showcases the potential of quantum-AGI drone systems in addressing global-scale climate challenges. The system's ability to provide both detailed local insights and global-scale modeling demonstrates its value in supporting international climate policy and forest management efforts.

2.5 Future Research Directions

2.5.1 Quantum-AGI Fusion for Climate Prediction

Future research should focus on developing more sophisticated quantum-AGI algorithms that can leverage the strengths of both quantum computing and artificial general intelligence for climate prediction. This could involve:-

(i)Developing quantum-enhanced neural networks that can process and analyze climate data at unprecedented scales and complexities.
(ii)Creating AGI systems that can interpret and explain the results of quantum climate simulations, bridging the gap between quantum computations and human understanding.
(iii)Exploring the potential of quantum-inspired algorithms in AGI systems for climate pattern recognition and anomaly detection.

Challenges:

Bridging the conceptual gap between quantum physics and AGI principles
Developing quantum hardware capable of supporting AGI-level computations
Ensuring the interpretability and reliability of quantum-AGI fusion models

2.5.2 Advanced Quantum Sensing for Environmental Monitoring

Future research should explore the development and integration of quantum sensors into drone-based monitoring systems. This could include:

(i)Quantum magnetometers for high-precision mapping of Earth's magnetic field changes related to climate phenomena.
(ii)Quantum gravimeters for monitoring ice mass changes and sea level rise with unprecedented accuracy.
(iii)Quantum-enhanced spectroscopy for ultra-sensitive detection of atmospheric pollutants and greenhouse gases.

Challenges:

(i)Miniaturizing quantum sensors for drone-based deployment.
(ii)Maintaining quantum coherence in variable environmental conditions.
(iii)Integrating quantum sensor data with classical monitoring systems.

2.5.3 Quantum-Secure Drone Swarm Communication

As drone swarms become larger and more complex, ensuring secure and efficient communication becomes crucial. Future research should focus on:

(i)Developing practical quantum key distribution protocols for dynamic drone networks.
(ii)Exploring the potential of entanglement-based communication for ultra-secure and efficient data transfer within drone swarms.
(iii)Creating quantum-resistant blockchain technologies for secure and transparent climate data management.

Challenges:

(i)Maintaining quantum entanglement over long distances in atmospheric conditions.

(ii)Scaling quantum key distribution to large, dynamic drone networks.
(iii)Balancing security requirements with communication efficiency and latency.

3. SATELITE ENABLED AGI DRONE

As we stand at the precipice of a new era in climate change monitoring and mitigation, the convergence of advanced technologies offers unprecedented opportunities to revolutionize our approach to environmental stewardship. This chapter explores the integration of Light Detection and Ranging (LIDAR) technology, Geostationary Earth Orbit (GEO) satellites, Low Earth Orbit (LEO) satellites, and Artificial General Intelligence (AGI)-enabled drones to create a comprehensive, multi-layered system for climate change monitoring and analysis.

The urgency of addressing climate change cannot be overstated. As global temperatures rise and extreme weather events become more frequent, the need for sophisticated, real-time monitoring and analysis tools has never been greater. Traditional methods of climate data collection and analysis, while valuable, are often limited in their scope, resolution, and ability to provide timely insights. Enter the era of AGI-enabled drones, enhanced by LIDAR capabilities and seamlessly integrated with both GEO and LEO satellite networks – a technological symphony that promises to transform our capacity to monitor and respond to environmental changes on a global scale.

We will explore how LIDAR technology can provide unprecedented detail in environmental mapping, how GEO and LEO satellites can offer complementary global coverage and data relay capabilities, and how AGI can tie these elements together to create a truly intelligent and adaptive climate monitoring network. Through case studies, we will illustrate the potential real-world applications of this technology, from tracking glacial retreats to monitoring urban heat islands and managing forest ecosystems.

As we look to the future, we will also consider the challenges that lie ahead and the exciting directions for further research and development. The integration of LIDAR, GEO and LEO satellites, and AGI in drone-based climate monitoring is not just a technological advancement – it represents a paradigm shift in our approach to understanding and protecting our planet.

3.1Architecture of LIDAR-Enabled AGI Drones Integrated with GEO and LEO Satellite Networks for Climate Change Monitoring

3.1.1 Overview

The proposed architecture integrates LIDAR-equipped AGI drones with GEO and LEO satellite networks to create a multi-layered, highly efficient, and intelligent climate monitoring system. This architecture aims to leverage the strengths of each technology to overcome the limitations of traditional monitoring systems and provide a comprehensive view of Earth's climate dynamics.

3.2 Key Components

3.2.1 LIDAR-Equipped AGI Drones

(i)Advanced LIDAR sensors for high-resolution 3D mapping
(ii)Onboard AGI systems for autonomous navigation and data analysis
(iii)Multi-spectral cameras and additional environmental sensors
(iv)Quantum-resistant communication modules

3.2.2 Geostationary Earth Orbit (GEO) Satellites

(i)Continuous, broad-scale Earth observation
(ii)High-bandwidth data relay capabilities
(iii)Advanced weather monitoring instruments

3.2.3 Low Earth Orbit (LEO) Satellites

(i)High-resolution Earth imaging
(ii)Frequent revisit times for dynamic phenomena monitoring
(iii)Specialized sensors for atmospheric composition analysis

3.2.4 Ground Control and Data Processing Centers

(i)AGI-powered central control systems
(ii)Quantum computing facilities for complex data analysis
(iii)Advanced visualization and simulation capabilities

3.2.5 Edge Computing Nodes

(i)Distributed processing units on drones and ground stations
(ii)Local data storage and analysis capabilities
(iii)AI-powered decision-making algorithms

3.3 Architectural Diagram

Figure 3. Satellite based AGI drone architecture

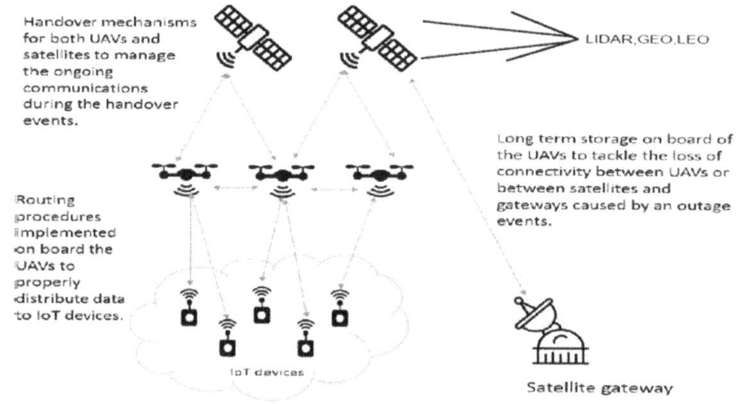

Fig 3:- Satelite Based AGI Drone Architecture

3.4 Data Flow and Processing

(i)LIDAR-equipped AGI drones collect high-resolution 3D data and environmental measurements.
(ii)Drones process data onboard using edge computing capabilities and AGI algorithms.
(iii)Processed data is transmitted to GEO satellites for global relay or directly to ground stations.
(iv)LEO satellites provide complementary high-resolution imagery and atmospheric data.
(v)Ground Control and Data Processing Centers integrate data from all sources using AGI and quantum computing.

(vi)Analyzed results inform drone swarm behavior and future data collection strategies.

3.5 Key Architectural Features

3.5.1 Multi-Scale Data Integration

The architecture seamlessly integrates data from multiple scales, from high-resolution LIDAR scans to broad-scale satellite imagery, providing a comprehensive view of climate phenomena.

3.5.2 Adaptive Sensing Strategies

AGI-powered drones and satellites dynamically adjust their sensing strategies based on real-time data analysis and emerging environmental patterns.

3.5.3 Global Real-Time Monitoring

The combination of GEO and LEO satellites with drone swarms enables continuous, global monitoring with the ability to focus on areas of particular interest or concern.

3.5.4 Quantum-Secure Communication

Utilizing quantum key distribution via satellite, the architecture ensures secure communication across the entire network, protecting sensitive climate data from interception or tampering.

3.5.5 Distributed Intelligence

Intelligence is distributed across the network, from edge computing on drones to centralized quantum computing facilities, enabling rapid local decision-making and complex global analysis.

3.6Key Features and Benefits

3.6.1 High-Resolution 3D Mapping with LIDAR

3.6.1.1 Feature: Centimeter-Level Accuracy in 3D Mapping

(i)LIDAR-equipped drones provide extremely detailed 3D maps of terrain, vegetation, and built environments.
(ii)Enables precise tracking of changes in land use, forest structure, and coastal erosion.

Benefit: This high-resolution data allows for unprecedented accuracy in monitoring small-scale changes that can have significant impacts on local and regional climate patterns.

3.6.1.2 Feature: Penetrative Sensing Capabilities

(i)LIDAR can penetrate vegetation canopies to map understory and ground surfaces.
(ii)Enables accurate biomass estimation and forest structure analysis.

Benefit: Improved understanding of forest ecosystems and their role in carbon sequestration, leading to more effective forest management strategies for climate change mitigation.

3.6.1.3 Feature: Multi-Temporal 3D Change Detection

(i)Repeated LIDAR scans allow for precise measurement of changes over time.
(ii)AGI algorithms can automatically detect and quantify changes in 3D structure.

Benefit: Early detection of subtle environmental changes, such as the onset of forest dieback or accelerated coastal erosion, allowing for timely intervention and adaptation strategies.

3.6.2 Global Coverage and Data Relay with GEO Satellites

3.6.2.1 Feature: Continuous Broad-Scale Monitoring

(i)GEO satellites provide constant coverage of large geographical areas.
(ii)Enables real-time monitoring of weather patterns and large-scale climate phenomena.

Benefit: Improved ability to track and predict extreme weather events and long-term climate trends on a global scale.

3.6.2.2 Feature: High-Bandwidth Data Relay

(i)GEO satellites act as communication hubs for the entire network.
(ii)Enables real-time data transmission from remote areas lacking ground infrastructure.

Benefit: Seamless integration of data from diverse sources, including remote drone swarms and ground stations, facilitating rapid response to emerging climate events.

3.6.2.3 Feature: Advanced Weather Monitoring Instruments

(i)GEO satellites equipped with state-of-the-art weather monitoring tools.
(ii)Provides continuous data on atmospheric conditions, cloud patterns, and storm systems.

Benefit: Enhanced weather forecasting capabilities and improved understanding of large-scale climate dynamics.

3.6.3 Dynamic Monitoring with LEO Satellites

3.6.3.1 Feature: High-Resolution Earth Imaging

(i)LEO satellites provide detailed imagery of Earth's surface.
(ii)Enables monitoring of land use changes, urban development, and ecosystem health.

Benefit: Comprehensive understanding of human impacts on the environment and the ability to track the effectiveness of climate change mitigation efforts.

3.6.3.2 Feature: Frequent Revisit Times

(i)LEO satellite constellations offer multiple daily passes over the same location.
(ii)Enables tracking of rapidly changing phenomena such as ice melt, flooding, and vegetation changes.

Benefit: Improved ability to monitor and respond to short-term environmental changes and extreme events.

3.6.3.3 Feature: Specialized Atmospheric Sensors

(i)LEO satellites are equipped with advanced sensors for measuring greenhouse gases and other atmospheric components.
(ii)Provides global data on atmospheric composition and pollution levels.

Benefit: Enhanced ability to track sources and sinks of greenhouse gases, informing climate models and policy decisions.

3.6.4 AGI-Driven Adaptive Monitoring and Response

3.6.4.1 Feature: Autonomous Mission Planning and Adaptation

(i)AGI systems dynamically adjust monitoring strategies based on collected data and environmental conditions.
(ii)Enables proactive repositioning of drones and satellites to areas of interest or emerging environmental phenomena.

Benefit: Maximizes the efficiency and effectiveness of the monitoring system, ensuring resources are always focused on the most critical areas.

3.6.4.2 Feature: Multi-Objective Optimization

(i)AGI balances multiple monitoring objectives (e.g., data quality, energy efficiency, coverage) in real time.
(ii)Enables sophisticated trade-off analysis and decision-making.

Benefit: Ensures that the monitoring system can adapt to changing priorities and constraints, optimizing its performance across a range of criteria simultaneously.

3.6.4.3 Feature: Predictive Analytics and Early Warning Systems

(i)AGI analyzes multi-source data to predict future climate trends and potential disasters.
(ii)Enables the development of advanced early warning systems for extreme weather events, sea-level rise, and other climate-related hazards.

Benefit: Improved preparedness and response capabilities, potentially saving lives and reducing economic impacts of climate-related disasters.

3.6.5 Integrated Multi-Scale Analysis

3.6.5.1 Feature: Seamless Integration of Drone, Satellite, and Ground-Based Data

(i)AGI-powered systems combine data from multiple sources and scales.
(ii)Enables comprehensive analysis of climate phenomena from local to global scales.

Benefit: Provides a more nuanced understanding of climate change impacts, revealing complex interactions between different environmental systems and improving the accuracy of climate models.

3.6.5.2 Feature: Cross-Domain Data Fusion

(i)Integrates data from multiple sensor types and domains (atmospheric, terrestrial, oceanic).
(ii)Enables holistic analysis of climate system interactions.

Benefit: Reveals previously undetected patterns and relationships in climate data, leading to new insights and more effective mitigation strategies.

3.7Case Studies

3.7.1 Case Study 1: Arctic Sea Ice Monitoring and Prediction

Background:

The rapid decline of Arctic sea ice is one of the most visible and concerning effects of global climate change. Traditional monitoring methods struggle to capture the complex dynamics of sea ice formation, movement, and melting across the vast Arctic region.

System Setup:

(i)500 LIDAR-equipped AGI drones deployed across key Arctic regions.
(ii)Constellation of LEO satellites with specialized ice-monitoring sensors.
(iii)GEO satellites for continuous broad-scale monitoring and data relay.
(iv)Ground stations in Arctic nations for local data processing and drone control.
(v)Central AGI-powered data analysis center.

Key Features Utilized:

(i)LIDAR 3D Mapping for Ice Thickness and Surface Characteristics
(ii)LEO Satellite High-Resolution Imagery for Ice Extent Tracking
(iii)GEO Satellite Continuous Monitoring of Arctic Weather Patterns
(iv)AGI-Driven Predictive Modeling of Sea Ice Dynamics

Outcomes:

(i)Creation of the most detailed and up-to-date 3D maps of Arctic sea ice ever produced.
(ii)Real-time tracking of ice formation, movement, and melting patterns.
(iii)Development of highly accurate short-term and seasonal sea ice forecasts.
(iv)Identification of key factors influencing Arctic amplification of global warming.
(v)Improved understanding of the impacts of sea ice loss on global climate patterns.

Challenges Encountered:

(i)Maintaining drone operations in extreme Arctic conditions
(ii)Ensuring consistent data quality across vast and remote areas
(iii)Integrating diverse data types (LIDAR, multispectral imagery, weather data) into coherent models

This case study demonstrates the power of integrating LIDAR-equipped AGI drones with satellite technologies for comprehensive monitoring of large-scale climate phenomena. The system's ability to provide both detailed, local-scale data and broad, continuous coverage showcases its potential for enhancing our understanding of complex climate systems like the Arctic.

3.7.2 Case Study 2: Global Forest Health and Carbon Sequestration Monitoring

Background:
Forests play a crucial role in the global carbon cycle and climate regulation. Accurate monitoring of forest health, carbon stocks, and the impacts of deforestation and degradation is essential for effective climate change mitigation strategies.
System Setup:

(i)10,000 LIDAR-equipped AGI drones deployed across major forest regions globally

(ii)LEO satellite constellation for high-resolution optical and radar imaging
(iii)GEO satellites for continuous monitoring and data relay
(iv)Network of ground-based monitoring stations in key forest areas
(v)Quantum computing facility for complex carbon cycle modeling

Key Features Utilized:

(i)LIDAR 3D Mapping for Precise Biomass Estimation and Forest Structure Analysis
(ii)LEO Satellite Multispectral Imaging for Vegetation Health Assessment
(iii)GEO Satellite Monitoring of Large-Scale Forest Disturbances (e.g., fires, storms)
(iv)AGI-Driven Predictive Modeling of Forest Carbon Dynamics

Outcomes:

(i)Creation of the most accurate global forest carbon stock map to date
(ii)Real-time detection and quantification of deforestation and forest degradation events
(iii)Development of high-resolution models of forest carbon sequestration potential
(iv)Identification of optimal areas for reforestation and forest conservation efforts
(v)Improved understanding of the impacts of climate change on forest ecosystems globally

Challenges Encountered:

(i)Managing and analyzing the massive volumes of data generated by the global monitoring system
(ii)Ensuring consistent drone coverage and data quality across diverse forest ecosystems
(iii)Developing accurate models to account for the complex interactions between forests and climate

This case study illustrates the potential of the integrated LIDAR-satellite-AGI system in addressing global-scale climate challenges. The ability to provide both detailed local insights and global-scale modeling demonstrates its value in supporting international climate policy and forest management efforts.

3.7.3 Case Study 3: Urban Climate Resilience and Heat Island Mitigation

Background:

Urban areas are particularly vulnerable to the impacts of climate change, with heat islands exacerbating temperature increases and posing significant health risks. Effective mitigation requires detailed, real-time monitoring of urban microclimates and the impacts of various interventions.

System Setup:

(i)1,000 LIDAR-equipped AGI drones deployed across 50 major cities worldwide.
(ii)LEO satellites for high-resolution thermal and multispectral imaging.
(iii)GEO satellites for continuous broad-scale monitoring and data relay.
(iv)Integration with city IoT networks and weather stations.
(v)Edge computing nodes distributed throughout urban areas.
(vi)Central AGI-powered urban climate modeling center.

Key Features Utilized:

(i)LIDAR 3D Mapping for Detailed Urban Morphology and Green Space Analysis.
(ii)LEO Satellite Thermal Imaging for Heat Island Mapping.
(iii)GEO Satellite Monitoring of Urban Weather Patterns.
(iv)AGI-Driven Simulation and Optimization of Urban Climate Interventions.

Outcomes:

(i)Creation of high-resolution, real-time 3D urban heat maps.
(ii)Identification of key contributors to heat island effects in different urban contexts.
(iii)Development of city-specific strategies for heat island mitigation (e.g., green infrastructure placement, building design optimization).
(iv)Real-time monitoring and prediction of urban air quality.
(v)Improved urban planning and design practices for climate resilience.

Challenges Encountered:

(i)Managing privacy concerns related to high-resolution urban monitoring.

(ii)Integrating the monitoring system with existing urban infrastructure and governance structures.
(iii)Developing models that can account for the complex interactions between urban form, climate, and human behavior.

This case study showcases the potential of the integrated LIDAR-satellite-AGI system in addressing complex urban climate challenges. The system's ability to provide detailed, real-time insights and predictive modeling capabilities demonstrates its value in supporting data-driven urban planning and climate adaptation strategies.

3.8 Future Research Directions

3.8.1 Quantum-Enhanced Climate Modeling and Simulation

Future research should focus on leveraging quantum computing capabilities to revolutionize climate modeling and simulation. This could involve:

(i)Developing quantum algorithms for solving complex climate system equations that are computationally intractable for classical computers. This could lead to more accurate and higher-resolution climate models.
(ii)Exploring quantum machine learning techniques for pattern recognition and anomaly detection in large-scale climate datasets. This could help identify subtle climate patterns and trends that might be missed by classical analysis methods.
(iii)Creating quantum-classical hybrid models that combine the strengths of both computing paradigms for climate simulation. This approach could optimize computational resources while pushing the boundaries of model complexity and accuracy.
(iv)Investigating the potential of quantum annealing and quantum simulation for specific climate-related optimization problems, such as optimizing renewable energy placement or carbon sequestration strategies.

Challenges:-

(i)Developing quantum algorithms that can effectively handle the vast scales and complexities of climate systems.
(ii)Ensuring the reliability and accuracy of quantum-enhanced climate models.
(iii)Bridging the gap between theoretical quantum advantages and practical implementations for climate science.

3.8.2 Advanced AI and Machine Learning for Data Integration and Analysis

As the volume and diversity of climate data continue to grow, advanced AI and machine learning techniques will be crucial for extracting meaningful insights. Future research should focus on:

(i)Developing sophisticated deep learning architectures capable of integrating and analyzing multi-modal data from LIDAR, satellites, and ground-based sensors.
(ii)Exploring reinforcement learning approaches for optimizing drone swarm behavior and adaptive sampling strategies in dynamic environmental conditions.
(iii)Investigating explainable AI techniques to enhance the interpretability of complex climate models and predictions, making them more accessible to policymakers and the public.
(iv)Advancing transfer learning methods to apply knowledge gained from well-studied regions to data-sparse areas, improving global coverage of climate insights.

Challenges:

(i)Managing the computational requirements of increasingly complex AI models.
(ii)Ensuring the reliability and generalizability of AI-driven climate predictions.
(iii)Addressing potential biases in AI models that could lead to misinterpretation of climate data.

3.8.3 Next-Generation LIDAR Technologies

Future research should focus on developing more advanced LIDAR technologies specifically tailored for climate monitoring applications. This could include:

(i)Exploring multi-spectral and hyperspectral LIDAR systems capable of simultaneously capturing 3D structure and detailed spectral information, enhancing our ability to monitor vegetation health, atmospheric composition, and other climate-relevant parameters.
(ii)Developing compact, energy-efficient LIDAR systems suitable for long-duration deployment on small drones and satellites, expanding the reach and persistence of 3D mapping capabilities.

(iii)Investigating adaptive LIDAR systems that can dynamically adjust their scanning patterns and resolutions based on real-time analysis of the environment, optimizing data collection for specific climate monitoring objectives.
(iv)Advancing LIDAR data processing algorithms to extract more nuanced information from point clouds, such as fine-scale changes in forest structure or subtle shifts in coastal morphology.

Challenges:-

(i)Miniaturizing advanced LIDAR systems while maintaining high performance.
(ii)Managing the massive data volumes generated by next-generation LIDAR systems.
(iii)Ensuring the reliability and accuracy of LIDAR measurements in diverse and challenging environmental conditions.

3.8.4Integrated Earth System Digital Twins

The concept of digital twins – highly detailed virtual replicas of physical systems – holds immense potential for climate science. Future research should focus on:

(i)Developing comprehensive Earth system digital twins that integrate data from multiple sources (LIDAR, satellites, ground sensors) with advanced climate models and AI algorithms.
(ii)Exploring the use of quantum computing to enhance the complexity and accuracy of Earth system digital twins, potentially enabling real-time simulation of global climate dynamics.
(iii)Investigating methods for continuous updating and refinement of digital twins based on real-time data inputs from the global monitoring network.
(iv)Developing user-friendly interfaces and visualization tools that allow scientists, policymakers, and the public to interact with and explore Earth system digital twins.

Challenges:-

(i)Managing the immense computational requirements of high-fidelity Earth system digital twins.
(ii)Ensuring the accuracy and reliability of digital twin predictions across multiple spatial and temporal scales.
(iii)Developing methods to validate and calibrate digital twins against real-world observations.

3.8.5 Autonomous Adaptive Monitoring Systems

Future research should focus on developing fully autonomous, self-optimizing climate monitoring systems. This could involve:

(i)Advancing AGI algorithms for autonomous mission planning and execution, allowing drone swarms and satellite constellations to adapt their monitoring strategies in real time based on emerging environmental patterns.
(ii)Exploring self-organizing sensor networks that can dynamically reconfigure themselves to optimize data collection and energy efficiency.
(iii)Developing predictive maintenance and self-healing capabilities for long-duration autonomous operations in challenging environments.
(iv)Investigating swarm intelligence approaches for coordinated monitoring and data collection across diverse platforms (drones, satellites, ground sensors).

Challenges:-

(i)Ensuring the reliability and safety of autonomous systems operating in diverse and unpredictable environments.
(ii)Developing robust decision-making algorithms that can handle the complexities of climate systems.
(iii)Addressing ethical and regulatory concerns related to highly autonomous climate monitoring systems.

By pursuing these research directions, we can push the boundaries of what's possible in climate change monitoring and analysis, leveraging the full potential of LIDAR, satellite technologies, and AGI to enhance our understanding of and response to global climate challenges.

CONCLUSION

The convergence of Artificial General Intelligence (AGI), Software-Defined Networking (SDN), quantum computing, edge computing, LIDAR technology, and satellite systems (both GEO and LEO) heralds a new era in climate change monitoring and mitigation. This integrated approach represents a paradigm shift in our ability to observe, understand, and respond to the complex dynamics of Earth's changing climate.

At the heart of this revolutionary system are AGI-enabled drones, serving as intelligent, autonomous platforms for data collection and analysis. These drones, equipped with advanced LIDAR sensors, provide unprecedented high-resolution 3D mapping capabilities. LIDAR technology allows for centimeter-level accuracy in measuring changes in terrain, vegetation structure, and urban landscapes, offering insights into subtle environmental shifts that were previously undetectable at scale.

The integration of SDN principles into the drone network architecture brings a new level of flexibility and efficiency to climate monitoring operations. SDN enables dynamic network slicing, allowing for the prioritization of critical data streams and optimal resource allocation. This adaptability is crucial in responding to rapidly evolving climate phenomena and ensuring that the most vital information is transmitted and processed without delay.

Quantum computing represents a quantum leap in our data processing capabilities. By leveraging quantum algorithms, we can tackle complex climate modeling problems that are intractable for classical computers. This opens up new possibilities for high-resolution climate simulations, enabling more accurate predictions of future climate scenarios and potential mitigation strategies. Quantum-inspired optimization techniques can also enhance the efficiency of drone swarm operations, maximizing coverage and data collection while minimizing energy consumption.

Edge computing brings intelligence to the field, allowing for real-time data processing and decision-making at the point of collection. This distributed approach reduces latency, conserves bandwidth, and enables rapid response to local environmental changes. Edge nodes on drones and ground stations can perform initial data analysis, identifying critical patterns or anomalies that require immediate attention, while filtering out less relevant information to optimize data transmission.

The integration of GEO and LEO satellite networks with this drone-based system creates a multi-layered, global monitoring infrastructure. GEO satellites provide continuous, broad-scale Earth observation and serve as high-bandwidth data relay points, ensuring global connectivity for the drone network. LEO satellites offer frequent revisit times and high-resolution imagery, complementing the drone-collected data with broader contextual information. This satellite-drone synergy enables comprehensive monitoring from local to global scales, capturing both fine-grained details and large-scale climate patterns.

The true power of this integrated system lies in its ability to adapt and respond to changing environmental conditions in real time. Powered by AGI, the network can autonomously adjust its monitoring strategies, redirecting resources to areas of particular interest or concern as they emerge. This adaptive capability ensures that we are always focusing our efforts where they are most needed, maximizing the value of the data collected and our ability to respond to emerging climate threats.

The applications of this advanced climate monitoring system are vast and transformative. In Arctic regions, it can provide detailed tracking of sea ice dynamics, crucial for understanding the impacts of global warming on polar ecosystems and global climate patterns. In forest ecosystems, it enables comprehensive monitoring of forest health, carbon stocks, and the impacts of deforestation, essential for effective climate change mitigation strategies. In urban environments, it can create high-resolution, real-time 3D urban heat maps, helping to identify and mitigate heat island effects and improve urban planning for climate resilience.

Moreover, this system has the potential to revolutionize our approach to climate modeling and prediction. By providing a constant stream of high-resolution, multi-scale data, it enables the development of more accurate and detailed climate models. The integration of quantum computing capabilities further enhances our ability to process and analyze this vast amount of data, potentially leading to breakthroughs in our understanding of complex climate systems and our ability to predict future changes.

However, realizing the full potential of this integrated system requires addressing several challenges and pursuing ambitious research directions. These include:

1. Advancing LIDAR technologies to capture even more detailed and diverse environmental data.
2. Developing more sophisticated quantum algorithms specifically tailored for climate data analysis and modeling.
3. Enhancing edge computing capabilities to handle increasingly complex on-site data processing tasks.
4. Improving the energy efficiency and autonomy of drone systems for long-duration missions.
5. Ensuring robust and secure communication protocols across the entire network, from drones to satellites.
6. Addressing ethical considerations and privacy concerns related to widespread environmental monitoring.

As we look to the future, the integration of AGI drones with SDN, quantum computing, edge computing, LIDAR, and satellite technologies represents not just a technological advancement, but a new paradigm in our relationship with Earth's climate system. It offers the potential for a level of understanding and responsiveness that was previously unimaginable, allowing us to monitor, predict, and respond to climate change with unprecedented precision and speed.

This integrated approach to climate monitoring has the potential to be a game-changer in our global response to climate change. It promises to provide us with the tools we need to better understand the complex dynamics of our changing climate and to develop more effective strategies for mitigation and adaptation. From tracking glacial retreat to monitoring urban heat islands and managing forest ecosystems, this system demonstrates its versatility in addressing diverse climate challenges across various scales and environments.

However, it is crucial to remember that technology alone cannot solve the climate crisis. The insights gained from this advanced monitoring system must be translated into concrete actions at all levels of society – from individual behavior changes to corporate practices and government policies. The global nature of climate change also underscores the need for international cooperation in the development and deployment of such advanced monitoring systems.

In conclusion, the future of AGI drones for climate change, integrated with SDN, quantum computing, edge computing, LIDAR, and satellite technologies, offers a powerful and comprehensive approach to one of the most pressing challenges of our time. As we continue to refine and expand these technologies, we move closer to a future where we can not only observe and predict climate change with extraordinary detail but also respond to it with the urgency and precision that the challenge demands. This integrated system represents not just hope, but a tangible path forward in our global efforts to understand, mitigate, and adapt to climate change, providing us with the tools we need to safeguard our planet for future generations.

REFERENCES

Alharthi, M., Taha, A. E. M., & Hassanein, H. S. (2019, December). Dynamic controller placement in software defined drone networks. In *2019 IEEE Global Communications Conference (GLOBECOM)* (pp. 1-6). IEEE. DOI: 10.1109/GLOBECOM38437.2019.9013799

Ateya, A. A., Algarni, A. D., Koucheryavy, A., & Soliman, N. (2022). Drone-based AI/IoT Framework for Monitoring, Tracking and Fighting Pandemics. *Computers, Materials & Continua*, *71*(3). Advance online publication. DOI: 10.32604/cmc.2022.021850

Ayan, O., Kutsevol, P., Özkan, H. Y., & Kellerer, W. (2022). Task-oriented scheduling for networked control systems: An age of information-aware implementation on software-defined radios. *arXiv preprint arXiv:2202.09189.*

Kasantikul, K., Yang, D., Wang, Q., & Lwin, A. (2018). A novel wind speed estimation based on the integration of an artificial neural network and a particle filter using BeiDou GEO reflectometry. *Sensors (Basel)*, *18*(10), 3350. DOI: 10.3390/s18103350 PMID: 30297601

Mendoza, F., Minardi, M., Chatzinotas, S., Lei, L., & Vu, T. X. (2021, September). An SDN based testbed for dynamic network slicing in satellite-terrestrial networks. In *2021 IEEE international mediterranean conference on communications and networking (MeditCom)* (pp. 36-41). IEEE. DOI: 10.1109/MeditCom49071.2021.9647537

Mertens, J. S., Milotta, G. M., Nagaradjane, P., & Morabito, G. (2020, August). SDN-(UAV) ISE: Applying software defined networking to wireless sensor networks with data mules. In *2020 IEEE 21st International Symposium on" A World of Wireless, Mobile and Multimedia Networks"(WoWMoM)* (pp. 323-328). IEEE.

Qin, C., & Pournaras, E. (2023). Short vs. long-term coordination of drones: When distributed optimization meets deep reinforcement learning. arXiv preprint arXiv:2311.09852.

Rajagopal, T., Balakrishnan, A., Valsalakumar, S., Rajagopal, T. K. R., & Sundaram, S. (2021). Application of MSVPC-5G multicast SDN network eminence video transmission in drone thermal imaging for solar farm monitoring. *Energies*, *14*(24), 8255. DOI: 10.3390/en14248255

Resop, J. P., Lehmann, L., & Hession, W. C. (2019). Drone laser scanning for modeling riverscape topography and vegetation: Comparison with traditional aerial lidar. *Drones (Basel)*, *3*(2), 35. DOI: 10.3390/drones3020035

Rozhon, J., Rezac, F., Safarik, J., Gresak, E., & Jalowiczor, J. (2019, May). Measuring and monitoring the QoS and QoE in software defined networking environments. In *Signal Processing, Sensor/Information Fusion, and Target Recognition XXVIII* (Vol. 11018, pp. 474–483). SPIE. DOI: 10.1117/12.2518838

Compilation of References

Abdel-Basset, M., Gamal, A., Moustafa, N., Abdel-Monem, A., & El-Saber, N. (2021). A Security-by-Design Decision-Making Model for Risk Management in Autonomous Vehicles. *IEEE Access : Practical Innovations, Open Solutions*, *9*, 107657–107679. DOI: 10.1109/ACCESS.2021.3098675

Abdelkader, M., & Koubaa, A. (2023). *Unmanned aerial vehicles applications: Challenges and trends*. Springer. DOI: 10.1007/978-3-031-32037-8

Abdulsattar, N. F., Abedi, F., Ghanimi, H. M., Kumar, S., Abbas, A. H., Abosinnee, A. S., Alkhayyat, A., Hassan, M. H., & Abbas, F. H. (2022). Botnet detection employing a dilated convolutional autoencoder classifier with the aid of hybrid shark and bear smell optimization algorithm-based feature selection in FANETs. *Big Data and Cognitive Computing*, *6*(4), 112. DOI: 10.3390/bdcc6040112

Aber, J. S., Marzolff, I., & Ries, J. B. (2010). Photogrammetry. In Aber, J. S., Marzolff, I., & Ries, J. B. (Eds.), *Small-format aerial photography* (pp. 23–39). Elsevier., DOI: 10.1016/B978-0-444-53260-2.10003-1

Abu Al-Haija, Q., & Al Badawi, A. (2022). High-performance intrusion detection system for networked UAVs via deep learning. *Neural Computing & Applications*, *34*(13), 10885–10900. DOI: 10.1007/s00521-022-07015-9

Abuljadayel, L., Yerima, S. Y., & Butt, U. (2024, December). Machine Learning-Based Intrusion Detection for Denial-of-Service De-Authentication Attacks on Drones. In 2024 IEEE 16th International Conference on Computational Intelligence and Communication Networks (CICN) (pp. 404-408). IEEE. DOI: 10.1109/CICN63059.2024.10847537

Adamopoulos, E., & Rinaudo, F. (2020). UAS-based archaeological remote sensing: Review, meta-analysis and state-of-the-art. *Drones (Basel)*, *4*(3), 46. DOI: 10.3390/drones4030046

Adams, S. M., & Friedland, C. J. (2011, September). A survey of unmanned aerial vehicle (UAV) usage for imagery collection in disaster research and management. In *9th international workshop on remote sensing for disaster response* (Vol. 8, pp. 1-8).

Ader, M., & Axelsson, D. (2017). *Drones in Arctic environments* (Master's thesis). KTH School of Industrial Engineering and Management, Stockholm, Sweden.

Agrawal, A., Abraham, S. J., Burger, B., Christine, C., Fraser, L., Hoeksema, J. M., & Cox, S. (2020, April). The next generation of human-drone partnerships: Co-designing an emergency response system. In *Proceedings of the 2020 CHI Conference on Human Factors in Computing Systems* (pp. 1-13).

Ahmad, W., Almaiah, M. A., Ali, A., & Al-Shareeda, M. A. (2024). Deep Learning Based Network intrusion detection for unmanned aerial vehicle (UAV). *2024 7th World Conference on Computing and Communication Technologies, WCCCT 2024*, 31–36. DOI: 10.1109/WCCCT60665.2024.10541665

Ahmad, W., Almaiah, M. A., Ali, A., & Al-Shareeda, M. A. (2024, April). Deep learning based network intrusion detection for unmanned aerial vehicle (uav). In *2024 7th World Conference on Computing and Communication Technologies (WC-CCT)* (pp. 31-36). IEEE.

Ahmed Fakhry, D., Lee, J., & Lee, J. T. (2025). *Drone Video Anomaly Detection by Future Segmentation Prediction and Spatio-Temporal Relational Modelling*. IEEE., DOI: 10.1109/ACCESS.2025.3536623

Ahmed, M., Ahmad, S., & Kheir, A. M. (2023). Climate Change: An Overview. Global Agricultural Production: Resilience to Climate Change, 1-30.

Ahmed, F., Mohanta, J. C., Keshari, A., & Yadav, P. S. (2022). Recent advances in unmanned aerial vehicles: A review. *Arabian Journal for Science and Engineering*, *47*(7), 7963–7984. DOI: 10.1007/s13369-022-06738-0 PMID: 35492958

Aissaoui, R., Deneuville, J. C., Guerber, C., & Pirovano, A. (2024). Evaluating Post-Quantum Key Exchange Mechanisms for UAV Communication Security. *AIAA/IEEE Digital Avionics Systems Conference - Proceedings*. DOI: 10.1109/DASC62030.2024.10749304

Akash, S. (March, 2023). What is the role of artificial intelligence in drone technology? Analytics Insight. https://www.analyticsinsight.net/artificial-intelligence/what-is-the-role-of-artificial-intelligence-in-drone-technology

Akib Uz Zaman, A., Abdelaty, A., & Jamal, M. S. (2024). Limitations and Barriers Affecting the Applications of UAVs in Transportation Infrastructure Systems Assessment. In Construction Research Congress 2024 (pp. 446-455).

Alam, K., Monir, M. F., Hassan, Z., & Habib, M. T. (2024). Optimizing IoT Network Intrusion Detection: A Deep Learning Approach. *7th Conference on Cloud and Internet of Things, CIoT 2024*. DOI: 10.1109/CIoT63799.2024.10757105

Alanezi, M. A., Shahriar, M. S., Hasan, B., Ahmed, S., Sha'aban, Y. A., & Bouchekara, H. R. E. H. (2022). Livestock management with unmanned aerial vehicles: A review. *IEEE Access : Practical Innovations, Open Solutions*, *10*, 45001–45028. DOI: 10.1109/ACCESS.2022.3168295

AL-Dosari, K., Hunaiti, Z., & Balachandran, W.AL-Dosari. (2023). Systematic review on civilian drones in safety and security applications. *Drones (Basel)*, *7*(3), 210. DOI: 10.3390/drones7030210

Alfeo, A. L., Cimino, M. G. C. A., De Francesco, N., Lega, M., & Vaglini, G. (2018). Design and simulation of the emergent behavior of small drones swarming for distributed target localization. *Journal of Computational Science*, *29*, 19–33. DOI: 10.1016/j.jocs.2018.09.014

Alharthi, M., Taha, A. E. M., & Hassanein, H. S. (2019, December). Dynamic controller placement in software defined drone networks. In *2019 IEEE Global Communications Conference (GLOBECOM)* (pp. 1-6). IEEE. DOI: 10.1109/GLOBECOM38437.2019.9013799

Ali, S. S., Kaur, R., & Abbas, H. (2022). Feasibility of drones as the next step in innovative solution for emerging society. Artificial Intelligence in Industry 4.0 and 5G Technology, 179-208.

Aliloo, J., Abbasi, E., Karamidehkordi, E., Parmehr, E. G., & Canavari, M. (2024). Dos and Don'ts of using drone technology in the crop fields. *Technology in Society*, *76*, 102456. DOI: 10.1016/j.techsoc.2024.102456

Allison, R. S., Johnston, J. M., Craig, G., & Jennings, S. (2016). Airborne optical and thermal remote sensing for wildfire detection and monitoring. *Sensors (Basel)*, *16*(8), 1310. DOI: 10.3390/s16081310 PMID: 27548174

Alquwayzani, A. A., & Albuali, A. A. (2024). A Systematic Literature Review of Zero Trust Architecture for Military UAV Security Systems. *IEEE Access : Practical Innovations, Open Solutions*, *12*, 176033–176056. DOI: 10.1109/ACCESS.2024.3503587

Alsamhi, S. H., Ma, O., Ansari, M. S., & Almalki, F. A. (2019). Survey on collaborative smart drones and internet of things for improving smartness of smart cities. *IEEE Access : Practical Innovations, Open Solutions*, *7*, 128125–128152. DOI: 10.1109/ACCESS.2019.2934998

Alyassi, R., Khonji, M., Karapetyan, A., Chau, S. C.-K., Elbassioni, K., & Tseng, C.-M. (2022). Autonomous recharging and flight mission planning for battery-operated autonomous drones. *IEEE Transactions on Automation Science and Engineering*, *20*(2), 1034–1046. DOI: 10.1109/TASE.2022.3175565

Amazon Monitoring. (n.d.). https://amazonmonitoring.com/

Amir, N. H. (2024). *Taking Flight for Nature's Plight: Framing Sustainable Reforestation with Unmanned Aerial Vehicles (UAVs)*. Hochschulbibliothek der Technischen Hochschule Köln.

Anagnostis, I., Kotzanikolaou, P., & Douligeris, C. (2024). Understanding and Securing Unmanned Aerial Vehicle (UAV) Services: A Comprehensive Tutorial. *Authorea Preprints*. DOI: 10.36227/techrxiv.170975064.43115762/v1

Anderson, K., & Gaston, K. J. (2013). Lightweight unmanned aerial vehicles will revolutionize spatial ecology. *Frontiers in Ecology and the Environment*, *11*(3), 138–146. DOI: 10.1890/120150

Anthoff, D., Nicholls, R. J., Tol, R. S. J., & Vafeidis, A. T. (2006). *Global and regional exposure to large rises in sea-level: A sensitivity analysis*. Tyndall Centre for Climate Change Research Working Paper 96.

Arthur, M. P. (2019). Detecting Signal Spoofing and Jamming Attacks in UAV Networks Using a Lightweight IDS. *Proceedings of the International Conference on Computing, Information and Telecommunication Systems (CITS)*, 2019. DOI: 10.1109/CITS.2019.8862148

Ashraf, S. N., Manickam, S., Zia, S. S., Abro, A. A., Obaidat, M., Uddin, M., Abdelhaq, M., & Alsaqour, R. (2023). IoT empowered smart cybersecurity framework for intrusion detection in internet of drones. *Scientific Reports*, *13*(1), 18422. DOI: 10.1038/s41598-023-45065-8 PMID: 37891186

Asif, M., Rahman, M. A., Akkaya, K., Shahriar, H., & Cuzzocrea, A. (2023, December). Adversarial data-augmented resilient intrusion detection system for unmanned aerial vehicles. In *2023 IEEE International Conference on Big Data (BigData)* (pp. 5428-5437). IEEE. DOI: 10.1109/BigData59044.2023.10386140

Askarzadeh, T., Bridgelall, R., & Tolliver, D. (2024). Monitoring Nodal Transportation Assets with Uncrewed Aerial Vehicles: A Comprehensive Review. *Drones (Basel)*, *8*(6), 233. DOI: 10.3390/drones8060233

Ateya, A. A., Algarni, A. D., Koucheryavy, A., & Soliman, N. (2022). Drone-based AI/IoT Framework for Monitoring, Tracking and Fighting Pandemics. *Computers, Materials & Continua, 71*(3). Advance online publication. DOI: 10.32604/cmc.2022.021850

Attaullah, H. M., Memon, S., Erkan, O. F., & Khawar, R. (2024). IoT Based Systems and Services: Recent Security Concerns and Feasible Solutions. *2024 IEEE 1st Karachi Section Humanitarian Technology Conference, Khi-HTC 2024*. DOI: 10.1109/KHI-HTC60760.2024.10481990

Attaullah, H. M., Memon, S., Erkan, O. F., & Khawar, R. (2024, January). IoT Based Systems and Services: Recent Security Concerns and Feasible Solutions. In 2024 IEEE 1st Karachi Section Humanitarian Technology Conference (KHI-HTC) (pp. 1-6). IEEE.

Attaullah, H. M., Khan, R. A., & Mughal, S. (2021). Cyber security for Industrial Control System–A Survey. *Iksp J. Emerg. Trends Basic Appl. Sci, 1*, 15–21.

Ausonio, E., Bagnerini, P., & Ghio, M. (2021). Drone swarms in fire suppression activities: A conceptual framework. *Drones (Basel), 5*(1), 17. DOI: 10.3390/drones5010017

Ayamga, M., Akaba, S., & Nyaaba, A. A. (2021). Multifaceted applicability of drones: A review. *Technological Forecasting and Social Change, 167*, 120677. DOI: 10.1016/j.techfore.2021.120677

Ayamga, M., Tekinerdogan, B., & Kassahun, A. (2021). Exploring the challenges posed by regulations for the use of drones in agriculture in the African context. *Land (Basel), 10*(2), 164. DOI: 10.3390/land10020164

Ayan, O., Kutsevol, P., Özkan, H. Y., & Kellerer, W. (2022). Task-oriented scheduling for networked control systems: An age of information-aware implementation on software-defined radios. *arXiv preprint arXiv:2202.09189.*

Azar, A. T., Koubaa, A., Mohamed, N. A., Ibrahim, H. A., Ibrahim, Z. F., Kazim, M., & Ammar, A.. (2021). Drone deep reinforcement learning: A review. *Electronics (Basel), 10*(9), 999. DOI: 10.3390/electronics10090999

Bachimanchi, H., Pinder, M. I. M., Robert, C., De Wit, P., Havenhand, J., Kinnby, A., Midtvedt, D., Selander, E., & Volpe, G. (2023). Deep-learning-powered data analysis in plankton ecology. *arXiv preprint arXiv:2309.08500.*

Baig, Z., Syed, N., & Mohammad, N. (2022). Securing the smart city airspace: Drone cyber attack detection through machine learning. *Future Internet, 14*(7), 205. DOI: 10.3390/fi14070205

Bajracharya, D. M., Shrestha, A. B., & Rajbhandari, R. (2020). Endangered species and habitat loss: A regional overview. *Journal of Environmental Protection*, *11*(3), 215–223.

Bakhsh, S. A., Khan, M. A., Ahmed, F., Alshehri, M. S., Ali, H., & Ahmad, J. (2023). Enhancing IoT network security through deep learning-powered Intrusion Detection System. *Internet of Things : Engineering Cyber Physical Human Systems*, *24*, 100936. Advance online publication. DOI: 10.1016/j.iot.2023.100936

Balasundram, S. K., Shamshiri, R. R., Sridhara, S., & Rizan, N. (2023). The role of digital agriculture in mitigating climate change and ensuring food security: An overview. *Sustainability (Basel)*, *15*(6), 5325. DOI: 10.3390/su15065325

Balážová, R., Hlinka, J., Belbachir, A., & Svoboda, J. (2024). Hyperlocal Weather Information for Drone Operations. *Proceedings of IEEE Southeastcon. IEEE Southeastcon*, *2024*, •••. https://cordis.europa.eu/project/id/101017643

Bansod, B., Singh, R., Thakur, R., & Singhal, G. (2017). A comparison between satellite-based and drone-based remote sensing technology to achieve sustainable development: A review. *Journal of Agriculture and Environment for International Development*, *111*, 383–407. DOI: 10.12895/jaeid.20171.657

Barbedo, J. G. A. (2013). Digital image processing techniques for detecting, quantifying and classifying plant diseases. *SpringerPlus*, *2*(1), 660. DOI: 10.1186/2193-1801-2-660 PMID: 24349961

Barka, E., Kerrache, C. A., Benkraouda, H., Shuaib, K., Ahmad, F., & Kurugollu, F. (2022). Towards a trusted unmanned aerial system using blockchain for the protection of critical infrastructure. *Transactions on Emerging Telecommunications Technologies*, *33*(8), e3706. DOI: 10.1002/ett.3706

Bashir, M. H., Ahmad, M., Rizvi, D. R., & El-Latif, A. A. A. (2024). Efficient CNN-based disaster events classification using UAV-aided images for emergency response application. *Neural Computing & Applications*, *36*(18), 1–14. DOI: 10.1007/s00521-024-09610-4

Bayomi, N., & Fernandez, J. E. (2023). Eyes in the sky: Drones applications in the built environment under climate change challenges. *Drones (Basel)*, *7*(10), 637. DOI: 10.3390/drones7100637

Beegum, T. R., Idris, M. Y. I., Ayub, M. N., & Shehadeh, H. A. (2023). Optimized Routing of UAVs Using Bio-Inspired Algorithm in FANET: A Systematic Review. *IEEE Access : Practical Innovations, Open Solutions*, *11*, 15588–15622. DOI: 10.1109/ACCESS.2023.3244067

Benameur, R., Dahane, A., Souihi, S., & Mellouk, A. (2024). A Novel Federated Learning Based Intrusion Detection System for IoT Networks. *IEEE International Conference on Communications*, 2402–2407. DOI: 10.1109/ICC51166.2024.10622538

Bendre, M. R., & Thool, R. C. (2016). Big data in precision agriculture: Weather forecasting for future farming. *2016 International Conference on Global Trends in Signal Processing, Information Computing and Communication (ICGTSPICC)*, 5–10. https://doi.org/DOI: 10.1109/ICGTSPICC.2016.7955252

Berguiga, A., Harchay, A., Massaoudi, A., Ben Ayed, M., & Belmabrouk, H. (2023). GMLP-IDS: A Novel Deep Learning-Based Intrusion Detection System for Smart Agriculture. *Computers, Materials & Continua*, *77*(1), 379–402. DOI: 10.32604/cmc.2023.041667

Bikkasani, D. C. (2024). Navigating artificial general intelligence (AGI): Societal implications, ethical considerations, and governance strategies. *AI and Ethics*, •••, 1–16. DOI: 10.1007/s43681-024-00642-z

Boccadoro, P., Striccoli, D., & Grieco, L. A. (2021). An extensive survey on the Internet of Drones. *Ad Hoc Networks*, *122*, 102600. DOI: 10.1016/j.adhoc.2021.102600

Bogucki, R., Cygan, M., Khan, C B., Klimek, M., Milczek, J K., & Mucha, M. (2018, November 28). Applying deep learning to right whale photo identification. Wiley-Blackwell, 33(3), 676-684. DOI: 10.1111/cobi.13226

Boogaard, F., Lima, R. L. P. D., & Graaf, R. D. (2020, April 23). Innovative Water Quality and Ecology Monitoring Using Underwater Unmanned Vehicles: Field Applications, Challenges and Feedback from Water Managers. *Water (Basel)*, *12*(4), 1196–1196. DOI: 10.3390/w12041196

Bouabdellah, K., Noureddine, H., & Larbi, S. (2013). Using wireless sensor networks for reliable forest fires detection. *Procedia Computer Science*, *19*, 794–801. DOI: 10.1016/j.procs.2013.06.104

Bouvry, P., Chaumette, S., Danoy, G., Guerrini, G., Jurquet, G., Kuwertz, A., Muller, W., Rosalie, M., & Sander, J. (2016, September 19–21). Using heterogeneous multilevel swarms of UAVs and high-level data fusion to support situation management in surveillance scenarios. In *Proceedings of the 2016 IEEE International Conference on Multisensor Fusion and Integration for Intelligent Systems (MFI)* (pp. 424–429). IEEE. DOI: 10.1109/MFI.2016.7849525

Brandsma, S., Lenzholzer, S., Carsjens, G. J., Brown, R. D., & Tavares, S. (2024). Implementation of urban climate-responsive design strategies: An international overview. *Journal of Urban Design*, *29*(5), 598–623. DOI: 10.1080/13574809.2024.2314760

Brown, A. (2022). AGI-powered drones: Applications and challenges. In *Proceedings of the International Conference on Robotics and Automation (ICRA)* (pp. 112–125).

Bu, F., & Gharajeh, M. S. (2019). Intelligent and vision-based fire detection systems: A survey. *Image and Vision Computing*, *91*, 103803. DOI: 10.1016/j.imavis.2019.08.007

Burgués, J., & Marco, S. (2020). Environmental chemical sensing using small drones: A review. *The Science of the Total Environment*, *748*, 141172. DOI: 10.1016/j.scitotenv.2020.141172 PMID: 32805561

Caballero-Martin, D., Lopez-Guede, J. M., Estevez, J., & Graña, M. (2024). Artificial intelligence applied to drone control: A state of the art. *Drones (Basel)*, *8*(7), 296. DOI: 10.3390/drones8070296

Cai, H., & Geng, Q. (2015). Research on the development process and trend of unmanned aerial vehicle. In *Proceedings of the 2015 International Industrial Informatics and Computer Engineering Conference* (pp. 37–40). Atlantis Press. DOI: 10.2991/iiicec-15.2015.418

Caillouet, C., Giroire, F., & Razafindralambo, T. (2019). Efficient data collection and tracking with flying drones. *Ad Hoc Networks*, *89*, 35–46. DOI: 10.1016/j.adhoc.2019.01.011

Callow, G., Watson, G., & Kalawsky, R. (2010). System Modelling for Run-time Verification and Validation of Autonomous Systems. *2010 5th International Conference on System of Systems Engineering (SoSE)*. https://doi.org/DOI: 10.1109/SOSE.2010.324

Camarretta, N. A., Harrison, P., Lucieer, A. M., Potts, B., Davidson, N., & Hunt, M. (2020). From drones to phenotype: Using UAV-LiDAR to detect species and provenance variation in tree productivity and structure. *Remote Sensing*, *12*(19), 3184. DOI: 10.3390/rs12193184

Carlucci, F., Campagna, L. M., & Fiorito, F. (2024). *Responsive Envelopes and Climate Change*. Springer. DOI: 10.1007/978-3-031-58101-4

Caruso, A., Chessa, S., Escolar, S., Barba, J., & López, J. C. (2021). Collection of data with drones in precision agriculture: Analytical model and LoRa case study. *IEEE Internet of Things Journal*, *8*(22), 16692–16704. DOI: 10.1109/JIOT.2021.3075561

Cengiz, K., Lipsa, S., Dash, R. K., Ivković, N., & Konecki, M. (2024). A novel intrusion detection system based on artificial neural network and genetic algorithm with a new dimensionality reduction technique for UAV communication. *IEEE Access : Practical Innovations, Open Solutions*, *12*, 4925–4937. DOI: 10.1109/ACCESS.2024.3349469

Ceviz, O., Sadioglu, P., & Sen, S. (2023). A survey of security in uavs and fanets: Issues, threats, analysis of attacks, and solutions. *arXiv preprint arXiv:2306.14281*.

Ceviz, O., Sadioglu, P., Sen, S., & Vassilakis, V. G. (2023). A Novel Federated Learning-Based IDS for Enhancing UAVs Privacy and Security. *arXiv preprint arXiv:2312.04135*.

Ceviz, O., Sadioglu, P., Sen, S., & Vassilakis, V. G. (2023). A novel federated learning-based intrusion detection system for flying ad hoc networks. *arXiv preprint arXiv:2312.04135*.

Ceviz, O., Sen, S., & Sadioglu, P. (2025). Distributed intrusion detection in dynamic networks of uavs using few-shot federated learning. *arXiv preprint arXiv:2501.13213*.

Ceviz, O., Sadioglu, P., & Sen, S. (2023). A Survey of Security in UAVs and FANETs: Issues, Threats, Analysis of Attacks, and Solutions. *IEEE Communications Surveys and Tutorials*. Advance online publication. DOI: 10.1109/COMST.2024.3515051

Chang, Z., Guo, W., Guo, X., Chen, T., Min, G., Abualnaja, K. M., & Mumtaz, S. (2021). Blockchain-empowered drone networks: Architecture, features, and future. *IEEE Network*, *35*(1), 86–93. DOI: 10.1109/MNET.011.2000202

Chapapría, V. E., Peris, J. S., & González-Escrivá, J. A. (2022). Coastal monitoring using unmanned aerial vehicles (UAVs) for the management of the Spanish Mediterranean coast: The case of almenara-Sagunto. *International Journal of Environmental Research and Public Health*, *19*(9), 5457. DOI: 10.3390/ijerph19095457 PMID: 35564850

Chatterjee, S. (2020). *Maritime Security Challenges in India: The Role of Emerging Technologies*. Maritime Institute.

Cheema, M. A., Ansari, R. I., Ashraf, N., Hassan, S. A., Qureshi, H. K., Bashir, A. K., & Politis, C. (2022). Blockchain-based secure delivery of medical supplies using drones. *Computer Networks*, *204*, 108706. DOI: 10.1016/j.comnet.2021.108706

Chen, W., Liu, J., Guo, H., & Kato, N. (2020). Toward robust and intelligent drone swarm: Challenges and future directions. *IEEE Network*, *34*(4), 278–283. DOI: 10.1109/MNET.001.1900521

Chisholm, R. A., Cui, J., Lum, S. K. Y., & Chen, B. M. (2013). UAV LiDAR for below-canopy forest surveys. *Journal of Unmanned Vehicle Systems*, *1*(1), 61–68. DOI: 10.1139/juvs-2013-0017

Chi, T., Ouyang, C., Wong, M. S., & Nichol, J. E. (2020). Smart sensing and artificial intelligence in the monitoring and management of biodiversity. *Environmental Science & Technology*, *54*(7), 3628–3640. DOI: 10.1021/acs.est.9b06279

Choudhary, A., & Kumar, R. (2022). Precision agriculture with AI-based drones: Applications and challenges. *International Journal of Mechanical Engineering*, *7*(1), 457–465.

Chou, P.-N. (2018). Smart technology for sustainable curriculum: Using drone to support young students' learning. *Sustainability (Basel)*, *10*(10), 3819. DOI: 10.3390/su10103819

Christin, S., Hervet, É., & Lecomte, N. (2019). Applications for deep learning in ecology. *Methods in Ecology and Evolution*, *10*(10), 1632–1644. DOI: 10.1111/2041-210X.13256

Chung, S.-J., Paranjape, A. A., Dames, P., Shen, S., & Kumar, V. (2016). A survey on aerial swarm robotics. *IEEE Transactions on Robotics*, *34*(4), 837–855. DOI: 10.1109/TRO.2018.2857475

CLARE Programme. (2023). *From sky to safety: How Uttaran is introducing UAV technology in coastal embankment mapping*. Retrieved from https://clareprogramme.org/update/from-sky-to-safety-how-uttaran-is-introducing-uav-technology-in-coastal-embankment-mapping/

Clarke, R. (2014). Understanding the drone epidemic. *Computer Law & Security Review*, *30*(3), 230–246. DOI: 10.1016/j.clsr.2014.03.002

Colomina, I., & Molina, P. (2014). Unmanned aerial systems for photogrammetry and remote sensing: A review. *ISPRS Journal of Photogrammetry and Remote Sensing*, *92*, 79–97. DOI: 10.1016/j.isprsjprs.2014.02.013

Cordeiro, T. A. O., Ferrão, R. C., & Cugnasca, P. S. (2024, September). Machine Learning Models for Intrusion Detection in Unmanned Aerial Vehicles: An Approach to Cybersecurity and Operational Safety. In *2024 IEEE 27th International Conference on Intelligent Transportation Systems (ITSC)* (pp. 4236-4241). IEEE.

Cortesão, J., Lenzholzer, S., Mülder, J., Klok, L., Jacobs, C., & Kluck, J. (2020). Visual guidelines for climate-responsive urban design. *Sustainable Cities and Society*, *60*, 102245. DOI: 10.1016/j.scs.2020.102245 d'Amato, G., Chong-Neto, H. J., Monge Ortega, O. P., Vitale, C., Ansotegui, I., Rosario, N., Haahtela, T., Galan, C., Pawankar, R., Murrieta-Aguttes, M., Cecchi, L., Bergmann, C., Ridolo, E., Ramon, G., Gonzalez Diaz, S., D'Amato, M., & Annesi-Maesano, I. (2020). The effects of climate change on respiratory allergy and asthma induced by pollen and mold allergens. *Allergy*, *75*(9), 2219–2228. DOI: 10.1111/all.14476 PMID: 32589303

d'Ambrosio, N., Perrone, G., Romano, S., & Urraro, A. (2025). A cyber-resilient open architecture for drone control. *Computers & Security*, *150*, 104205. Advance online publication. DOI: 10.1016/j.cose.2024.104205

Da Silva, L. M., Ferrão, I. G., Dezan, C., Espes, D., & Branco, K. R. (2023, June). Anomaly-based intrusion detection system for in-flight and network security in uav swarm. In *2023 International Conference on Unmanned Aircraft Systems (ICUAS)* (pp. 812-819). IEEE. DOI: 10.1109/ICUAS57906.2023.10155873

Damaševičius, R., Bacanin, N., & Misra, S. (2023). From sensors to safety: Internet of Emergency Services (IoES) for emergency response and disaster management. *Journal of Sensor and Actuator Networks*, *12*(3), 41. DOI: 10.3390/jsan12030041

Danda, A. A., Ghosh, N., Bandyopadhyay, J., & Hazra, S. (2019). Managed retreat: Adaptation to climate change in the Sundarbans ecoregion in the Bengal Delta. *Journal of the Indian Ocean Region*, *15*(3), 317–335. DOI: 10.1080/19480881.2019.1652974

Dandois, J. P., & Ellis, E. C. (2010). Remote sensing of vegetation structure using computer vision. *Remote Sensing (Basel)*, *2*(4), 1157–1176. DOI: 10.3390/rs2041157

Dario, J., & Millian, R. (2019). *Towards the application of UAS for road maintenance at the Norvik Port* (Master's thesis). KTH Royal Institute of Technology, Stockholm, Sweden.

De Almeida, D. R. A., Broadbent, E. N., Ferreira, M. P., Meli, P., Zambrano, A. M. A., Gorgens, E. B., & Brancalion, P. H. (2021). Monitoring restored tropical forest diversity and structure through UAV-borne hyperspectral and lidar fusion. *Remote Sensing of Environment*, *264*, 112582. DOI: 10.1016/j.rse.2021.112582

De Schiller, S., Bentley, I., & Watson, G. B. (2016). Sustainable urban form: environment and climate responsive design. In *Designing Sustainable Cities in the Developing World* (pp. 163–178). Routledge.

De Vries, R., & Mann, Z. Á. (2023, December 4). Secure Neural Network Inference as a Service with Resource-Constrained Clients. *16th IEEE/ACM International Conference on Utility and Cloud Computing, UCC 2023*. DOI: 10.1145/3603166.3632132

Deane, S., Avdelidis, N. P., Ibarra-Castanedo, C., Williamson, A. A., Withers, S., Zolotas, A., Maldague, X. P. V., Ahmadi, M., Pant, S., & Genest, M.. (2022). Development of a thermal excitation source used in an active thermographic UAV platform. *Quantitative Infrared Thermography Journal*, **•••**, 1–32. DOI: 10.1080/17686733.2022.2102523

Dhara, C. (2019, March 21). West Bengal's climate change conundrum part III: Extraordinarily rapid sea-level rise in Sundarbans turns families into refugees. *Acclimatise News*.

Dharmarathne, G., Waduge, A. O., Bogahawaththa, M., Rathnayake, U., & Meddage, D. P. P. (2024). Adapting cities to the surge: A comprehensive review of climate-induced urban flooding. *Results in Engineering*, *22*, 102123. DOI: 10.1016/j.rineng.2024.102123

Díaz-Báñez, J.-M., & Fabila-Monroy, R. (2024). Filming runners with drones is hard. *arXiv preprint arXiv:2403.14033*.

Dileep, M., Navaneeth, A., Ullagaddi, S., & Danti, A. (2020). A study and analysis on various types of agricultural drones and its applications. 2020 fifth international conference on research in computational intelligence and communication networks (ICRCICN), Enemark, C. (2020). On the responsible use of armed drones: The prospective moral responsibilities of states. *International Journal of Human Rights*, *24*(6), 868–888.

Dou, F., Ye, J., Yuan, G., Lu, Q., Niu, W., Sun, H., Guan, L., Lu, G., Mai, G., & Liu, N. (2023). Towards artificial general intelligence (agi) in the internet of things (iot): Opportunities and challenges. *ArXiv Preprint ArXiv:2309.07438*.

DrDrone. (n.d.). *DJI Phantom 4 RTK*.https://drdrone.ca/pages/dji-phantom-4-rtk

Drone Skyz. (2023). *The integration of artificial intelligence in drone technology. Drone Skyze*.https://droneskyz.com/the-integration-of-artificial-intelligence-in-drone-technology/

Eastaff, M. S., & Priya, M. A. S. (n.d.). Artificial General Intelligence (AGI)–Balancing Promise and Peril. Advances in Future Trends and Technology, 186.

El Faouz, S. A., Souri, A., & Inanc, N. (2024). LSTM-based Deep Learning Model for Cyber-attack Detection Systems in the Internet of Drones. *2024 15th International Conference on Computing Communication and Networking Technologies, ICCCNT 2024*. DOI: 10.1109/ICCCNT61001.2024.10726101

Elghaish, F., Matarneh, S., Talebi, S., Kagioglou, M., Hosseini, M. R., & Abrishami, S. (2021). Toward digitalization in the construction industry with immersive and drones technologies: A critical literature review. *Smart and Sustainable Built Environment, 10*(3), 345–363. DOI: 10.1108/SASBE-06-2020-0077

Emimi, M., Khaleel, M., & Alkrash, A. (2023). The current opportunities and challenges in drone technology. *Int. J. Electr. Eng. and Sustain.*, 74-89.

Erdelj, M., Natalizio, E., Chowdhury, K. R., & Akyildiz, I. F. (2017). Help from the sky: Leveraging UAVs for disaster management. *IEEE Pervasive Computing, 16*(1), 24–32. DOI: 10.1109/MPRV.2017.11

European Commission. (2023). *Biodiversity policies at the global level.* https://biodiversity.europa.eu/policy/biodiversity-policies-at-the-global-level

Famili, A., Stavrou, A., Wang, H., Park, J.-M., & Gerdes, R. (2024). Securing your airspace: Detection of drones trespassing protected areas. *Sensors (Basel), 24*(7), 2028. DOI: 10.3390/s24072028 PMID: 38610239

Fang, Z., & Hong, Z. (2024). Research on Deep Learning Models Combining Transformer and BERT: Application in Network Intrusion Detection. *Proceedings - 2024 4th International Signal Processing, Communications and Engineering Management Conference, ISPCEM 2024*, 218–223. DOI: 10.1109/ISPCEM64498.2024.00045

Fas-Millán, M. Á., Soro, F., Jung, O., & Shaaban, A. (2023, September). Cybersecurity Analysis in the UAV Domain: the Practical Approach of the Labyrinth Project. In *Proceedings of the 2023 ACM Conference on Information Technology for Social Good* (pp. 446-454). DOI: 10.1145/3582515.3609566

FEDS. (2024). Drone innovation for environmental conservation. Retrieved from https://feds.ae/industry/environment-and-ecology/

Fernandez, A., Li, Q., & Banerjee, D. (2024). Long-term trends in biodiversity under climate stress. *The Science of the Total Environment, 902*, 167221. DOI: 10.1016/j.scitotenv.2023.167221

Gabrielle Huet, V. (2020). Infrastructure Projects and Climate Change Adaption in the Era of Grassroots Movement Resurgence: Suggestions fro Transformational Actions.

Gadewar, A. B., Patil, R. V., Mahajan, S. A., & Patil, L. V. (2024). Automated Network Vulnerability Detection in IoT: Lightweight IDS Approaches for Enhanced Security. *2024 Second International Conference on Advanced Computing & Communication Technologies (ICACCTech)*, 612–618. DOI: 10.1109/ICACCTech65084.2024.00103

Gardezi, H., Ikrama, M., Usama, M., Iqbal, M., Jalal, F. E., Hussain, A., & Li, X. (2024). Predictive modeling of rutting depth in modified asphalt mixes using gene-expression programming (GEP): A sustainable use of RAP, fly ash, and plastic waste. *Construction & Building Materials*, *443*, 137809. DOI: 10.1016/j.conbuildmat.2024.137809

Gaur, A., Singh, A., Kumar, A., Kulkarni, K. S., Lala, S., Kapoor, K., Srivastava, V., Kumar, A., & Mukhopadhyay, S. C. (2019). Fire sensing technologies: A review. *IEEE Sensors Journal*, *19*(9), 3191–3202. DOI: 10.1109/JSEN.2019.2894665

Gaur, A., Singh, A., Kumar, A., Kumar, A., & Kapoor, K. (2020). Video flame and smoke based fire detection algorithms: A literature review. *Fire Technology*, *56*(5), 1943–1980. DOI: 10.1007/s10694-020-00986-y

Ghosh, A., Mukherjee, S., & Dutta, S. (2022). Tidal inundation and mangrove health assessment in the Sundarbans using remote sensing technologies. *Environmental Monitoring and Assessment*, *194*(7), 4682. DOI: 10.1007/s10661-022-09876-2

Giles, A., Ren, K., Davies, J., Abrego, D., & Kelaher, B. P. (2023, April 23). Combining Drones and Deep Learning to Automate Coral Reef Assessment with RGB Imagery. *Remote Sensing (Basel)*, *15*(9), 2238–2238. DOI: 10.3390/rs15092238

GIM International. (2023). *Monitoring coastal erosion with UAV LiDAR*. Retrieved from https://www.gim-international.com/content/article/monitoring-coastal-erosion-with-uav-lidar

Glira, P., Pfeifer, N., & Mandlburger, G. (2019). Hybrid orientation of airborne LiDAR point clouds and aerial images. *ISPRS Annals of the Photogrammetry, Remote Sensing and Spatial Information Sciences*, *IV-2*(W5), 567–574. DOI: 10.5194/isprs-annals-IV-2-W5-567-2019

Godfrey, I., & Brenes, J. P. S. (2021, April). Practical Applications for UAS Designed to Assist Climatologists in Studying Toxic Gas Emissions Relative to Climate Change. In *International Conference on Unmanned Aerial System in Geomatics* (pp. 457-484). Cham: Springer International Publishing.

Goertzel, B., Goertzel, I. F., & Heljakka, A. (2010). Artificial general intelligence: Concept, state of the art, and future prospects. *Journal of Artificial General Intelligence*, *1*(1), 1–48. DOI: 10.2478/jagi-2014-0001

Gogoll, J., & Müller, J. F. (2017). Autonomous cars: In favor of a mandatory ethics setting. *Science and Engineering Ethics*, *23*(3), 681–700. DOI: 10.1007/s11948-016-9806-x PMID: 27417644

Goodfellow, I., Bengio, Y., & Courville, A. (2016). *Deep learning*. MIT Press.

Greco, C., Pace, P., Basagni, S., & Fortino, G. (2021). Jamming detection at the edge of drone networks using Multi-layer Perceptrons and Decision Trees. *Applied Soft Computing*, *111*, 107806. Advance online publication. DOI: 10.1016/j.asoc.2021.107806

Grieves, M. (2019). Virtually intelligent product systems: Digital and physical twins. In Flumerfelt, S., Schwartz, K. G., Mavris, D., & Briceno, S. (Eds.), *Complex systems engineering: Theory and practice* (pp. 175–200). American Institute of Aeronautics and Astronautics. DOI: 10.2514/5.9781624105654.0175.0200

Guarnera, C. (2024). The future of drones: Top trends to watch in 2024. Blue Falcon Aerial. https://www.bluefalconaerial.com/the-future-of-drones-top-trends-to-watch-in-2024/

Gundu, R., & Maleki, M. (2022). Securing CAN Bus in Connected and Autonomous Vehicles Using Supervised Machine Learning Approaches. *IEEE International Conference on Electro Information Technology, 2022-May*, 42–46. DOI: 10.1109/eIT53891.2022.9813985

Gupta, A., Afrin, T., Scully, E., & Yodo, N. (2021). Advances of UAVs toward future transportation: The state-of-the-art, challenges, and opportunities. *Future Transportation*, *1*(2), 326–350. DOI: 10.3390/futuretransp1020019

Guvenc, I., Koohifar, F., Singh, S., Sichitiu, M. L., & Matolak, D. (2018). Detection, tracking, and interdiction for amateur drones. *IEEE Communications Magazine*, *56*(4), 75–81. DOI: 10.1109/MCOM.2018.1700455

Hadi, H. J., Cao, Y., Li, S., Hu, Y., Wang, J., & Wang, S. (2024). Real-Time Collaborative Intrusion Detection System in UAV Networks Using Deep Learning. *IEEE Internet of Things Journal*, *11*(20), 33371–33391. Advance online publication. DOI: 10.1109/JIOT.2024.3426511

Hadi, H. J., Cao, Y., Li, S., Hu, Y., Wang, J., & Wang, S. (2024). Real-time collaborative intrusion detection system in uav networks using deep learning. *IEEE Internet of Things Journal*.

Hafiza, L., Wahyuningrum, R. D., Maliah Rachmawati, S., & Salshabila, S. (2023). Network Intrusion Detection using Wide Tuned Convolutional Neural Network on IoT Network. *2023 International Conference on Advanced Mechatronics, Intelligent Manufacture and Industrial Automation, ICAMIMIA 2023 - Proceedings*, 788–792. DOI: 10.1109/ICAMIMIA60881.2023.10427676

Haider, M., Ahmed, I., & Rawat, D. B. (2022). Cyber Threats and Cybersecurity Reassessed in UAV-assisted Cyber Physical Systems. *International Conference on Ubiquitous and Future Networks, ICUFN, 2022-July*, 222–227. DOI: 10.1109/ICUFN55119.2022.9829584

Halofsky, J. E., Peterson, D. L., & Harvey, B. J. (2020). Changing wildfire, changing forests: The effects of climate change on fire regimes and vegetation in the Pacific Northwest, USA. *Fire Ecology*, *16*(1), 1–26. DOI: 10.1186/s42408-019-0062-8

Hanover, D., Loquercio, A., Bauersfeld, L., Romero, A., Penicka, R., Song, Y., Cioffi, G., Kaufmann, E., & Scaramuzza, D. (2024). Autonomous drone racing: A survey. *IEEE Transactions on Robotics*, *40*, 3044–3067. DOI: 10.1109/TRO.2024.3400838

Haque, E., Hasan, K., Ahmed, I., Alam, M. S., & Islam, T. (2024). Enhancing UAV Security Through Zero Trust Architecture: An Advanced Deep Learning and Explainable AI Analysis. *2024 International Conference on Computing, Networking and Communications, ICNC 2024*, 463–467. DOI: 10.1109/ICNC59896.2024.10556279

Harbi, Y., Medani, K., Gherbi, C., Senouci, O., Aliouat, Z., & Harous, S. (2023). A Systematic Literature Review of Blockchain Technology for Internet of Drones Security. *Arabian Journal for Science and Engineering*, *48*(2), 1053–1074. DOI: 10.1007/s13369-022-07380-6 PMID: 36337772

Hart, J. K., & Martinez, K. (2006). Environmental sensor networks: A revolution in the earth system science? *Earth-Science Reviews*, *78*(3-4), 177–191. DOI: 10.1016/j.earscirev.2006.05.001

Hartsell, C., Ramakrishna, S., Dubey, A., Stojcsics, D., Mahadevan, N., & Karsai, G. (2021). ReSonAte: A Runtime Risk Assessment Framework for Autonomous Systems. *Proceedings - 2021 International Symposium on Software Engineering for Adaptive and Self-Managing Systems, SEAMS 2021*, 118–129. DOI: 10.1109/SEAMS51251.2021.00025

Hasna, A. M. (2009). Climate change, contemporary society and engineering practice: A sustainability journey. *The International Journal of Climate Change*, *1*(4), 139–150. DOI: 10.18848/1835-7156/CGP/v01i04/37278

Hassanalian, M., Rice, D., & Abdelkefi, A. (2018). Evolution of space drones for planetary exploration: A review. *Progress in Aerospace Sciences*, *97*, 61–105. DOI: 10.1016/j.paerosci.2018.01.003

Hassija, V., Chamola, V., Agrawal, A., Goyal, A., Luong, N. C., Niyato, D., Yu, F. R., & Guizani, M. (2021). Fast, Reliable, and Secure Drone Communication: A Comprehensive Survey. *IEEE Communications Surveys and Tutorials*, *23*(4), 2802–2832. DOI: 10.1109/COMST.2021.3097916

Hassler, S. C., Mughal, U. A., & Ismail, M. (2023). Cyber-physical intrusion detection system for unmanned aerial vehicles. *IEEE Transactions on Intelligent Transportation Systems*, *25*(6), 6106–6117. DOI: 10.1109/TITS.2023.3339728

He, X., Chen, Q., Wang, W., Liu, T., Li, L., Tang, L., & Liu, Q. (2023, December). Stacked broad learning system empowered FCL assisted by dtn for intrusion detection in UAV networks. In *GLOBECOM 2023-2023 IEEE Global Communications Conference* (pp. 5372-5377). IEEE. DOI: 10.1109/GLOBECOM54140.2023.10436983

Henrion, L., Zhang, D., Li, V., & Sick, V. (2021). Built infrastructure renewal and climate change mitigation can both find solutions in CO2. *Frontiers in Sustainability*, *2*, 733133. DOI: 10.3389/frsus.2021.733133

Hensel, E., Wenclawski, S., & Layman, C. A. (2018, November 10). Using a small, consumer grade drone to identify and count marine megafauna in shallow habitats. Escuela de Ciencias del Mar, Facultad de Recursos Naturales. *Latin American Journal of Aquatic Research*, *46*(5), 1025–1033. DOI: 10.3856/vol46-issue5-fulltext-15

Hermanus, D. R., Supangkat, S. H., & Hidayat, F. (2023). Web-based anomaly detection for smart urban living: Drone photography and videography. *Jurnal Sistem Cerdas*, *6*(2), 144–152. DOI: 10.37396/jsc.v6i2.330

Hernández, D., Cano, J.-C., Silla, F., Calafate, C. T., & Cecilia, J. M. (2021). AI-enabled autonomous drones for fast climate change crisis assessment. *IEEE Internet of Things Journal*, *9*(10), 7286–7297. DOI: 10.1109/JIOT.2021.3098379

Hobbs, A., & Lyall, B. (2016). Human factors guidelines for unmanned aircraft systems. *Ergonomics in Design*, *24*(4), 23–28. DOI: 10.1177/1064804616640632

Hodgson, J. C., Baylis, S. M., Mott, R., Herrod, A., & Clarke, R. H. (2016). Precision wildlife monitoring using unmanned aerial vehicles. *Scientific Reports*, *6*(1), 22574. DOI: 10.1038/srep22574 PMID: 26986721

Holden, P. B., Rebelo, A. J., & New, M. G. (2021). Mapping invasive alien trees in water towers: A combined approach using satellite data fusion, drone technology and expert engagement. *Remote Sensing Applications: Society and Environment*, *21*, 100448. DOI: 10.1016/j.rsase.2020.100448

Hosseini, N., Jamal, H., Matolak, D. W., Haque, J., & Magesacher, T. (2019). UAV Command and Control, Navigation and Surveillance: A Review of Potential 5G and Satellite Systems. *2019 IEEE/AIAA 38th Digital Avionics Systems Conference (DASC).*

Ilinca, V., andric, I., Chi u, Z., Irimia, R., & Gheuca, I. (2022). UAV applications to assess short-term dynamics of slow-moving landslides under dense forest cover. *Landslides*, *19*(7), 1717–1734. DOI: 10.1007/s10346-022-01877-9

Indian Maritime Foundation. (2023). *Blue Economy and Environmental Sustainability: The Indian Perspective*. IM Foundation.

International Rhino Foundation. (2023). A New Poaching Problem in South Africa. International Rhino Foundation. https://rhinos.org/blog/a-new-poaching-problem-in-south-africa/

Iost Filho, F. H., Heldens, W. B., Kong, Z., & de Lange, E. S. (2020). Drones: Innovative technology for use in precision pest management. *Journal of Economic Entomology*, *113*(1), 1–25. DOI: 10.1093/jee/toz268 PMID: 31811713

Iqbal, N. (2021). *Reverse Engineering and CFD analysis of a BWB drone system* (Master's thesis, Universitat Politècnica de Catalunya).

Irshad, M. A., Sattar, S., Nawaz, R., Al-Hussain, S. A., Rizwan, M., Bukhari, A., Waseem, M., Irfan, A., Inam, A., & Zaki, M. E. A. (2023). Enhancing chromium removal and recovery from industrial wastewater using sustainable and efficient nanomaterial: A review. *Ecotoxicology and Environmental Safety*, *263*, 115231. DOI: 10.1016/j.ecoenv.2023.115231 PMID: 37429088

Isibue, E. W., & Pingel, T. J. (2019). Unmanned aerial vehicle based measurement of urban forests. *Urban Forestry & Urban Greening*, *48*, 126574. DOI: 10.1016/j.ufug.2019.126574

Janke, C., & de Haag, M. U. (2022). Implementation of European Drone Regulations - Status Quo and Assessment. *Journal of Intelligent & Robotic Systems*, *106*(2), 33. Advance online publication. DOI: 10.1007/s10846-022-01714-0

Javed, S., Khan, M. A., Abdullah, A. M., Alsirhani, A., Alomari, A., Noor, F., & Ullah, I. (2022). An efficient authentication scheme using blockchain as a certificate authority for the internet of drones. *Drones (Basel)*, *6*(10), 264. DOI: 10.3390/drones6100264

Jiang, W., Liu, L., Xiao, H., Zhu, S., Li, W., & Liu, Y. (2021). Composition and distribution of vegetation in the water level fluctuating zone of the Lantsang cascade reservoir system using UAV multispectral imagery. *PLoS One*, *16*(2), e0247682. DOI: 10.1371/journal.pone.0247682 PMID: 33780473

Jiménez López, J., & Mulero-Pázmány, M. (2019). Drones for conservation in protected areas: Present and future. *Drones (Basel)*, *3*(1), 10. DOI: 10.3390/drones3010010

Johnson, A. M., Cunningham, C. J., Arnold, E., Rosamond, W. D., & Zègre-Hemsey, J. K. (2021). Impact of using drones in emergency medicine: What does the future hold? *Open Access Emergency Medicine : OAEM*, *13*, 487–498. DOI: 10.2147/OAEM.S247020 PMID: 34815722

Johnson, J. (2020). Artificial intelligence, drone swarming and escalation risks in future warfare. *RUSI Journal*, *165*(2), 26–36. DOI: 10.1080/03071847.2020.1752026

Jumaah, H. J., Kalantar, B., Halin, A. A., Mansor, S., Ueda, N., & Jumaah, S. J. (2021). Development of UAV-based PM2. 5 monitoring system. *Drones (Basel)*, *5*(3), 60. DOI: 10.3390/drones5030060

Júnior, M. R. B., de Almeida Moreira, B. R., dos Santos Carreira, V., de Brito Filho, A. L., Trentin, C., de Souza, F. L. P., Tedesco, D., Setiyono, T., Flores, J. P., & Ampatzidis, Y. (2024). Precision agriculture in the United States: A comprehensive meta-review inspiring further research, innovation, and adoption. *Computers and Electronics in Agriculture*, *221*, 108993. DOI: 10.1016/j.compag.2024.108993

Kandrot, S., & Holloway, P. (2020). Applications of drone technology for sustainable development of the coastal zone: A literature review. *Project report.*

Kandrot, S., Hayes, S., & Holloway, P. (2022). Applications of uncrewed aerial vehicles (UAV) technology to support integrated coastal zone management and the UN sustainable development goals at the coast. *Estuaries and Coasts*, *45*(5), 1230–1249. DOI: 10.1007/s12237-021-01001-5 PMID: 34690615

Kar, G., Mustafa, H., Wang, Y., Chen, Y., Xu, W., Gruteser, M., & Vu, T. (2014, November). Detection of on-road vehicles emanating GPS interference. In *Proceedings of the 2014 ACM SIGSAC conference on computer and communications security* (pp. 621-632). DOI: 10.1145/2660267.2660336

Kasantikul, K., Yang, D., Wang, Q., & Lwin, A. (2018). A novel wind speed estimation based on the integration of an artificial neural network and a particle filter using BeiDou GEO reflectometry. *Sensors (Basel)*, *18*(10), 3350. DOI: 10.3390/s18103350 PMID: 30297601

Kaul, M., Hill, R. L., & Walthall, C. (2015). Artificial intelligence for precision agriculture: Applications and limitations. *Agricultural Research*, *4*(2), 75–86.

Kellenberger, B., Marcos, D., & Tuia, D. (2018). Detecting mammals in UAV images: Best practices to address a substantially imbalanced dataset with deep learning. *Remote Sensing of Environment*, *216*, 139–153. DOI: 10.1016/j.rse.2018.06.028

Kesavan, P C., & Swaminathan, M S. (2006, June 27). Managing extreme natural disasters in coastal areas. Royal Society, 364(1845), 2191-2216. DOI: 10.1098/rsta.2006.1822

Khalil, A. A., & Rahman, M. A. (2023). Adaptive Neuro-Fuzzy Inference System-based Lightweight Intrusion Detection System for UAVs. *Proceedings of the IEEE Conference on Local Computer Networks (LCN)*, 2023. DOI: 10.1109/LCN58197.2023.10223340

Khamooshi, M., & Mohammadpour, J. (2020). *Advances in unmanned aerial vehicles: State of the art and the road to autonomy*. Springer.

Khan, I. U., Ayub, M. Y., Abdollahi, A., & Dutta, A. (2023). A Hybrid Deep Learning Model-Based Intrusion Detection System for Emergency Planning Using IoT-Network. *Proceedings - 2023 IEEE 8th International Conference on Information and Communication Technologies for Disaster Management, ICT-DM 2023*. DOI: 10.1109/ICT-DM58371.2023.10286954

Khan, A., Gupta, S., & Gupta, S. K. (2022). Emerging UAV technology for disaster detection, mitigation, response, and preparedness. *Journal of Field Robotics*, *39*(6), 905–955. DOI: 10.1002/rob.22075

Khan, I. U., Abdollahi, A., Alturki, R., Alshehri, M. D., Ikram, M. A., Alyamani, H. J., & Khan, S. (2021). Intelligent detection system enabled attack probability using Markov chain in aerial networks. *Wireless Communications and Mobile Computing*, *2021*(1), 1–9. DOI: 10.1155/2021/1542657

Khan, I. U., Abdollahi, A., Jamil, A., Baig, B., Aziz, M. A., & Subhan, F. (2022). A Novel Design of FANET Routing Protocol Aided 5G Communication Using IoT. J. *Mobile Multimedia*, *18*(5), 1333–1354.

Khan, I. U., Nain Zukhraf, S. Z., Abdollahi, A., Imran, S. A., Qureshi, I. M., Aziz, M. A., & Hussian Shah, S. B. (2020, November). Reinforce based optimization in wireless communication technologies and routing techniques using internet of flying vehicles. In *Proceedings of the 4th International Conference on Future Networks and Distributed Systems* (pp. 1-6).

Khan, I. U., Qureshi, I. M., Aziz, M. A., Cheema, T. A., & Shah, S. B. H. (2020). Smart IoT control-based nature inspired energy efficient routing protocol for flying ad hoc network (FANET). *IEEE Access : Practical Innovations, Open Solutions*, *8*, 56371–56378. DOI: 10.1109/ACCESS.2020.2981531

Khan, I. U., Shah, S. B. H., Wang, L., Aziz, M. A., Stephan, T., & Kumar, N. (2021). Routing protocols & unmanned aerial vehicles autonomous localization in flying networks. *International Journal of Communication Systems*, •••, 4885.

Khan, M. A., Alvi, B. A., Safi, A., & Khan, I. U. (2018). Drones for good in smart cities: a review. *Proc. Int. Conf. Elect., Electron., Comput., Commun., Mech. Comput.(EECCMC)*, 1–6.

Khan, M. A., Javaid, S., Mohsan, S. A. H., Tanveer, M., & Ullah, I. (2024). Future-Proofing Security for UAVs With Post-Quantum Cryptography: A Review. In *IEEE Open Journal of the Communications Society*. Institute of Electrical and Electronics Engineers Inc., DOI: 10.1109/OJCOMS.2024.3486649

Khan, M. A., Menouar, H., Eldeeb, A., Abu-Dayya, A., & Salim, F. D. (2022). On the detection of unauthorized drones—Techniques and future perspectives: A review. *IEEE Sensors Journal*, *22*(12), 11439–11455. DOI: 10.1109/JSEN.2022.3171293

Khosiawan, Y., Park, Y., Moon, I., Nilakantan, J. M., & Nielsen, I. (2018). Task scheduling system for UAV operations in indoor environment. *Neural Computing & Applications*, *31*(9), 5431–5459. DOI: 10.1007/s00521-018-3373-9

Kim, K., & Kang, Y. (2020). Drone Security Module for UAV Data Encryption. *ICTC 2020 (International Conference on Information and Communication Technology Convergence).*

Kim, K. T., Kim, H., Jeong, S., Lee, Y. S., Choi, E., & Kim, J. Y. (2025). Quantification of Greenhouse Gas Emissions from a Municipal Solid Waste Incinerator Using an Uncrewed Aerial Vehicle. *Environment International*, *198*, 109396. Advance online publication. DOI: 10.1016/j.envint.2025.109396 PMID: 40184891

Kim, Y., Eisenberg, D. A., Bondank, E. N., Chester, M. V., Mascaro, G., & Underwood, B. S. (2017). Fail-safe and safe-to-fail adaptation: Decision-making for urban flooding under climate change. *Climatic Change*, *145*(3-4), 397–412. DOI: 10.1007/s10584-017-2090-1

Koh, L. P., & Wich, S. A. (2012). Dawn of drone ecology: Low-cost autonomous aerial vehicles for conservation. *Tropical Conservation Science*, *5*(2), 121–132. DOI: 10.1177/194008291200500202

Koranteng, C., Simons, B., & Essel, C. (2019). Climate responsive buildings: A comfort assessment of buildings on KNUST campus, Kumasi. *Journal of Engineering. Defence Technology*, *17*(5), 862–877.

Koubaa, A., Ammar, A., Abdelkader, M., Alhabashi, Y., & Ghouti, L. (2023). AERO: AI-enabled remote sensing observation with onboard edge computing in UAVs. *Remote Sensing (Basel)*, *15*(7), 1873. DOI: 10.3390/rs15071873

Koulianos, A., & Litke, A. (2023). Blockchain technology for secure communication and formation control in smart drone swarms. *Future Internet*, *15*(10), 344. DOI: 10.3390/fi15100344

Krichen, M., Adoni, W. Y. H., Mihoub, A., Alzahrani, M. Y., & Nahhal, T. (2022). Security Challenges for Drone Communications: Possible Threats, Attacks and Countermeasures. *Proceedings - 2022 2nd International Conference of Smart Systems and Emerging Technologies, SMARTTECH 2022*, 184–189. DOI: 10.1109/SMARTTECH54121.2022.00048

Kucharczyk, M., & Hugenholtz, C. H. (2021). Remote sensing of natural hazard-related disasters with small drones: Global trends, biases, and research opportunities. *Remote Sensing of Environment*, *264*, 112577. DOI: 10.1016/j.rse.2021.112577

Kumar, S., & Kanwar, S. R. (2022). Community Participatory Approach to Reverse Climate Change.

Kumar, A. (2020). Drone proliferation and security threats. *Indian Journal of Asian Affairs*, *33*(1/2), 43–62.

Kumar, A., de Jesus Pacheco, D. A., Kaushik, K., & Rodrigues, J. J. (2022). Futuristic view of the internet of quantum drones: Review, challenges and research agenda. *Vehicular Communications*, *36*, 100487. DOI: 10.1016/j.vehcom.2022.100487

Kumar, A., & Jain, S. (2021). Drone-based monitoring and redirecting system. In *Development and Future of Internet of Drones (IoD): Insights, Trends and Road Ahead* (pp. 163–183). Springer International Publishing.

Kumar, R., & Mehta, A. (2024). Monitoring biodiversity and ecosystem resilience: Impacts of urban expansion. *Global Ecology and Conservation*, *42*, e02543. DOI: 10.1016/j.gecco.2024.e02543

Kumar, V., Ranjan, D., & Verma, K. (2021). Global climate change: the loop between cause and impact. In *Global Climate Change* (pp. 187–211). Elsevier. DOI: 10.1016/B978-0-12-822928-6.00002-2

Kunertova, D. (2023). The war in Ukraine shows the game-changing effect of drones depends on the game. *Bulletin of the Atomic Scientists, 79*(2), 95–102. DOI: 10.1080/00963402.2023.2178180

La Roche, P., & Liggett, R. (2001). Very simple design tools: A web based assistant for the design of climate responsive buildings. *Architectural Science Review, 44*(4), 437–448. DOI: 10.1080/00038628.2001.9696924

Labib, N. S., Brust, M. R., Danoy, G., & Bouvry, P. (2021). The rise of drones in internet of things: A survey on the evolution, prospects and challenges of unmanned aerial vehicles. *IEEE Access : Practical Innovations, Open Solutions, 9*, 115466–115487. DOI: 10.1109/ACCESS.2021.3104963

Lambeck, K., Esat, T., & Potter, E. K. (2002). Links between climate and sea levels for the past three million years. *Nature, 419*(6903), 199–206. DOI: 10.1038/nature01089 PMID: 12226674

Lamoureux, S., Forbes, D. L., Bell, T., Manson, G., Rudy, A. C. A., Lalonde, J., & Fraser, P. R. (2015). In Stern, G. A., & Gaden, A. (Eds.), *The impact of climate change on infrastructure in the western and central Canadian Arctic. From Science to Policy in the Western and Central Canadian Arctic: An Integrated Regional Impact Study (IRIS) of climate change and modernization* (pp. 301–341). ArcticNet.

Lan, Y., & Chen, S. (2018). Current status and trends of plant protection UAV and its spraying technology in China. *International Journal of Precision Agricultural Aviation, 1*(1), 1–9. DOI: 10.33440/j.ijpaa.20180101.0002

Laukys, J., Maršalka, B., Daugėla, I., & Stankūnavičius, G. (2023). Drone-Based Vertical Atmospheric Temperature Profiling in Urban Environments. *Drones (Basel), 7*(11), 645. Advance online publication. DOI: 10.3390/drones7110645

Li, Y., & Pu, C. (2020, December). Lightweight digital signature solution to defend micro aerial vehicles against man-in-the-middle attack. In *2020 IEEE 23rd international conference on computational science and engineering (CSE)* (pp. 92-97). IEEE. DOI: 10.1109/CSE50738.2020.00020

Liang, H., & Yin, X. (2023). Self-Healing Control: Review, Framework, and Prospect. *IEEE Access : Practical Innovations, Open Solutions, 11*, 79495–79512. DOI: 10.1109/ACCESS.2023.3298554

Lichocki, P., Canonge, S., & Breazeal, C. (2011). The impact of robot swarms in environmental monitoring. *Proceedings of the International Workshop on Swarm Robotics*, 121–130.

Lin, J., Sedigh, S., & Miller, A. (2012). Self-healing autonomous systems: A literature survey. In 2012 24th International Conference on Microelectronics (ICM) (pp. 1–4). IEEE. https://doi.org/DOI: 10.1109/ICM.2012.6413372

Lindsey, R. (2022, April 19). Climate change: Global sea level. *Climate.gov.* Reviewed by Lumpkin, R., Johnson, G., Thompson, P., & Sweet, W. Retrieved from https://www.climate.gov/news-features/understanding-climate/climate-change-global-sea-level

Liu, Y., Paris, J. D., Vrekoussis, M., Antoniou, P., Constantinides, C., Desservettaz, M., Keleshis, C., Laurent, O., Leonidou, A., Philippon, C., Vouterakos, P., Quéhé, P.-Y., Bousquet, P., & Sciare, J. (2022). Improvements of a low-cost CO 2 commercial nondispersive near-infrared (NDIR) sensor for unmanned aerial vehicle (UAV) atmospheric mapping applications. *Atmospheric Measurement Techniques*, *15*(15), 4431–4442. DOI: 10.5194/amt-15-4431-2022

Looman, R. (2017). Climate-responsive design: A framework for an energy concept design-decision support tool for architects using principles of climate-responsive design. A+ BEl Architecture and the Built Environment, (1), 1-282.

Lounis, K., Ding, S. H. H., & Zulkernine, M. (2023). D2D-MAP: A Drone to Drone Authentication Protocol Using Physical Unclonable Functions. *IEEE Transactions on Vehicular Technology*, *72*(4), 5079–5093. DOI: 10.1109/TVT.2022.3224611

Lucas, M. (2024). The Role of AI in Climate Change Mitigation and Environmental Monitoring. DOI: 10.13140/RG.2.2.21153.38247

Luo, J., & Oracevic, A. (2024). Adaptive Machine Learning for Efficient Anomaly Detection in Autonomous UAVs Swarm Operations. *Proceedings of the International Symposium on Networks, Computers and Communications (ISNCC)*, 2024. DOI: 10.1109/ISNCC62547.2024.10758945

Luu, T. M. N., Yadav, M., Srivastava, A. P., & Gopal, K. (2025). Innovation and Technology to Address the Challenges of Climate Change. In *Effects of Climate Change on Social and Economic Factors* (pp. 423–464). IGI Global.

Lv, Z. (2019). The security of Internet of drones. *Computer Communications*, *148*, 208–214. DOI: 10.1016/j.comcom.2019.09.018

Lv, Z., Chen, D., Feng, H., Lou, R., & Wang, H. (2021). Beyond 5G for digital twins of UAVs. *Computer Networks*, *197*, 108366. Advance online publication. DOI: 10.1016/j.comnet.2021.108366

Machida, F., & Andrade, E. (2022). Modeling and Analysis of Deforestation Prevention by Uncrewed Aerial Vehicles-based monitoring systems. *Environmental Modelling & Software, 158*, 105540. Advance online publication. DOI: 10.1016/j.envsoft.2022.105540

Mader, D., Blaskow, R., Westfeld, P., & Maas, H.-G. (2015). UAV-based acquisition of 3D point cloud—A comparison of a low-cost laser scanner and SfM-tools. *ISPRS International Archives of the Photogrammetry. Remote Sensing and Spatial Information Sciences, XL-3*(W3), 335–341. DOI: 10.5194/isprsarchives-XL-3-W3-335-2015

Maghazei, O., & Netland, T. (2020). Drones in manufacturing: Exploring opportunities for research and practice. *Journal of Manufacturing Technology Management, 31*(6), 1237–1259. DOI: 10.1108/JMTM-03-2019-0099

Magistretti, S., & Dell'Era, C. (2019). Unveiling opportunities afforded by emerging technologies: Evidences from the drone industry. *Technology Analysis and Strategic Management, 31*(5), 606–623. DOI: 10.1080/09537325.2018.1538497

Mahar, W. A. (2021). *Methodology for the design of climate-responsive houses for improved thermal comfort in cold semi-arid climates*. Universite de Liege.

Mairaj, A., Baba, A. I., & Javaid, A. Y. (2019). Application specific drone simulators: Recent advances and challenges. *Simulation Modelling Practice and Theory, 94*, 100–117. DOI: 10.1016/j.simpat.2019.01.004

Majeed, R., Abdullah, N. A., Faheem Mushtaq, M., Umer, M., & Nappi, M. (2021). Intelligent cyber-security system for iot-aided drones using voting classifier. *Electronics (Basel), 10*(23), 2926. DOI: 10.3390/electronics10232926

Manfreda, S., McCabe, M. F., Miller, P. E., Lucas, R., Madrigal, V. P., Mallinis, G., Ben Dor, E., Helman, D., Estes, L., & Ciraolo, G.. (2018). On the use of unmanned aerial systems for environmental monitoring. *Remote Sensing (Basel), 10*(5), 641. DOI: 10.3390/rs10040641

Mangewa, L. J., Ndakidemi, P. A., & Munishi, L. K. (2019). Integrating UAV technology in an ecological monitoring system for community wildlife management areas in Tanzania. *Sustainability (Basel), 11*(21), 6116. DOI: 10.3390/su11216116

MarketsandMarkets. (2022). Drone services market by type, duration, application, industry, and region—Global forecast to 2030. https://www.marketsandmarkets.com/Market-Reports/drone-service-market-80726041.html

Marr, B. (n.d.). Drones and Robots in the Vineyard: Revolutionizing Viticulture Through Technology. BM Wine Cellar. https://bmwineguide.co.uk/drones-and-robots-in-the-vineyard-revolutionizing-viticulture-through-technology/

Maseer, Z. K., Yusof, R., Mostafa, S. A., Bahaman, N., Musa, O., & Al-Rimy, B. A. S. (2021). DeepIoT.IDS: Hybrid deep learning for enhancing IoT network intrusion detection. *Computers, Materials & Continua*, *69*(3), 3946–3967. DOI: 10.32604/cmc.2021.016074

McClelland, G. T. W., Bond, A. L., Sardana, A., & Glass, T. (2016). Rapid population estimate of a surface-nesting seabird on a remote island using a low-cost unmanned aerial vehicle. *Marine Ornithology*, *44*, 215–220.

Mckinsey. (2024). "What is Artificial General Intelligence (AGI)?" available at: https://www.mckinsey.com/featured-insights/mckinsey-explainers/what-is-artificial-general-intelligence-agi

MDPI. (2023). Advancements in air quality monitoring: The role of drone technology. *Proceedings*, *105*(1), 19.

Measurusa. (n.d.). https://drones.measurusa.com/pages/drones-for-environmental-monitoring-and-conservation

Memane, S. E., & Kulkarni, V. S. (2015). A review on flame and smoke detection techniques in video's. *Int. J. Adv. Res. Electrical, Electronics Instrumen.Eng*, *4*(2), 885–889.

Mendoza, F., Minardi, M., Chatzinotas, S., Lei, L., & Vu, T. X. (2021, September). An SDN based testbed for dynamic network slicing in satellite-terrestrial networks. In *2021 IEEE international mediterranean conference on communications and networking (MeditCom)* (pp. 36-41). IEEE. DOI: 10.1109/MeditCom49071.2021.9647537

Merkert, R., & Bushell, J. (2020). Managing the drone revolution: A systematic literature review into the current use of airborne drones and future strategic directions for their effective control. *Journal of Air Transport Management*, *89*, 101929. DOI: 10.1016/j.jairtraman.2020.101929 PMID: 32952321

Mertens, J. S., Milotta, G. M., Nagaradjane, P., & Morabito, G. (2020, August). SDN-(UAV) ISE: Applying software defined networking to wireless sensor networks with data mules. In *2020 IEEE 21st International Symposium on" A World of Wireless, Mobile and Multimedia Networks"(WoWMoM)* (pp. 323-328). IEEE.

Mhlanga, D., & Dzingirai, M. (2024). Smart technologies in agriculture: Advancements and benefits for sustainable development. In *Responsible Business and Sustainable Development* (pp. 57–72). Routledge.

Minkina, W. (2020). Theoretical basics of radiant heat transfer—Practical examples of calculation for the infrared (IR) used in infrared thermography measurements. *Quantitative Infrared Thermography Journal, 18*(3), 269–282. DOI: 10.1080/17686733.2020.1739260

Minn, W., Tun, Y. N., Shar, L. K., & Jiang, L. (2024). DronLomaly: Runtime Log-based Anomaly Detector for DJI Drones. *Proceedings - International Conference on Software Engineering*, 6–10. DOI: 10.1145/3639478.3640042

Mishra, D., Vegni, A. M., Loscrí, V., & Natalizio, E. (2022). Drone networking in the 6G era: A technology overview. *IEEE Communications Standards Magazine, 5*(4), 88–95. DOI: 10.1109/MCOMSTD.0001.2100016

Mitra, D., Bhattacharya, A., & Banerjee, K. (2021). Mapping coastal vulnerability in the Indian Sundarbans using geospatial techniques. *Journal of Coastal Research, 37*(2), 377–389. DOI: 10.2112/JCOASTRES-D-20-00085.1

Modern Technology For Enhancing Agricultural Productivity. (2020, August 29). https://practicalagrculture1214.blogspot.com/search/label/%23Agriculture

Mohamed, N., Al-Jaroodi, J., Jawhar, I., Idries, A., & Mohammed, F. (2020). Unmanned aerial vehicles applications in future smart cities. *Technological Forecasting and Social Change, 153*, 119293. DOI: 10.1016/j.techfore.2018.05.004

Mohan, M., Richardson, G., Gopan, G., Aghai, M. M., Bajaj, S., Galgamuwa, G. P., Vastaranta, M., Arachchige, P. S. P., Amorós, L., Corte, A. P. D., de-Miguel, S., Leite, R. V., Kganyago, M., Broadbent, E. N., Doaemo, W., Shorab, M. A. B., & Cardil, A. (2021). UAV-supported forest regeneration: Current trends, challenges and implications. *Remote Sensing (Basel), 13*(13), 2596. DOI: 10.3390/rs13132596

Mohan, M., Rue, H. A., Bajaj, S., Galgamuwa, G. P., Adrah, E., Aghai, M. M., Broadbent, E. N., Khadamkar, O., Sasmito, S. D., Roise, J., Doaemo, W., & Cardil, A. (2021). Afforestation, reforestation and new challenges from COVID-19: Thirty-three recommendations to support Civil Society Organizations (CSOs). [Niazy, M. M. The role of drones in detecting cadmium pollution.]. *Journal of Environmental Management, 287*, 112277. DOI: 10.1016/j.jenvman.2021.112277 PMID: 33756214

Mohsan, S. A. H., Othman, N. Q. H., Li, Y., Alsharif, M. H., & Khan, M. A. (2023). Unmanned aerial vehicles (UAVs): Practical aspects, applications, open challenges, security issues, and future trends. *Intelligent Service Robotics, 16*(1), 1–29. DOI: 10.1007/s11370-022-00452-4 PMID: 36687780

Moi, T., Cibicik, A., & Rølvåg, T. (2020). Digital twin based condition monitoring of a knuckle boom crane: An experimental study. *Engineering Failure Analysis*, *112*, 104517. DOI: 10.1016/j.engfailanal.2020.104517

Montalbano, E. (2022). Drone Technology Soars into 2023. Design News. https://www.designnews.com/automation/drone-technology-soars-into-2023

Morshedi, R., & Matinkhah, S. M. *, & Sadeghi, M. T. (2024). Intrusion Detection for IoT Network Security with Deep Learning. [JAIDM]. *Journal of Artificial Intelligence and Data Mining*.

Mowla, N. I., Tran, N. H., Doh, I., & Chae, K. (2019). Federated learning-based cognitive detection of jamming attack in flying ad-hoc network. *IEEE Access : Practical Innovations, Open Solutions*, *8*, 4338–4350. DOI: 10.1109/ACCESS.2019.2962873

Mughal, U. A., Atat, R., & Ismail, M. (2024). Graph Neural Network-Based Intrusion Detection System for a Swarm of UAVs. *Proceedings - IEEE Military Communications Conference MILCOM*, 578–583. DOI: 10.1109/MILCOM61039.2024.10773671

Mughal, U. A., Hassler, S. C., & Ismail, M. (2023, October). Machine Learning-Based Intrusion Detection for Swarm of Unmanned Aerial Vehicles. In *2023 IEEE Conference on Communications and Network Security (CNS)* (pp. 1-9). IEEE. DOI: 10.1109/CNS59707.2023.10288962

Mulla, D. J. (2013). Twenty five years of remote sensing in precision agriculture: Key advances and remaining knowledge gaps. *Biosystems Engineering*, *114*(4), 358–371. DOI: 10.1016/j.biosystemseng.2012.08.009

Myagmar-Ochir, Y., & Kim, W. (2023, August 23). A Survey of Video Surveillance Systems in Smart City. *Electronics (Basel)*, *12*(17), 3567–3567. DOI: 10.3390/electronics12173567

Myburgh, A., Botha, H., Downs, C. T., & Woodborne, S. M. (2021). The application and limitations of a low-cost UAV platform and open-source software combination for ecological mapping and monitoring. *African Journal of Wildlife Research*, *51*(1), 166–177. DOI: 10.3957/056.051.0166

Nabi, F., Zhou, X., Iftikhar, U., & Attaullah, H. M. (2024). A Case Study of Cyber Subversion Attack based Design Flaw in Service Oriented Component Application Logic. *Journal of Cyber Security Technology*, *8*(3), 204–228. DOI: 10.1080/23742917.2023.2261169

Naqvi, R. A., Hussain, D., & Loh, W.-K. (2020). Artificial intelligence-based semantic segmentation of ocular regions for biometrics and healthcare applications. *Computer Modeling in Engineering & Sciences*, *66*, 715–732. DOI: 10.32604/cmc.2020.011016

Narejo, I. H., Matloob, M. J., Attaullah, H. M., & Khawar, R. (2024). Designing and Evaluating a Mobile Social Robot as an Intervention Tool for Autism Spectrum Disorder. *2024 IEEE 1st Karachi Section Humanitarian Technology Conference, Khi-HTC 2024*. DOI: 10.1109/KHI-HTC60760.2024.10482306

National Aeronautics and Space Administration. (n.d.). Are sea levels rising the same all over the world, as if we're filling a giant bathtub? *Sea Level Change: Observations from Space*. Retrieved from https://sealevel.nasa.gov/faq/9/are-sea-levels-rising-the-same-all-over-the-world-as-if-were-filling-a-giant-bathtub/

Nawaz, M., Mehmood, Z., Nazir, T., Naqvi, R. A., Rehman, A., Iqbal, M., & Saba, T. (2021). Skin cancer detection from dermoscopic images using deep learning and fuzzy k-means clustering. *Microscopy Research and Technique*, *85*(2), 339–351. DOI: 10.1002/jemt.23846 PMID: 34448519

Nayyar, A., Nguyen, B. L., & Nguyen, N. G. (2020). The internet of drone things (IoDT): future envision of smart drones. In *First International Conference on Sustainable Technologies* DOI: 10.1007/978-981-15-0029-9_45

Neumann, J. E., & Price, J. C. (2009). Adapting to Climate Change: The Public Policy Response–Public Infrastructure.

Neumann, J. (2009). Adaptation to climate change: Revisiting infrastructure norms. *Issue Brief*, •••, 9–15.

Newbold, T. (2018). Future effects of climate and land-use change on terrestrial vertebrate community diversity under different scenarios. Proceedings of the Royal Society B: Biological Sciences, 285(1881), 20180792. https://doi.org/DOI: 10.1098/rspb.2018.0792

Newswire, P. R. (2023). *XAG drones in vineyards make wine growing safer and easier.* https://www.prnewswire.com/news-releases/xag-drones-in-vineyards-make-wine-growing-safer-and-easier-301755238.html

Nex, F., & Remondino, F. (2014). UAV for 3D mapping applications: A review. *Applied Geomatics*, *6*(1), 1–15. DOI: 10.1007/s12518-013-0120-x

Ngo, R. (2020). *AGI Safety From First Principles. Future of Humanity Institute (FHI)*. University of Oxford.

Nguyen, A.-T., Tran, Q.-B., Tran, D.-Q., & Reiter, S. (2011). An investigation on climate responsive design strategies of vernacular housing in Vietnam. *Building and Environment*, *46*(10), 2088–2106. DOI: 10.1016/j.buildenv.2011.04.019

Niyonsaba, S., Konate, K., & Soidridine, M. M. (2024). Deep Learning Based Intrusion Detection for Cybersecurity in Unmanned Aerial Vehicles Network. *International Conference on Electrical, Computer, and Energy Technologies, ICECET 2024*. DOI: 10.1109/ICECET61485.2024.10698453

Noor, M., Rehman, N. U., Jalil, A., Fahad, S., Adnan, M., Wahid, F., ... & Hassan, S. (2020). Climate change and costal plant lives. Environment, climate, plant and vegetation growth, 93-108.

Nourmohammadi, A., Jafari, M., & Zander, T. O. (2018). A survey on unmanned aerial vehicle remote control using brain–computer interface. *IEEE Transactions on Human-Machine Systems*, *48*(4), 337–348. DOI: 10.1109/THMS.2018.2830647

Ntizikira, E., Lei, W., Alblehai, F., Saleem, K., & Lodhi, M. A. (2023). Secure and privacy-preserving intrusion detection and prevention in the internet of unmanned aerial vehicles. *Sensors (Basel)*, *23*(19), 8077. DOI: 10.3390/s23198077 PMID: 37836907

Obaid, O. I. (2023). From machine learning to artificial general intelligence: A roadmap and implications. *Mesopotamian Journal of Big Data*, *2023*, 81–91. DOI: 10.58496/MJBD/2023/012

Ojo, B. (2024). Strategies for the optimization of critical infrastructure projects to enhance urban resilience to climate change. *The Journal of Scientific and Engineering Research*, *11*, 107–123.

Olson, D., & Anderson, J. (2021). Review on unmanned aerial vehicles, remote sensors, imagery processing, and their applications in agriculture. *Agronomy Journal*, *113*(2), 971–992. DOI: 10.1002/agj2.20595

Omolara, A. E., Alawida, M., & Abiodun, O. I. (2023). Drone cybersecurity issues, solutions, trend insights and future perspectives: A survey. *Neural Computing & Applications*, *35*(31), 23063–23101. DOI: 10.1007/s00521-023-08857-7

Osman, O., Yehya, A., & Maalouf, E. (2025). New perspective on designing climate-responsive latent energy storage systems. *Renewable & Sustainable Energy Reviews*, *211*, 115286. DOI: 10.1016/j.rser.2024.115286

Pajares, G. (2015). Overview and current status of remote sensing applications based on unmanned aerial vehicles (UAVs). *Photogrammetric Engineering and Remote Sensing*, *81*(4), 281–329. DOI: 10.14358/PERS.81.4.281

Panarelli, G. (2015). Climate responsive design. Research, strategies and assessment criteria. Technological innovation and Climate Responsive Architecture Approach in the project. *Housing Policies and Urban Economics*, *3*, 92–94.

Panday, U. S., Pratihast, A. K., Aryal, J., & Kayastha, R. B. (2020). A review on drone-based data solutions for cereal crops. *Drones (Basel)*, *4*(3), 41. DOI: 10.3390/drones4030041

Pande, A. K., Brantley, P., Tanveer, M. H., & Voicu, R. C. (2024). From AI to AGI - The Evolution of Real-Time Systems with GPT Integration. *SoutheastCon*, *2024*, 699–707. DOI: 10.1109/SoutheastCon52093.2024.10500172

Pandey, G. K., Gurjar, D. S., Nguyen, H. H., & Yadav, S. (2022). Security Threats and Mitigation Techniques in UAV Communications: A Comprehensive Survey. *IEEE Access : Practical Innovations, Open Solutions*, *10*, 112858–112897. DOI: 10.1109/ACCESS.2022.3215975

Pang, X., Zhao, N., Tang, J., Niyato, D., & Wong, K. K. (2021). Joint Trajectory and Beamforming Optimization for Secure UAV Transmission Aided by IRS. *13th International Conference on Wireless Communications and Signal Processing, WCSP 2021*. DOI: 10.1109/WCSP52459.2021.9613308

Park, K. H., Park, E., & Kim, H. K. (2020). Unsupervised intrusion detection system for unmanned aerial vehicle with less labeling effort. In *Information Security Applications:21st International Conference, WISA 2020,Jeju Island, South Korea,August 26–28, 2020, Revised Selected Papers 21* (pp. 45-58). Springer International Publishing. DOI: 10.1007/978-3-030-65299-9_4

Park, S., Kim, H. T., Lee, S., Joo, H., & Kim, H. (2021). Survey on anti-drone systems: Components, designs, and challenges. *IEEE Access : Practical Innovations, Open Solutions*, *9*, 42635–42659. DOI: 10.1109/ACCESS.2021.3065926

Parra-López, C., Ben Abdallah, S., Garcia-Garcia, G., Hassoun, A., Sánchez-Zamora, P., Trollman, H., Jagtap, S., & Carmona-Torres, C. (2024). Integrating digital technologies in agriculture for climate change adaptation and mitigation: State of the art and future perspectives. In *Computers and Electronics in Agriculture* (Vol. 226). Elsevier B.V., DOI: 10.1016/j.compag.2024.109412

Paterson, M. (2020). Climate change and international political economy: Between collapse and transformation. *Review of International Political Economy*, *28*(2), 394–405. DOI: 10.1080/09692290.2020.1830829

Pathak, H., Kumar, G., Mohapatra, S., Gaikwad, B., & Rane, J. (2020). Use of drones in agriculture: Potentials, Problems and Policy Needs. *ICAR-National Institute of Abiotic Stress Management*, *300*, 4–15.

Pettorelli, N., Laurance, W. F., O'Brien, T. G., Wegmann, M., Nagendra, H., & Turner, W. (2014). Satellite remote sensing for applied ecologists: Opportunities and challenges. *Journal of Applied Ecology*, *51*(4), 839–848. DOI: 10.1111/1365-2664.12261

Pichler, P., & Hartig, F. (2022). Deep learning as a tool for ecology and evolution. *Methods in Ecology and Evolution*, *13*(9), 1903–1920.

Pichler, P., Hartig, F., & Dormann, C. F. (2022). Machine learning and deep learning—A review for ecologists. *Methods in Ecology and Evolution*, *13*(7), 1640–1656.

Piekkoontod, T., Pachana, B., Hrimpeng, K., & Charoenjit, K. (2020). Assessments of Nipa forest using Landsat imagery enhanced with unmanned aerial vehicle photography. *Applied Environmental Research*, *42*(3), 49–59. DOI: 10.35762/AER.2020.42.3.5

Pliatsios, D., Goudos, S. K., Lagkas, T., Argyriou, V., Boulogeorgos, A. A. A., & Sarigiannidis, P. (2022). Drone-Base-Station for Next-Generation Internet-of-Things: A Comparison of Swarm Intelligence Approaches. *IEEE Open Journal of Antennas and Propagation*, *3*, 32–47. DOI: 10.1109/OJAP.2021.3133459

Politi, E., Varlamis, I., Tserpes, K., Larsen, M., & Dimitrakopoulos, G. (2022). The future of safe BVLOS drone operations with respect to system and service engineering. *Proceedings - 16th IEEE International Conference on Service-Oriented System Engineering, SOSE 2022*, 133–140. DOI: 10.1109/SOSE55356.2022.00022

Półka, M., Ptak, S., & Kuziora, Ł. (2017). The Use of UAV's for Search and Rescue Operations. *Procedia Engineering*, *192*, 748–752. DOI: 10.1016/j.proeng.2017.06.129

Power, E. E. (2023). *Sophisticated drone technology applications in 2023*.https://eepower.com/tech-insights/sophisticated-drone-technology-applications-in-2023/

Pradhan, B., Dini Hairi Bin Suliman, M., & Arshad Bin Awang, M. (2007). Forest fire susceptibility and risk mapping using remote sensing and geographical information systems (GIS). *Disaster Prevention and Management*, *16*(3), 344–352. DOI: 10.1108/09653560710758297

Prasetio, E. A., & Nurliyana, C. (2023). Evaluating perceived safety of autonomous vehicle: The influence of privacy and cybersecurity to cognitive and emotional safety. *IATSS Research*, *47*(2), 160–170. DOI: 10.1016/j.iatssr.2023.06.001

Proag, V., & Proag, V. (2021). Climate change and infrastructure. Infrastructure Planning and Management: An Integrated Approach, 279-304.

Qin, C., & Pournaras, E. (2023). Short vs. long-term coordination of drones: When distributed optimization meets deep reinforcement learning. arXiv preprint arXiv:2311.09852.

Qi, N., Wang, W., Ye, D., Wang, M., Tsiftsis, T. A., & Yao, R. (2021). Energy-efficient full-duplex UAV relaying networks: Trajectory design for channel-model-free scenarios. *ETRI Journal*, *43*(3), 436–446. DOI: 10.4218/etrij.2020-0060

Quinn, J., & Bouk, S. (2023). Intrusion Detection Systems for Flying Ad-hoc Networks. arXiv preprint.

Radmanesh, M., Kumar, M., Nemati, A., & Sarim, M. (2015). Dynamic optimal UAV trajectory planning in the national airspace system via mixed integer linear programming. *Proceedings of the Institution of Mechanical Engineers. Part G, Journal of Aerospace Engineering*, *230*(9), 1668–1682. DOI: 10.1177/0954410015609361

Radoglou-Grammatikis, P., Sarigiannidis, P., Lagkas, T., & Moscholios, I. (2020). A compilation of UAV applications for precision agriculture. *Computer Networks*, *172*, 107148. DOI: 10.1016/j.comnet.2020.107148

Rafael Aditya Cahyo, W., Jonarta, H., Clarissa, A. C., Syafalni, I., Sutisna, N., & Hasanuddin, M. O. (2023). Securing UAV Communication with TinyJAMBU Cryptography. *Proceedings of the International Conference on Electrical Engineering and Informatics*. DOI: 10.1109/ICEEI59426.2023.10346669

Rahman, H., Siregar, V. P., & Putra, R. S. (2022). Coastal inundation risk assessment in Indonesia using AGI-powered drones. *Marine Geodesy*, *45*(1), 29–47. DOI: 10.1080/01490419.2022.2045678

Rahman, K., Aziz, M. A., Usman, N., Kiren, T., Cheema, T. A., Shoukat, H., Bhatia, T. K., Abdollahi, A., & Sajid, A. (2023). Cognitive Lightweight Logistic Regression-Based IDS for IoT-Enabled FANET to Detect Cyberattacks. *Mobile Information Systems*, *2023*(1), 7690322. DOI: 10.1155/2023/7690322

Raja, G., Anbalagan, S., Ganapathisubramaniyan, A., Selvakumar, M. S., Bashir, A. K., & Mumtaz, S. (2021). Efficient and Secured Swarm Pattern Multi-UAV Communication. *IEEE Transactions on Vehicular Technology*, *70*(7), 7050–7058. DOI: 10.1109/TVT.2021.3082308

Rajagopal, T., Balakrishnan, A., Valsalakumar, S., Rajagopal, T. K. R., & Sundaram, S. (2021). Application of MSVPC-5G multicast SDN network eminence video transmission in drone thermal imaging for solar farm monitoring. *Energies*, *14*(24), 8255. DOI: 10.3390/en14248255

Rajen, G. (1999). *Cooperative environmental monitoring in the coastal regions of India and Pakistan*. Cooperative Monitoring Center Occasional Paper/11, Sandia National Laboratories. Retrieved from http://www.cmc.sandia.gov/issues/papers/coastal/index.html

Rakha, T., Liberty, A., Gorodetsky, A., Kakillioglu, B., & Velipasalar, S. (2018). Heat mapping drones: An autonomous computer-vision-based procedure for building envelope inspection using unmanned aerial systems (UAS). *Technology | Architecture + Design, 2*(1), 30–44. https://doi.org/DOI: 10.1080/24751448.2018.1428137

Ramadan, R. A., Emara, A. H., Al-Sarem, M., & Elhamahmy, M. (2021). Internet of drones intrusion detection using deep learning. *Electronics (Basel), 10*(21), 2633. Advance online publication. DOI: 10.3390/electronics10212633

Ramesh, K., & Kumar, V. (2023). *Sea Level Rise Monitoring in India: The Role of UAVs and Remote Sensing Technologies*. Oceanographic Research Institute.

Rammer, W., & Seidl, R. (2019). Harnessing deep learning in ecology: An example predicting bark beetle outbreaks. *Frontiers in Plant Science, 10*, 1327. DOI: 10.3389/fpls.2019.01327 PMID: 31719829

Rani, C., Modares, H., Sriram, R., Mikulski, D., & Lewis, F. L. (2016). Security of unmanned aerial vehicle systems against cyber-physical attacks. *The Journal of Defense Modeling and Simulation, 13*(3), 331–342. DOI: 10.1177/1548512915617252

Ranyal, E., & Jain, K. (2020). Unmanned aerial vehicle's vulnerability to GPS spoofing: A review. *Photonirvachak (Dehra Dun), 49*(3), 585–591. DOI: 10.1007/s12524-020-01225-1

Rasmussen, K. (2011). On the requirements for successful GPS spoofing attacks.

Rault, T., Bouabdallah, A., & Challal, Y. (2014). Energy efficiency in wireless sensor networks: A top-down survey. *Computer Networks, 67*, 104–122. DOI: 10.1016/j.comnet.2014.03.027

Rault, T., Bouabdallah, A., & Challal, Y. (2023). Advances and applications of machine learning and deep learning in environmental ecology and health: A review. *Environmental Pollution, 306*, 119347.

Raya, A. M., & Esteve, J. S. (2024). Challenges of Civil Application of 'Drones' for Commercial Purposes in the Logistics and Transportation Sector. In *Applying Drones to Current Societal and Industrial Challenges* (pp. 235–253). Springer. DOI: 10.1007/978-3-031-55571-8_9

Reiche, J., Lucas, R., Mitchell, A. L., Verbesselt, J., Hoekman, D., Haarpaintner, J., Kellndorfer, J. M., Rosenqvist, A., Lehmann, E. A., Woodcock, C. E., Seifert, F. M., & Herold, M. (2016). Combining satellite data for better tropical forest monitoring. *Nature Climate Change*, *6*(2), 120–122. DOI: 10.1038/nclimate2919

Rele, M., & Patil, D. Examining the Impact of Artificial Intelligence on Cybersecurity within the Internet of Things.

Research, K. B. V. (n.d.). *Drone data services market.*https://www.kbvresearch.com/drone-data-services-market/

ResearchGate. (2019). *Types of UAV communication [Figure].*https://www.researchgate.net/figure/types-of-uav-communication_fig1_331471106

Resop, J. P., Lehmann, L., & Hession, W. C. (2019). Drone laser scanning for modeling riverscape topography and vegetation: Comparison with traditional aerial lidar. *Drones (Basel)*, *3*(2), 35. DOI: 10.3390/drones3020035

Restas, A. (2015). Drone applications for supporting disaster management. *World Journal of Engineering and Technology*, *3*(3), 316–321. DOI: 10.4236/wjet.2015.33C047

Restituyo, R., & Hayajneh, T. (2018, November). Vulnerabilities and attacks analysis for military and commercial iot drones. In *2018 9th IEEE Annual Ubiquitous Computing, Electronics & Mobile Communication Conference (UEMCON)* (pp. 26-32). IEEE. DOI: 10.1109/UEMCON.2018.8796596

Rezwan, S., & Choi, W. (2022). Artificial intelligence approaches for UAV navigation: Recent advances and future challenges. *IEEE Access : Practical Innovations, Open Solutions*, *10*, 26320–26339. DOI: 10.1109/ACCESS.2022.3157626

Ripple, W. J., Wolf, C., Newsome, T. M., Barnard, P., & Moomaw, W. R. (2020). World scientists' warning of a climate emergency. *Bioscience*, *70*(1), 8–100. DOI: 10.1093/biosci/biz152

Rocque, R. J., Beaudoin, C., Ndjaboue, R., Cameron, L., Poirier-Bergeron, L., Poulin-Rheault, R.-A., Fallon, C., Tricco, A. C., & Witteman, H. O. (2021). Health effects of climate change: An overview of systematic reviews. *BMJ Open*, *11*(6), e046333. DOI: 10.1136/bmjopen-2020-046333 PMID: 34108165

Rodríguez, D. A., Tafur, C. L., Daza, P. F. M., Vidales, J. A. V., & Rincón, J. C. D. (2024). Inspection of aircrafts and airports using UAS: A review. *Results in Engineering*, *22*, 102330. DOI: 10.1016/j.rineng.2024.102330

Roldán-Gómez, J. J., González-Gironda, E., & Barrientos, A. (2021). A survey on robotic technologies for forest firefighting: Applying drone swarms to improve firefighters' efficiency and safety. *Applied Sciences (Basel, Switzerland), 11*(1), 363. DOI: 10.3390/app11010363 PMID: 36660242

Roldán, J. J., Joossen, G., Sanz, D., Del Cerro, J., & Barrientos, A. (2015). Mini-UAV based sensory system for measuring environmental variables in greenhouses. *Sensors (Basel), 15*(2), 3334–3350. DOI: 10.3390/s150203334 PMID: 25648713

Rozhon, J., Rezac, F., Safarik, J., Gresak, E., & Jalowiczor, J. (2019, May). Measuring and monitoring the QoS and QoE in software defined networking environments. In *Signal Processing, Sensor/Information Fusion, and Target Recognition XXVIII* (Vol. 11018, pp. 474–483). SPIE. DOI: 10.1117/12.2518838

Ruckelshaus, M., Reguero, B. G., Arkema, K., Compeán, R. G., Weekes, K., Bailey, A., & Silver, J. (2020). Harnessing new data technologies for nature-based solutions in assessing and managing risk in coastal zones. *International Journal of Disaster Risk Reduction, 51*, 101795. DOI: 10.1016/j.ijdrr.2020.101795

Ryan, M., Nuhoff-Isakhanyan, G., & Tekinerdoğan, B. (2023, January 30). An interdisciplinary approach to artificial intelligence in agriculture. Taylor & Francis, 95(1). DOI: 10.1080/27685241.2023.2168568

Saad, W., Hashash, O., Thomas, C. K., Chaccour, C., Debbah, M., Mandayam, N., & Han, Z. (2025). Artificial General Intelligence (AGI)-Native Wireless Systems: A Journey Beyond 6G. *Proceedings of the IEEE*, 1–39. DOI: 10.1109/JPROC.2025.3526887

Sachs, J. D., Schmidt-Traub, G., & Williams, J. (2023). *Biodiversity and ecosystem services: Monitoring for sustainability*. Springer.

Saharan, S., & Yadav, B. (2022). Digital and cyber forensics: A contemporary evolution in forensic sciences. In *Crime Scene Management within Forensic Science: Forensic Techniques for Criminal Investigations* (pp. 267-294). Springer. DOI: 10.1007/978-981-16-6683-4_11

Saharan, S., Singh, S., Bora, D., & Saxena, G. (2024). Cyber Forensics: A Boon to Cybersecurity. In *Digital Technologies in Modeling and Management: Insights in Education and Industry* (pp. 93-115). IGI Global.

Saharan, S., Bhandari, A. K., & Yadav, B. (2024). Use of Advanced Technologies for Drones in the Context of Security Issues and Challenges. In *Cybersecurity Issues and Challenges in the Drone Industry* (pp. 196–216). IGI Global. DOI: 10.4018/979-8-3693-0774-8.ch008

Saharan, S., Singh, S., Bhandari, A. K., & Yadav, B. (2024). The Future of Cyber-Crimes and Cyber War in the Metaverse. In *Forecasting Cyber Crimes in the Age of the Metaverse* (pp. 126–148). IGI Global.

Sajid, J., Hayawi, K., Malik, A. W., Anwar, Z., & Trabelsi, Z. (2023). A Fog Computing Framework for Intrusion Detection of Energy-Based Attacks on UAV-Assisted Smart Farming. *Applied Sciences (Basel, Switzerland)*, *13*(6), 3857. Advance online publication. DOI: 10.3390/app13063857

Salmon, P. M., McLean, S., Carden, T., King, B. J., Thompson, J., Baber, C., Stanton, N. A., & Read, G. J. M. (2024). When tomorrow comes: A prospective risk assessment of a future artificial general intelligence-based uncrewed combat aerial vehicle system. *Applied Ergonomics*, *117*, 104245. DOI: 10.1016/j.apergo.2024.104245 PMID: 38320386

Samadzadegan, F., Dadrass Javan, F., Ashtari Mahini, F., & Gholamshahi, M. (2022). Detection and recognition of drones based on a deep convolutional neural network using visible imagery. *Aerospace (Basel, Switzerland)*, *9*(1), 31. DOI: 10.3390/aerospace9010031

Samanta, S., Hazra, S., Mondal, P. P., Chanda, A., Giri, S., French, J. R., & Nicholls, R. J. (2021). Assessment and attribution of mangrove forest changes in the Indian Sundarbans from 2000 to 2020. *Remote Sensing (Basel)*, *13*(24), 4957. DOI: 10.3390/rs13244957

Sandalow, D. B., McCormick, C., Kucukelbir, A., Friedman, J., Nagrani, T., Fan, Z., ... & Ruane, A. (2023). Artificial Intelligence for Climate Change Mitigation Roadmap.

Sanyal, N., & Shuvo, S. R. S. (2020, October 29). Coastal Hazards in Asian Countries in the Context of Climate Change., 20. http://www.rericjournal.ait.ac.th/index.php/reric/article/download/2531/pdf

Sarim, M., Radmanesh, M., Dechering, M., Kumar, M., Pragada, R., & Cohen, K. (2019). Distributed detect-and-avoid for multiple unmanned aerial vehicles in national air space. *Journal of Dynamic Systems, Measurement, and Control*, *141*(7), 071014. DOI: 10.1115/1.4043190

Sarkar, S., Basak, S., & Sinha, A. (2020). Applications of AI in traffic monitoring and urban planning. *International Journal of Emerging Technology and Advanced Engineering*, *10*(7), 214–219.

Schönbein, R., Krause, J., & Brunsch, L. (2021). Artificial general intelligence and drones: Ethical and technological implications of merging autonomy and aerial robotics. *AI & Society*, *37*, 135–146. DOI: 10.1007/s00146-021-01203-6

Scotese, C. R., Song, H., Mills, B. J., & van der Meer, D. G. (2021). Phanerozoic paleotemperatures: The earth's changing climate during the last 540 million years. *Earth-Science Reviews*, *215*, 103503. DOI: 10.1016/j.earscirev.2021.103503

Secretariat of the Convention on Biological Diversity. (2022). *Biodiversity-related conventions*. https://www.cbd.int/brc

Seelam, D. R., Kidiyur, M. D., Whig, P., Gupta, S. K., & Balantrapu, S. S. (2025). Integrating Artificial Intelligence in Blue-Green Infrastructure: Enhancing Sustainability and Resilience. In Integrating Blue-Green Infrastructure Into Urban Development (pp. 347-372). IGI Global Scientific Publishing.

Semieniuk, G., Taylor, L., Rezai, A., & Foley, D. K. (2021). Plausible energy demand patterns in a growing global economy with climate policy. *Nature Climate Change*, *11*(4), 313–318. DOI: 10.1038/s41558-020-00975-7

Sentrycs. (2024). *The 2024 forecast in drone technology & drone security: A Sentrycs perspective*. https://sentrycs.com/the-counter-drone-blog/the-2024-forecast-in-drone-technology-drone-security-a-sentrycs-perspective/

Serafinelli, E. (2022). Imagining the social future of drones. *Convergence (London)*, *28*(6), 1376–1391. DOI: 10.1177/13548565211054904

Serafinelli, E., & O'Hagan, L. A. (2024). Drone views: A multimodal ethnographic perspective. *Visual Communication*, *23*(2), 223–243. DOI: 10.1177/14703572211065093

Shafique, A., Mehmood, A., & Elhadef, M. (2021). Survey of Security Protocols and Vulnerabilities in Unmanned Aerial Vehicles. *IEEE Access : Practical Innovations, Open Solutions*, *9*, 46927–46948. DOI: 10.1109/ACCESS.2021.3066778

Shahmoradi, J., Talebi, E., Roghanchi, P., & Hassanalian, M. (2020). A comprehensive review of applications of drone technology in the mining industry. *Drones (Basel)*, *4*(3), 34. DOI: 10.3390/drones4030034

Shakhatreh, H., Sawalmeh, A. H., Al-Fuqaha, A., Dou, Z., Almaita, E., Khalil, I., Othman, N. S., Khreishah, A., & Guizani, M. (2019). Unmanned aerial vehicles (UAVs): A survey on civil applications and key research challenges. *IEEE Access : Practical Innovations, Open Solutions*, *7*, 48572–48634. DOI: 10.1109/ACCESS.2019.2909530

Sharma, D. D., & Sarojwal, A. (2024). Secure and Resilient Threat Modelling Strategy for Drone Delivery System. *2024 IEEE Region 10 Symposium, TENSYMP 2024*. DOI: 10.1109/TENSYMP61132.2024.10752205

Sharma, R. (2021). *The Future of Maritime Surveillance: Integrating UAVs into India's Coastal Defense*. National Defense Review.

Sharma, R., & Chatterjee, A. (2023). Advancements in drone technology for environmental monitoring in India: A case study of the Sundarbans. *Sustainability*, *15*(4), 2341. DOI: 10.3390/su15042341

Sharma, V. (2019). Advances in drone communications, state-of-the-art and architectures. *Drones (Basel)*, *3*(1), 21. DOI: 10.3390/drones3010021

Shaw, J. T., Shah, A., Yong, H., & Allen, G. (2021). Methods for quantifying methane emissions using unmanned aerial vehicles: A review. *Philosophical Transactions. Series A, Mathematical, Physical, and Engineering Sciences*, *379*(2210), 20200450. DOI: 10.1098/rsta.2020.0450 PMID: 34565219

Sheng, T., Jin, R., Yang, C., Qiu, K., Wang, M., Shi, J., Zhang, J., Gao, Y., Wu, Q., Zhou, X., Wang, H., Zhang, J., Fang, Q., Pan, N., Xue, Y., Wang, Y., Xiong, R., Gao, F., Zhang, Y., & Gu, Z. (2023). Unmanned aerial vehicle mediated drug delivery for first aid. *Advanced Materials*, *35*(10), 2208648. DOI: 10.1002/adma.202208648 PMID: 36563167

Sherwani, F., Asad, M. M., & Ibrahim, B. S. K. K. (2020, March). Collaborative robots and industrial revolution 4.0 (ir 4.0). In *2020 International Conference on Emerging Trends in Smart Technologies (ICETST)* (pp. 1-5). IEEE. DOI: 10.1109/ICETST49965.2020.9080724

Sibanda, M., Mutanga, O., Chimonyo, V. G., Clulow, A. D., Shoko, C., Mazvimavi, D., Dube, T., & Mabhaudhi, T. (2021). Application of drone technologies in surface water resources monitoring and assessment: A systematic review of progress, challenges, and opportunities in the global south. *Drones (Basel)*, *5*(3), 84. DOI: 10.3390/drones5030084

Singh, A. (2023). Artificial Intelligence-Based Autonomous Vehicle for Making Real-Time Decisions, and Navigating Safely. *Proceedings of the 2023 International Conference on Innovative Computing, Intelligent Communication and Smart Electrical Systems, ICSES 2023*. DOI: 10.1109/ICSES60034.2023.10465299

Singh, S. (2020). *Biodiversity hotspots in India and the challenges of conservation*. Oxford Research Encyclopedia of Environmental Science., DOI: 10.1093/acrefore/9780199389414.013.95

Siregar, V. P., Rahman, H., & Lubis, R. F. (2023). AI-driven disaster management using drones: Case studies from Indonesia. *International Journal of Disaster Risk Reduction*, *84*, 103456. DOI: 10.1016/j.ijdrr.2023.103456

Skendžić, S., Zovko, M., Živković, I. P., Lešić, V., & Lemić, D. (2021). The impact of climate change on agricultural insect pests. *Insects*, *12*(5), 440. DOI: 10.3390/insects12050440 PMID: 34066138

Slater, T. F., Biggs, C. N., & Sanchez, R. L. (2021). Positive influence of education partnerships for teaching integrated STEM through drone competition. *Journal of Astronomy and Earth Sciences Education*, *8*(2), 113–124. DOI: 10.19030/jaese.v8i2.10409

Sliusar, N., Filkin, T., Huber-Humer, M., & Ritzkowski, M. (2022). Drone technology in municipal solid waste management and landfilling: A comprehensive review. *Waste Management (New York, N.Y.)*, *139*, 1–16. DOI: 10.1016/j.wasman.2021.12.006 PMID: 34923184

Smith, J., Doe, A., & Chen, L. (2023). Artificial general intelligence: Current developments and future prospects. *Journal of AI Research*, *45*(2), 210–225.

Solodov, A., Williams, A., Al Hanaei, S., & Goddard, B. (2018). Analyzing the threat of unmanned aerial vehicles (UAV) to nuclear facilities. *Security Journal*, *31*(1), 305–324. DOI: 10.1057/s41284-017-0102-5

Sonko, S., Adewusi, A. O., Obi, O. C., Onwusinkwue, S., & Atadoga, A. (2024). A critical review towards artificial general intelligence: Challenges, ethical considerations, and the path forward. *World Journal of Advanced Research and Reviews*, *21*(3), 1262–1268. DOI: 10.30574/wjarr.2024.21.3.0817

St. Clair, P., & Hyde, R. (2009). Towards a new model for climate responsive design at the university of the sunshine coast chancellery. *Journal of Green Building*, *4*(3), 3–20. DOI: 10.3992/jgb.4.3.3

Staneva, J., Pinardi, N., Coppini, G., Jacob, B., Chen, W., Jayson-Quashigah, P.-N., Alessandri, J., Mentaschi, L., & Drillet, Y. (2024). Advancing marine sustainability through digital twin what-if scenarios in nature-based solutions. *EGU General Assembly 2024*, Vienna, Austria, 14–19 Apr 2024, EGU24-14266. DOI: 10.5194/egusphere-egu24-14266

Subramanian, K., Pazhanivelan, S., Srinivasan, G., Santhi, R., & Sathiah, N. (2021). Drones in insect pest management. *Frontiers in Agronomy*, *3*, 640885. DOI: 10.3389/fagro.2021.640885

Sumra, I. A., Sellappan, P., Abdullah, A., & Ali, A. (2018). Security issues and Challenges in MANET-VANET-FANET: A Survey. *EAI Endorsed Transactions on Energy Web*, *5*(17), 155884. Advance online publication. DOI: 10.4108/eai.10-4-2018.155884

Suran, N. A., Shafri, H. Z. M., Shaharum, N. S. N., Radzali, N. A. W. M., & Kumar, V. (2019). UAV-based hyperspectral data analysis for urban area mapping. *ISPRS International Archives of the Photogrammetry. The International Archives of the Photogrammetry, Remote Sensing and Spatial Information Sciences*, *XLII-4*(W16), 621–626. DOI: 10.5194/isprs-archives-XLII-4-W16-621-2019

Szalanczi-Orban, V., & Vaczi, D. (2022). Use of drones in logistics: Options in inventory control systems. *Interdisciplinary Description of Complex Systems*, *20*(3), 295–303. DOI: 10.7906/indecs.20.3.9

Tanase, M. A., Aponte, C., Mermoz, S., Bouvet, A., Le Toan, T., & Heurich, M. (2018). Detection of windthrows and insect outbreaks by L-band SAR: A case study in the Bavarian Forest National Park. *Remote Sensing of Environment*, *209*, 700–711. DOI: 10.1016/j.rse.2018.03.009

Tarr, A. A., Tarr, J.-A., Thompson, M., & Ellis, J. (2021). *Drone law and policy: Global development, risks, regulation and insurance*. Routledge. DOI: 10.4324/9781003028031

Thompson, A. K., Patel, D. D., & Ilhan Akbas, M. (2024). Formulating a Comprehensive Cybersecurity Framework for Uncrewed Aerial Vehicles. *Conference Proceedings of the IEEE International Performance, Computing, and Communications Conference*. DOI: 10.1109/IPCCC59868.2024.10850308

Tiansawat, P., & Elliott, S. (2020). *Unmanned aerial vehicles for automated forest restoration*. Faculty of Science, Chiang Mai University.

Tlili, F., & Ayed, S., & Chaari Fourati, L. (2023, August). Dynamic Intrusion Detection Framework for UAVCAN Protocol Using AI. In *Proceedings of the 18th International Conference on Availability, Reliability and Security* (pp. 1-10). DOI: 10.1145/3600160.3605071

Tlili, F., Ayed, S., & Chaari Fourati, L. (2024). Advancing UAV security with artificial intelligence: A comprehensive survey of techniques and future directions. In *Internet of Things (Netherlands)* (Vol. 27). Elsevier B.V. DOI: 10.1016/j.iot.2024.101281

Tmušić, G., Manfreda, S., Aasen, H., James, M. R., Gonçalves, G., Ben-Dor, E., Brook, A., Polinova, M., Arranz, J. J., Mészáros, J., Zhuang, R., Johansen, K., Malbeteau, Y., de Lima, I. P., Davids, C., Herban, S., & McCabe, M. F. (2020). Current practices in UAS-based environmental monitoring. *Remote Sensing (Basel), 12*(6), 1001. DOI: 10.3390/rs12061001

Toh, I. (2020). Drone localisation with machine learning.

TOPODRONE. (2023). *TOPODRONE-RASA Surveying partnership augments coastal mapping, serves resilient communities*. LIDAR Magazine. Retrieved from https://lidarmag.com/2023/04/04/topodrone-rasa-surveying-partnership-augments-coastal-mapping-serves-resilient-communities/

Tout, S., Abualkibash, M., & Patil, P. (2018). Emerging Research in the Security of Modern and Autonomous Vehicles. *IEEE International Conference on Electro/Information Technology (EIT)*, 199. DOI: 10.1109/EIT.2018.8500204

Tsiamis, N., Efthymiou, L., & Tsagarakis, K. P. (2019). A comparative analysis of the legislation evolution for drone use in OECD countries. *Drones (Basel), 3*(4), 75. DOI: 10.3390/drones3040075

Tucker, C., Smith, J., & Wilson, R. (2023). Mapping intertidal estuaries using UAV LiDAR: A case study from South Carolina. *Sustainability, 15*(22), 15823. https://www.mdpi.com/2071-1050/15/22/15823

Turner, W. (2014). Sensing biodiversity. *Science, 346*(6207), 301–302. DOI: 10.1126/science.1256014 PMID: 25324372

Turner, W., Spector, S., Gardiner, N., Fladeland, M., Sterling, E., & Steininger, M. (2003). Remote sensing for biodiversity science and conservation. *Trends in Ecology & Evolution, 18*(6), 306–314. DOI: 10.1016/S0169-5347(03)00070-3

Tyagi, P., & Dembla, D. (2014). A taxonomy of security attacks and issues in vehicular ad-hoc networks (vanets). *International Journal of Computer Applications, 91*(7), 22–29. DOI: 10.5120/15893-5040

UNESCO. (2023). World Network of Biosphere Reserves. Retrieved from https://www.unesco.org

Upadhyaya, S., Gyawali, P., Shrestha, S., & Bhandari, S. (n.d.). *KC, S*. The Fundamental Role of GNSS in Modern Surveying and Mapping to Support Climate Responsive Land Governance and to Enhance Disaster Resilience.

Usama, M. M., Gardezi, H., Jalal, M. F. E., Rehman, M. M. A., Javed, M. N., Janjua, D. S., & Iqbal, M. M. (2023). Predictive Modelling of Compression Strength of Waste Gp/Fa Blended Expansive Soils Using Multi-Expression Programming. *Construction & Building Materials*, *392*(May), 131956. DOI: 10.1016/j.conbuildmat.2023.131956

Varela, M. R., Patrício, A. R., Anderson, K., Broderick, A. C., DeBell, L., Hawkes, L. A., Tilley, D., Snape, R. T. E., Westoby, M. J., & Godley, B. J. (2018). Assessing climate change associated sea-level rise impacts on sea turtle nesting beaches using drones, photogrammetry and a novel GPS system. *Global Change Biology*. Advance online publication. DOI: 10.1111/gcb.14526 PMID: 30430701

Vasconcelos, G., Carrijo, G., Miani, R., Souza, J., & Guizilini, V. (2016, October). The impact of DoS attacks on the AR. Drone 2.0. In *2016 XIII Latin American Robotics Symposium and IV Brazilian Robotics Symposium (LARS/SBR)* (pp. 127-132). IEEE. DOI: 10.1109/LARS-SBR.2016.28

Vasconcelos, G., Miani, R. S., Guizilini, V. C., & Souza, J. R. (2019). Evaluation of dos attacks on commercial wi-fi-based uavs. *International Journal of Communication Networks and Information Security*, *11*(1), 212–223.

Vatan Kaptan, M. (2020). Climate-responsive design strategy for Erbil city. *Archnet-IJAR: International Journal of Architectural Research*, *14*(1), 90–111. DOI: 10.1108/ARCH-04-2019-0089

Ventura, D., Grosso, L., Pensa, D., Casoli, E., Mancini, G., Valente, T., Scardi, M., & Rakaj, A. (2023). Coastal benthic habitat mapping and monitoring by integrating aerial and water surface low-cost drones. *Frontiers in Marine Science*, *9*, 1096594. DOI: 10.3389/fmars.2022.1096594

Villa, T. F., Gonzalez, F., Miljievic, B., Ristovski, Z. D., & Morawska, L. (2016). An overview of small unmanned aerial vehicles for air quality measurements: Present applications and future prospectives. *Sensors (Basel)*, *16*(7), 1072. DOI: 10.3390/s16071072 PMID: 27420065

Vinuesa, R., Azizpour, H., Leite, I., Balaam, M., Dignum, V., Domisch, S., Felländer, A., Langhans, S. D., Tegmark, M., & Fuso Nerini, F. (2020). The role of artificial intelligence in achieving the Sustainable Development Goals. *Nature Communications*, *11*(1), 233. DOI: 10.1038/s41467-019-14108-y PMID: 31932590

Vujičić, M. D., Kennell, J., Stankov, U., Gretzel, U., Vasiljević, Đ. A., & Morrison, A. M. (2022). Keeping up with the drones! Techno-social dimensions of tourist drone videography. *Technology in Society*, *68*, 101838. DOI: 10.1016/j.techsoc.2021.101838

Wagner, B., & Egerer, M. (2022). Application of UAV remote sensing and machine learning to model and map land use in urban gardens. *Journal of Urban Ecology*, *8*(1), juac008. Advance online publication. DOI: 10.1093/jue/juac008

Wakefield, L. (2024). How drones monitor environmental changes. Cover Drone. https://www.coverdrone.com/how-drones-monitor-environmental-changes/

Wäldchen, J., & Mäder, P. (2018). Plant species identification using computer vision techniques: A systematic literature review. *Archives of Computational Methods in Engineering*, *25*(2), 507–543. DOI: 10.1007/s11831-016-9206-z PMID: 29962832

Wang, J., Zhang, S., Guo, H., Tian, Y., Liu, S., Du, C., & Wu, J. (2024). Stereoscopic monitoring of transportation infrastructure. *Automation in Construction*, *164*, 105472. DOI: 10.1016/j.autcon.2024.105472

Wang, N., Christen, M., & Hunt, M. (2021). Ethical considerations associated with "humanitarian drones": A scoping literature review. *Science and Engineering Ethics*, *27*(4), 51. DOI: 10.1007/s11948-021-00327-4 PMID: 34342721

Wang, Q., Mao, B., Stoliarov, S. I., & Sun, J. (2019). A review of lithium-ion battery failure mechanisms and fire prevention strategies. *Progress in Energy and Combustion Science*, *73*, 95–131. DOI: 10.1016/j.pecs.2019.03.002

Wang, Y. (2022). A Formal Theory of AI Trustworthiness for Evaluating Autonomous AI Systems. *Conference Proceedings - IEEE International Conference on Systems, Man and Cybernetics, 2022-October*, 137–142. DOI: 10.1109/SMC53654.2022.9945351

Wanner, D., Hashim, H. A., Srivastava, S., & Steinhauer, A. (2024). UAV avionics safety, certification, accidents, redundancy, integrity, and reliability: a comprehensive review and future trends. In *Drone Systems and Applications* (Vol. 12, pp. 1–23). Canadian Science Publishing., DOI: 10.1139/dsa-2023-0091

Watkins, S., Burry, J., Mohamed, A., Marino, M., Prudden, S., Fisher, A., Kloet, N., Jakobi, T., & Clothier, R. (2020). Ten questions concerning the use of drones in urban environments. *Building and Environment*, *167*, 106458. DOI: 10.1016/j.buildenv.2019.106458

Waymo. "The World's Most Experienced Driver". Available at: https://waymo.com/

Webber, S., Nelson, S., Millington, N., Bryant, G., & Bigger, P. (2022). Financing reparative climate infrastructures: Capital switching, repair, and decommodification. *Antipode*, *54*(3), 934–958. DOI: 10.1111/anti.12806

Weiskopf, S. R., Rubenstein, M. A., Crozier, L. G., Gaichas, S., Griffis, R., Halofsky, J. E., Hyde, K. J., Morelli, T. L., Morisette, J. T., Muñoz, R. C., Pershing, A. J., Peterson, D. L., Poudel, R., Staudinger, M. D., Sutton-Grier, A. E., Thompson, L., Vose, J., Weltzin, J. F., & Whyte, K. P. (2020). Climate change effects on biodiversity, ecosystems, ecosystem services, and natural resource management in the United States. *The Science of the Total Environment, 733*, 137782. DOI: 10.1016/j.scitotenv.2020.137782 PMID: 32209235

Weiss, S. (2024). Environmental Sensing is Here, Tracking Everything from Forest Fires to Threatened Species. Wired. https://www.wired.com/story/environmental-sensing-is-here-tracking-everything-from-forest-fires-to-threatened-species/#:~:text=7%3A30%20AM-,Environmental%20Sensing%20Is%20Here%2C%20Tracking%20Everything%20from%20Forest%20Fires%20to,listen%20out%20for%20threatened%20animals.&text=You%20are%20in%20a%20lush%20forest

Whelan, J., Almehmadi, A., Braverman, J., & El-Khatib, K. (2020, September). Threat Analysis of a long range autonomous unmanned aerial system. In *2020 international conference on computing and information technology (ICCIT-1441)* (pp. 1-5). IEEE. DOI: 10.1109/ICCIT-144147971.2020.9213789

White, B. (2021). The impact of AGI on drone technology. *IEEE Transactions on Robotics, 37*(4), 567–580.

Wich, S. A., Hudson, M., Andrianandrasana, H., & Longmore, S. N. (2021). Drones for conservation. *Conservation technology, 35.*

Wich, S. A., Spaan, D., Pintea, L., Delahunty, R., & Kerby, J. 4.8 UAVs in conservation research. *UAVs for the Environmental Sciences*, 403.

Wilson, R. J., Thomas, C. D., & Maclean, I. M. D. (2020). Ecological monitoring for biodiversity conservation. *Biological Conservation, 248*, 108658. DOI: 10.1016/j.biocon.2020.108658

World Health Organization. (2020). *Biodiversity and health.* https://www.who.int/news-room/fact-sheets/detail/biodiversity

Wu, J., Snell, G., & Samji, H. (2020). Climate anxiety in young people: A call to action. *The Lancet. Planetary Health, 4*(10), e435–e436. DOI: 10.1016/S2542-5196(20)30223-0 PMID: 32918865

Wu, Y., Yang, L., Zhang, L., Nie, L., & Zheng, L. (2024). Intrusion detection for unmanned aerial vehicles security: A tiny machine learning model. *IEEE Internet of Things Journal, 11*(12), 20970–20982. DOI: 10.1109/JIOT.2024.3360231

XDynamics. (n.d.). *Unmanned and intelligent: The rise of AI in drone technology.* https://www.xdynamics.com/blog/unmanned-and-intelligent-the-rise-of-ai-in-drone-technology/

Yaacoub, J. P., Noura, H., Salman, O., & Chehab, A. (2020). Security analysis of drones systems: Attacks, limitations, and recommendations. *Internet of Things : Engineering Cyber Physical Human Systems, 11*, 100218. DOI: 10.1016/j.iot.2020.100218 PMID: 38620271

Yampolskiy, R., & Fox, J. (2013). Safety Engineering for Artificial General Intelligence. *Topoi, 32*(2), 217–226. DOI: 10.1007/s11245-012-9128-9

Yan, S., Han, H., Dong, X., & Xu, Z. (2024). Lightweight Deep Learning Method based on Group Convolution: Detecting DDoS Attacks in IoT Environments. *Proceedings - 2024 10th International Symposium on System Security, Safety, and Reliability, ISSSR 2024*, 181–189. DOI: 10.1109/ISSSR61934.2024.00027

Yan, C., Xu, W., & Liu, J. (2016). Can you trust autonomous vehicles: Contactless attacks against sensors of self-driving vehicle. *Def Con, 24*(8), 109.

Yang, L., Fu, R., He, W., He, Q., & Liu, Y. (2020). Adaptive thermal comfort and climate responsive building design strategies in dry–hot and dry–cold areas: Case study in Turpan, China. *Energy and Building, 209*, 109678. DOI: 10.1016/j.enbuild.2019.109678

Yao, A., Jiang, F., Li, X., Dong, C., Xu, J., Xu, Y., Li, G., & Liu, X. (2021). A Novel Security Framework for Edge Computing based UAV Delivery System. *Proceedings - 2021 IEEE 20th International Conference on Trust, Security and Privacy in Computing and Communications, TrustCom 2021*, 1031–1038. DOI: 10.1109/TrustCom53373.2021.00142

Yazdinejad, A., Parizi, R. M., Dehghantanha, A., & Karimipour, H. (2021). Federated learning for drone authentication. *Ad Hoc Networks, 120*, 102574. DOI: 10.1016/j.adhoc.2021.102574

Yigitcanlar, T., Desouza, K. C., Butler, L., & Roozkhosh, F. (2020). Contributions and risks of artificial intelligence (AI) in building smarter cities: Insights from a systematic review of the literature. *Energies, 13*(6), 1473. DOI: 10.3390/en13061473

Young, J C., Arthur, R., Spruce, M., & Williams, H T P. (2022, May 1). Social sensing of flood impacts in India: A case study of Kerala 2018. Elsevier BV, 74, 102908-102908. https://doi.org/DOI: 10.1016/j.ijdrr.2022.102908

Zailani, M. A. H., Sabudin, R. Z. A. R., Rahman, R. A., Saiboon, I. M., Ismail, A., & Mahdy, Z. A. (2020). Drone for medical products transportation in maternal healthcare: A systematic review and framework for future research. *Medicine*, *99*(36), e21967. DOI: 10.1097/MD.0000000000021967 PMID: 32899033

Zaki, A. M., Abdelhamid, A. A., Ibrahim, A., Eid, M. M., & El-Kenawy, E. S. M. (2024). Securing the Skies: A Study of Cybersecurity Measures in Unmanned Aerial Vehicles. *Full Length Article*, *8*(1), 51–1. DOI: 10.54216/IJWAC.080106

Zenadrone. (2023). Drones for a greener future: Innovations in environmental monitoring. Retrieved from https://www.zenadrone.com/drones-for-a-greener-future-innovations-in-environmental-monitoring/

Zeng, K. C., Liu, S., Shu, Y., Wang, D., Li, H., Dou, Y., . . . Yang, Y. (2018). All your {GPS} are belong to us: Towards stealthy manipulation of road navigation systems. In 27th USENIX security symposium (USENIX security 18) (pp. 1527-1544).

Zeng, Q., Barnt, K., Ragan, L., & Nait-Abdesselam, F. (2023, December). Realtime intrusion detection in unmanned aerial vehicles using active learning and generative adversarial networks. In *2023 IEEE 29th International Conference on Parallel and Distributed Systems (ICPADS)* (pp. 2802-2803). IEEE. DOI: 10.1109/ICPADS60453.2023.00386

Zeng, Q., Olatunde-Salawu, S., & Nait-Abdesselam, F. (2024, October). FGA-IDS: A Federated Learning and GAN-Augmented Intrusion Detection System for UAV Networks. In *2024 IEEE 10th International Conference on Collaboration and Internet Computing (CIC)* (pp. 50-59). IEEE. DOI: 10.1109/CIC62241.2024.00017

Zhang, Z., Si, X., Li, L., Gao, Y., Li, X., Yuan, J., & Xing, G. (2023). An Intrusion Detection Method Based on Transformer-LSTM Model. *2023 3rd International Conference on Neural Networks, Information and Communication Engineering, NNICE 2023*, 352–355. DOI: 10.1109/NNICE58320.2023.10105733

Zhang, J., Jung, J., Sohn, G., & Cohen, M. (2015). Thermal infrared inspection of roof insulation using unmanned aerial vehicles. *ISPRS International Archives of the Photogrammetry. The International Archives of the Photogrammetry, Remote Sensing and Spatial Information Sciences*, *XL-1*(W4), 381–386. DOI: 10.5194/isprsarchives-XL-1-W4-381-2015

Zhi, Y., Fu, Z., Sun, X., & Yu, J. (2020). Security and Privacy Issues of UAV: A Survey. *Mobile Networks and Applications*, *25*(1), 95–101. DOI: 10.1007/s11036-018-1193-x

Zhou, Y., Pan, C., Yeoh, P. L., Wang, K., Elkashlan, M., Vucetic, B., & Li, Y. (2020). Secure Communications for UAV-Enabled Mobile Edge Computing Systems. *IEEE Transactions on Communications*, *68*(1), 376–388. DOI: 10.1109/TCOMM.2019.2947921

Zhou, Y., Wang, H., & Tang, J. (2020). Ethical design in autonomous drones: Policy and challenges. *AI & Society*, *35*, 729–740. DOI: 10.1007/s00146-019-00885-2

Zohdi, T. I. (2018). Multiple UAVs for mapping: A review of basic modeling, simulation, and applications. *Annual Review of Environment and Resources*, *43*(1), 523–543. DOI: 10.1146/annurev-environ-102017-025912

About the Contributors

Inam Ullah Khan is a distinguished academic and industry professional, widely recognized for his contributions to Artificial Intelligence, Artificial General Intelligence, Unmanned Aerial Vehicles, Routing Protocols, Intrusion Detection Systems, Machine Learning, Deep Learning, and Evolutionary Computing. He is the Founder of AI-Explain Your Science (AI-EYS) and a Senior Member of IEEE, with active memberships in prestigious organizations such as the International Association of Engineers (IAENG), IEEE Young Professionals, IEEE Systems Council, and various IAENG societies focused on Artificial Intelligence, Computer Science, Internet Computing & Web Services, Information System Engineering, Scientific Computing, Software Engineering, and Wireless Networks. Dr. Khan currently serves as a Mentor of Artificial Intelligence at Corvit Systems, Rawalpindi, Pakistan, and as a Global Mentor and Guest Lecturer at Impact Xcelerator, School of Science & Technology, IE University, Madrid, Spain. He is also a Trainer at the National Vocational and Technical Training Commission (NAVTTC), Pakistan. Additionally, he is a Co-Supervisor in the Lincoln Global Postdoc and Research Associate Programme (LGPR) at Lincoln University College, Malaysia, and an Advisor for Cisco Community Pakistan. Previously, Dr. Khan was a Visiting Researcher at King's College London, UK, and has held faculty positions at several esteemed institutions in Pakistan, including the National University of Technology (NUTECH), Islamabad; Center for Emerging Sciences, Engineering & Technology (CESET), Islamabad; Abdul Wali Khan University (Garden Campus and Timergara Campus); University of Swat; and Shaheed Zulfikar Ali Bhutto Institute of Science and Technology (SZABIST), Islamabad Campus. Academically, Dr. Khan earned his Ph.D. in Electronics Engineering from Isra University, Islamabad Campus, where he also completed his M.S. in Electronic Engineering. He holds a Bachelor's degree in Computer Science from Abdul Wali Khan University, Mardan, Pakistan. His Master's thesis, titled "Route Optimization with Ant Colony Optimization (ACO)," was published as a book in Germany and is available on Amazon. Additionally, he has completed the Huawei Technologies Pakistan Train the Trainer (TTT) Program. With a prolific research portfolio, Dr. Khan has authored and co-authored over 100 research articles published in leading journals, conferences, and book chapters. He is also an editor for approximately 20 books covering various advanced topics. His expertise has been recognized at the national level, as he has been featured multiple times as a technology expert on Pakistan National Television, showcasing his thought leadership in Artificial Intelligence and emerging technologies. Currently, Dr. Khan is a Postdoctoral Research Fellow at Multimedia University, Cyberjaya, Malaysia, and also serves as Visiting Faculty at Mechatronics Engineering Lab at the NAVTTC Center of Excellence, Islamabad, Pakistan. Additionally, he is an Adjunct Faculty Member at PSGR Krishnammal College for Women, Coimbatore, India.

Oroos Arshi (Member IEEE) is a Research Scientist at AI-EYS, where she is shaping the future through innovation and academic excellence. She was awarded the Young Researcher Award at the International Conference on Emerging Trends and Innovations (ICETI) 2024. She is an active member of IEEE Women in Engineering, IEEE Robotics & Automation Society, and IEEE Young Professionals. Additionally, she is a member of the International Association of Engineers (IAENG). She completed her postgraduate degree, M.Tech in Cyber Security and Forensics, from the School of Computer Science, University of Petroleum and Energy Studies, Dehradun, India. She earned her undergraduate degree, B.Tech in Computer Science and Engineering, from Integral University, Lucknow, India. She has around 15 international publications in reputed journals, conferences, books, and book chapters. Recently, Oroos published her book titled Unmanned Aerial Vehicle Swarms for Protecting Smart Cities: Future Trends & Challenges. In addition to this, six of her books have been accepted as an editor by prestigious publishers, including Wiley, IGI Global, Springer, Apple Academic Press, and Taylor & Francis. She served as a session chair for AIoT's Emerging Technologies and Future Applications at the 4th International Conference on Advances in Computation Technology, Computing, and Engineering in Morocco. Additionally, she was the chief organizer and publication chair at the 2nd International Conference on Emerging Trends & Innovation (ICETI) in July 2024. More notably, she has served as a speaker at numerous conferences. Moreover, she has earned both a gold and silver badge for problem-solving in Design Analysis and Algorithm (DAA) from Hacker Rank. Her expertise has been demonstrated on multiple platforms, reflecting her commitment to sharing insights and contributing to advancements in her field globally. Her research interests include Artificial Intelligence, Unmanned Aerial Vehicles, the Internet of Things, Computer Vision, and Natural Language Processing.

Keshav Kaushik is a distinguished cybersecurity and digital forensics expert, currently serving as an Associate Professor at the Sharda School of Computer Science & Engineering, Sharda University, Greater Noida, India. A key member of the Cybersecurity Centre of Excellence, he has been instrumental in advancing research and education in cybersecurity, AI-driven security solutions, and digital forensics. Recognized among the World's Top 2% Scientists by Stanford University and Elsevier (2024), he has made significant contributions to academia and research. His academic journey includes a prestigious faculty internship at IIT Ropar during the Summer Faculty Research Fellow Programme 2016, demonstrating his commitment to continuous learning. With over 160 publications, including 30 peer-reviewed SCI/SCIE/Scopus-indexed journal articles and 50+ Scopus-indexed conference papers, he has established himself as a leading researcher. Additionally, he has authored 35 books and 30 book chapters, reinforcing his expertise in cybersecurity and digital forensics. An innovator in his field, Dr. Kaushik holds five granted patents, six published patents, and five granted copyrights. His editorial leadership includes serving as Guest Editor for the IEEE Journal of Biomedical and Health Informatics (SCIE, IF: 7.7) and Associate Editor for journals such as IECE Transactions on Emerging Trends in Network Systems, IECE Transactions on Sustainable Computing, International Journal of Sensors, Wireless Communications and Control (Scopus-indexed, Bentham Science), and International Journal of Information Security and Privacy (IJISP, ESCI, Scopus-indexed). He is also an Editorial Board Member of Springer's International Cybersecurity Law Review. Dr. Kaushik is a Certified Ethical Hacker (CEH v11), CQI & IRCA Certified ISO/IEC 27001:2013 Lead Auditor, Quick Heal Academy Certified Cyber Security Professional (QCSP), and IBM Cybersecurity Analyst, further validating his expertise. He is a Vice-Chairperson of the Meerut ACM Professional Chapter, an appointed Bentham Ambassador by Bentham Science Publishers, and a member of the International Association of Engineers (IAENG). A dynamic speaker and mentor, he has delivered over 50 national and international talks on cybersecurity, AI security, and digital forensics. His mentorship was recognized in the Smart India Hackathon 2017, earning appreciation from AICTE, MHRD, and ISRO. He has also contributed to cybercrime investigation training and has been honored by the Uttarakhand Police for his efforts in cybersecurity education. A two-time GATE qualifier with a 96.07 percentile (2012 & 2016), he continues to drive impactful research, innovation, and education in cybersecurity, making significant contributions to academia and the industry.

Hamed Taherdoost is an award-winning leader in research and development, known for his contributions across both industry and academia. He is the founder of Hamta Business Corporation, Associate Professor and Chair of RSAC at University

Canada West, & Director of R&D at Q Minded | Quark Minded Technology Inc. He has over 20 years of experience in both industry and academic sectors. He has worked at international companies from Cyprus, the UK, Malta, Iran, Malaysia, and Canada and has been highly involved in development of several projects in different industries, healthcare, transportation, residential, oil and gas and IT. Additionally, he has served as a trusted technical and technology consultant for multiple companies, providing advisory and mentorship. In academia, Dr. Taherdoost has held teaching positions in Southeast Asia, the Middle East, Europe and North America since 2009. He began his academic career as a lecturer at AU & PNU and later served as an adjunct professor, dissertation chair and faculty fellow at Westcliff University, USA, & adjunct professor and thesis supervisor at Gisma University of Applied Sciences, Germany. His research tenure at IAU lasted over eight years, during which he supervised numerous students. Dr. Taherdoost has a prolific publishing record, with over 350 scientific articles in top-tier journals and conference proceedings. His work is widely recognized, with a strong citation impact reflecting in a high h-index (>50), and his contributions span book chapters, edited volumes, and authored books focusing on technology and research methodology. His textbooks on E-Business and Digital Transformation are teaching at global institutions, including the University of Malta, the University of Macau, and University Canada West. His research accomplishments have earned him a place among the top 10 SSRN Business Authors from 2021 to 2024, as well as being featured in the Stanford-Elsevier list of the world's top 2% of Scientists in 2021, 2022 2023 and 2024. Dr. Taherdoost has organized and chaired many workshops and conferences and has frequently been invited as a keynote speaker. He is an active member of the editorial, reviewer, and advisory boards for several prestigious journals published by Taylor & Francis, Springer, Emerald, Elsevier, MDPI, EAI, IGI Publishing, and Inderscience. He has also participated as an organising, scientific and technical committee member in over 400 conferences held across Europe, America, Australia, Asia, and Africa. Dr. Taherdoost's leadership and innovation have earned him numerous accolades, including THE BIZZ Business Excellence Award, PeerJ Award, and recognition at the Asia Corporate Excellence & Sustainability Awards. He was a finalist for the Innovation in Teaching of Research Methodology Excellence Awards and Southeast Asian Startup Awards by Global Startup Awards. Additionally, his research achievements also include winning several best paper awards, outstanding reviewer awards and best presentation awards like MLIS awards, CSA award, CSTFM Lecture award, the Outstanding Editorial Board Member award from Bilingual Publishing, Best Paper of the Year of Computers MDPI, Editor's Choice Articles of Algorithm Journal, and Best Interview Award of Encyclopedia. Dr. Taherdoost is the Editor of a book series with Routledge (Taylor & Francis Group) titled Mastering Academic Excellence: Research, Teaching, Learning, and Publishing & Editor of a book series entitled "Quantum Computing" with Jenny Stanford Publishing Pte Ltd. He serves on the editorial boards of several high-impact journals, including the International Journal of Information Technology Project Management (IGI, IF: 0.8), Transactions on Scalable Information Systems (EAI, Q2), International Journal of Electronic Government Research (IGI, IF: 1.2), Information Resources Management Journal (IGI, IF: 1.4), Journal of Blockchain (MDPI), and International Journal of Data Mining, Modelling, and Management (InderScience, IF: 0.5, CS: 0.9). He is also an Associate Editor of Frontiers in Research Metrics and Analytics (Scopus) and Academic Editor of PLOS ONE. He's been a guest editor of special issues in Results in Engineering, Elsevier (IF: 5 & CS:4.5), Electronics MDPI (IF: 1.9 & CS: 4.7), Computers MDPI (IF:2.8 & CS: 4.7), Discover Computing by Springer. He is a Certified Cyber Security Professional and Certified Graduate Technologist. He is a GUS Fellow - GUS Institute I Global University Systems, senior member of IEEE, and Working Group Member of International Federation for Information Processing - IFIP TC 11 - Human Aspects of Information Security and Assurance and Information Security Management. Currently, he is involved in several multidisciplinary research projects, including studying innovation in information technology, blockchain, cybersecurity, and technology acceptance.

Khadija Slimani completed her Ph.D., which was jointly supervised by the University of Ibn Tofail in Morocco and the University of Technology of Belfort Montbeliard (UTBM) in France. This collaborative international effort showcased her commitment to academic excellence and cross-cultural research. During her doctoral studies, Dr. Slimani delved into the realms of machine learning, deep learning, pattern recognition, and computer vision, with a specific focus on academic emotion recognition. This interdisciplinary approach, spanning both Moroccan and French academic environments, enriched her research experience and contributed to the global perspective of her work. After completing her Ph.D. thesis, Dr. Slimani pursued a postdoctoral position at the University of Poitiers, France, demonstrating her versatility in research by delving into Objects DRI (Detection, Recognition, and Identification). Her application of AI methodologies aimed at enhancing video content filtering for optimal security reflected the international scope of her expertise. Dr. Slimani's academic contributions extended to various engineering schools in Paris, France, where she took on teaching responsibilities across multiple modules, including data science, deep learning, machine Learning, data bases, and computer vision. Currently holding the esteemed title of Associate Professor at the Graduate School of Automatic Electronic Computing in Paris, France. Her scholarly achievements include over 18 published articles in esteemed journals and conferences, supported by an impressive h-index of 10 in Scopus. Notably, she played a key role as a guest editor for the special issue "Integrating Intelligence:

Advancements in Applied Research with Smart Technologies" in the Journal of Autonomous Intelligence (eISSN: 2630-5046) and served as a guest editor for the proceedings published in E3S Web of Conference 2022 and 2023, both Scopus/web of science indexed. Through her diverse roles, Dr. Slimani consistently inspires and contributes to shaping the next generation of researchers.

***

Yasir khan is currently serving as Information Security Analyst at the department of Science and Technology & Information Technology (ST&IT). Previously, he worked as a Young Professional Officer (YPO) at the Department of Finance, Government of Khyber Pakhtunkhwa, Pakistan. He also served as Offensive Security Engineer at Virtually Testing Foundation (VTF). He completed his undergraduate degree in Bachelor's of Studies in computer science from Abdul Wali Khan University Mardan, Pakistan. His main areas of research include information security, network security, machine learning, cryptography, network intrusion and prevention, deep learning, security of IoT, and blockchain.

Meha Khiria is an Assistant Professor of Law at Mody University of Science and Technology, Rajasthan, India, with over 3 years of teaching experience. As an advocate enrolled in the Bar Council of India, Ms. Meha excels in both academia and legal practice. Currently completing a Ph.D. thesis on women's adoption rights in India, her research focuses on gender equality and child welfare. Passionate about contributing to global legal advancements, Ms. Meha aims to enhance women's rights and legal education internationally. With a Master's degree in Criminal Law, she is committed to shaping legal discourse and promoting justice worldwide.

Gabriela Michael is a Research Scholar pursuing her PhD from Indian Maritime University. She holds an LLM in Coastal and Maritime Security Law (Gold medalist) from Rashtriya Raksha University, Gujarat, and an award of distinction in her BA.LLB (Honors) from VIT School of Law, Chennai. She has a teaching experience in Symbiosis Law School, Nagpur. Her research interest lies in multi-disciplinary and inter-disciplinary research that has addressed critical contemporary issues, especially in criminology, maritime law, technology, and international law to societal challenges and a focused exploration on climate change.

Sameer Saharan is a dedicated researcher in Forensic Science, specializing in fingerprint analysis, cyber forensics, and DNA fingerprinting. He holds a Ph.D. in Forensic Science from Amity University, Haryana, and has authored several research papers and book chapters. With a passion for continuous learning, Dr. Saharan actively contributes to the advancement of forensic science through his expertise and knowledge. He is committed to nurturing the next generation of forensic scientists and making a positive impact in the field.

Anurag Sarthi is an alumnus of University of Delhi and an aspirational academician who is currently pursuing his PhD in law at Faculty of Law, University of Delhi, India. He was previously working as an Assistant Professor of Law at Raffles University, Neemrana, Rajasthan, India. He had secured 2nd rank in his Master's Degree in Coastal and Maritime Security Law and Governance from Rashtriya Raksha University, India and a PG Diploma in Environmental Law and Policy from National Law University, Delhi. His area of research interest lies in environmental law, climate change, maritime law, international governance, and security. .

Mridu Sharma holds a PhD, Master's, and Bachelor's degree in Forensic Science. Her research interests lie in the areas of forensic toxicology, fingerprint and questioned documents examination, and cyber forensics. Her work has been recognised with several awards for best presentations at international conferences and has been published in Scopus-indexed journals of national and international repute. In addition to her research, she is committed to teaching and mentoring students at the undergraduate and postgraduate levels.

Baqar Ali Zardari received a (B.S) degree in Computer Science, (M.S) in Information Technology and (PhD) From QUEST Nawab shah His Area Of Interest Includes HCI, Distributed Databases, User Experience, Artificial Intelligence, DDOS And Shallow Neural Network.

Index

D

E

G

I

L

M

P

R

S

T

U

Z

www.ingramcontent.com/pod-product-compliance
Lightning Source LLC
LaVergne TN
LVHW080342011026
849161LV00009B/2165